MC68000 Assembly Language Programming

Brian Bramer B.Tech., Ph.D., C.Eng., M.I.E.E., M.I.E.E.E
Principal Lecturer in Computing Science,
Department of Computing Science,
Leicester Polytechnic.

Susan Bramer B.A.
Lecturer in Computing Science,
Department of Computing Science,
Leicester Polytechnic.

Edward Arnold
A division of Hodder & Stoughton
LONDON MELBOURNE AUCKLAND

First published in Great Britain 1991
Reprinted 1993 with corrections

British Library Cataloguing in Publication Data

Bramer, B.
MC 68000 assembly language programming.—2nd ed.
I. Title II. Bramer, S. M. S.
005.1

ISBN 0-340-54451-1

Printed and bound in Great Britain for Edward Arnold, a division of Hodder and Stoughton Limited, Mill Road, Dunton Green, Sevenoaks, Kent TN13 2YA by St Edmundsbury Press Limited, Bury St Edmunds, Suffolk and by Hartnolls Ltd, Bodmin, Cornwall.

Preface

This book is intended as a learning text for students of Computer Science, Electronic Engineering, Information Technology and other courses, who are attending a module which introduces assembly language programming of the Motorola MC68000 microprocessor. In addition, the book may also be used for self-instruction by experienced programmers who wish to program MC68000 based systems in assembly language.

This learning text would normally accompany a course of practical/tutorial sessions which may be backed by lectures. Each chapter is a self contained unit that can be read by the student and contains exercises to be attempted during tutorial/practical sessions (answers to the exercises are given in Appendix G). To complete each chapter is a problem that should be attempted and may be used by instructors as a means of assessment. At critical points in the text reviews of topics covered in the previous few chapters are presented.

Good programming practice is encouraged throughout the book by the use of modular and structured programming techniques. In particular, as an aid to modular programming, the use and writing of subroutines is described early in the text and a library of MC68000 assembly code subroutines is provided which enables student programs to read and write numeric and other data via the computer console. This enables the use of subroutines in the exercises and problems before the operation of the subroutine call and return instructions are described in detail.

Although this book is a self contained text, it is expected that a computer architecture and assembly language programming module that uses it, as part of a formal course, would follow a module on high-level language programming (e.g. Modula 2). This would serve to reinforce the modular and structured programming techniques outlined in this book, which could then be formalised in a Software Engineering module at a later stage.

This book has not been written with any particular MC68000 microcomputer system in mind and the only assumption made is that single character input/output routines are available which can be accessed from assembly language programs (Appendix C discusses some of the problems that may arise when using different 68000 based microcomputers).

Outline of the Book

An introduction to computer systems, information representation, and computer hardware and software fundamentals is presented in Chapters 1, 2 and 3 (these initial chapters may be omitted if these topics have already been covered in other modules). While these chapters are being covered, the students could be carrying out practical exercises on number systems (Appendix A describes the binary and hexadecimal number systems and the ASCII character code). The register and memory architecture of the MC68000 is introduced in Chapter 4 which should be accompanied by practical/tutorial sessions using the monitor/operating system of the target microcomputer.

When MC68000 assembly language programming is introduced in Chapters 5, 6, 7 and 8 the student should be familiar with the use of the microcomputer and be ready to attempt the programming exercises and problems. The use of modular programming techniques is introduced in Chapter 9 (this could be omitted if the student has already covered this topic in another module) followed by an introduction to subroutine programming (covered in detail in Chapter 14) and the use of the assembly language subroutine library of Appendix D. Subroutines can then be implemented and used in the exercises and problems of Chapter 10 to 16 which cover advanced assembly language programming. Chapter 17 discusses the implementation of medium to large software systems and discusses Software Engineering techniques.

Basic input/output programming is introduced in Chapter 18, followed in Chapter 19 by an introduction to parallel and serial communications. Parallel interface chips are then introduced in Chapter 20 (MC6821 PIA) and 21/22 (MC68230 PIT). The exercises of these chapters may be carried out using ready built experiments which are described in Appendix F. For students who have studied basic digital sequential and combinational logical circuits, extra exercises and problems can be provided in which the students build circuits to be attached to the parallel interfaces and tested with suitable programs.

Chapter 23 gives details of MC68000 exceptions (interrupts are a type of exception) and Chapter 24 then describes the use of interrupts with the MC6821 PIA and MC68230 PIT. Input/output device programming concludes with serial interface chips, introduced in Chapters 25 (MC6850 ACIA) and 26 (MC68681 DUART), and the programming of time intervals (using the MC6840 PTM and MC68230 PIT) in Chapter 27.

Chapter 28 introduces the use of macros and conditional assembly, Chapter 29 provides examples of interfacing 68000 assembly language to a high-level language (C in this case) and the book concludes with a overview of the MC68000 family in Chapter 30.

Recommendations to Tutors

It is recommended that the sample programs and exercises are entered and tested (see below). This should give an indication of any problems associated with the particular microcomputer and assembler being used. Students can then be provided with information to enable them to overcome any problems (e.g. the use of a non-standard assembler), and other details such as I/O device register addresses, etc.

If the students are to attempt to program input/output devices (MC6821 PIA, MC68320 PIT) plug-in cards and other external equipment will have to be provided (sample circuits that can be built for use with the PIA or PIT are shown in Appendix F).

To obtain copies of the sample programs and exercises call email bb@dmu.ac.uk

or send a floppy disk to: Dr Brian Bramer
Department of Computing Science,
De Montfort University,
The Gateway,
Leicester LE1 9BH,
U.K.

Acknowledgements and dedications

We wish to acknowledge the assistance of colleagues who helped with the preparation of this book. In particular Ian Sexton of Leicester Polytechnic and Clive Beckwith of Coventry Technical College for their help with the chapters on the MC68230 PIT.

This book is dedicated to our lovely daughters Elinor and Isobel without whose assistance it would probably have been finished three months earlier !

Brian Bramer, Susan Bramer, 1993

Contents

1

Introduction to Computer Systems

Until the early 1980's computer systems were large, expensive and required many expert staff to maintain, operate and program them. The users of such systems were generally restricted to staff within large organisations that could afford the high purchase and operating costs. The advent of high-powered microcomputers has changed this, providing systems which are not only sufficiently powerful to be used throughout commerce and industry but cheap enough to be used in many every-day appliances, e.g. cars, video recorders, etc.

1.1 Users of computer systems

1.1.1 An end-user's viewpoint of a computer system

The vast majority of users of computer systems are *end-users*, in that they use computer systems as a tool to aid their everyday work. The computer system can be considered as a 'black box' into which information is fed for processing and then results are produced. The information input can range from text (a wordprocessor) or numerical data (a spreadsheet), entered via a keyboard, to diagrams (a CAD design system) entered via a mouse or a digitising tablet. Similarly, information output may be text, numbers or diagrams and pictures presented on a display screen or printer. To communicate with the system, end-users employ terms which are common to their everyday working environment, e.g. accountants use columns of numbers and artists use pictures.

In general, an end-user would purchase a complete computer system consisting of hardware (the physical components) and software (the programs that tell the hardware what to do) to suit their application environment. They do not need to have knowledge of how a computer system works or skills in the design, implementation and testing of programs. In fact, attempts by end-users to learn these skills would divert them from gaining more important knowledge, e.g.:

1. The applications of computer systems to their specialty with consideration of limitations due to accuracy, size of problem which can be solved, etc.
2. Purchase and installation of a computer system (Bramer 1989a):
   - (a) carrying out a feasibility study to determine the advantages and disadvantages of installing a computer system;
   - (b) drawing up a specification of a computer system to suit their requirements;
   - (c) from the specification draw up an ITT (Invitation to Tender) which will be sent to prospective vendors;
   - (d) processing tenders and shortlisting vendors systems using various criteria, e.g. cost, functionality, usability;
   - (e) system installation and acceptance tests;
   - (f) training, system management, maintenance, etc.

1.1.2 A computer programmer's viewpoint

1.1.2.1 Application programmers

In many application areas there is still a need for specialists with a knowledge of computer programming. These specialists may be professional computer scientists implementing software packages in particular application areas, or scientists or engineers writing programs for their own applications. Such programs would generally be written in a high-level problem solving language (Hunt 1982) that, apart from some appreciation of information storage and accuracy limitations, requires little knowledge of the internal workings of computer systems.

1.1.2.2 Low-level programmers and hardware designers

Computer scientists and electronic engineers working in the areas of real-time systems, hardware design, process control, etc., are expected to implement low-level programs, e.g. to control input and output devices. As such they require a knowledge of computer architecture and skills in low-level programming techniques. In the past, for reasons of efficiency and speed, such programs would have been written in an assembly language. However, programming in assembly language is difficult, requiring highly skilled programmers, and large programs implemented in it are very difficult to update and maintain. Today, the majority of low-level programs would be written in a systems implementation language (a language which has many of the facilities of an assembly language, e.g. bit-manipulation, ability to access to I/O registers) such as C, C++, Modula 2 or Ada. Assembly language would be used for time critical modules and modules which cannot be coded in the high-level language. In educational establishments the teaching of assembly languages is still regarded as important (Loui 1988) for a number of reasons, including:

(a) they are still used in time critical applications in industry;

(b) they provide an understanding of instruction execution (arithmetic, logical, I/O) and the representation of information (simple and complex data structures, pointers);

and (c) they assist in the understanding of the principles of processor architecture (control unit, ALU, memory, I/O) and the effect of these on system performance.

This book introduces the assembly language of the Motorola MC68000 microprocessor family and how it can be applied to control input/output devices (for information on 68000 hardware and interfacing refer to Clements 1987 and Wilcox 1987). It is expected that readers will already have used a modern high-level language and gained experience in program design, implementation and testing.

Although the MC68000 processor is over ten years old (Stritter & Gunter 1979, Toong & Gupta 1981) a range of upward compatible successors, the 68010 (MacGregor et al 1983), 68020 (Farrell 1984, MacGregor et al 1984), 68030 and 68040 (Edenfield et al 1990), have preserved the fundamental architecture and the family is extensively used in commercial and industrial applications. The MC68000 itself is still used in cheaper low-end computer systems which do not need the power and functionality of its successors, e.g. in simple control applications and for educational use (Coats 1985/86).

1.2 A microcomputer system

At a superficial level a computer system can be considered as consisting of three components, namely *hardware, software* and *data.*

1.2.1 Hardware

The term hardware embraces the physical components of the system:

1 The box which contains the printed circuit boards, power supply, etc.
2 The display screen and keyboard for user interaction.
3 The peripheral devices such as disks and printers.

The internal electronic circuits of modern computers are made up from a number of integrated circuit chips and other components. An integrated circuit chip is a small packaged device a few centimetres square which contains complex electronic circuits. The heart of the modern microcomputer is the microprocessor which is an integrated circuit chip containing the central processing unit (the basic control and processing circuits) of a small computer system. A complete microcomputer system contains a microprocessor plus memory, input/output devices, power supplies, etc.

However, before the computer hardware can perform a task (for example add numbers or read a character from a keyboard), it requires a program to tell it what to do.

1.2.2 Software

Software comprises the programs that tell the hardware what to do. A program is a sequence of instructions stored in the memory of the computer system. The central processing unit fetches an instruction, decodes it and then executes the required operation (e.g. to add two numbers). When an instruction has been executed the next instruction is fetched, decoded and executed, etc. A program may be very simple, for example, to calculate the average of ten numbers, or very complex, as would be required to draw a television quality picture on a display screen.

1.2.3 Data

The data is the information to be processed by the computer system. Data may be simple numbers for mathematical calculations, text such as names and addresses or more complex structures such as pictures or drawings. The instructions that make up the program define what data is to be processed, in what form and at what time.

1.3 Instruction and data storage

Within the computer hardware there must be **memory** to store the instructions of the program to be executed and the data to be processed.

1.3.1 Representation of integer numbers

Within modern computer systems the basic element of storage is the *binary digit (or bit)* which can represent a 0 or a 1. The reason for this is that it is very easy to build electronic switches where an off/on condition is used to represent a 0/1 binary value. Although a single bit can only have two states, 0 or 1, a sequence of bits can be used to represent a larger range of values. Such a sequence is called a word of storage and is usually 8, 16, 32, 64 or 128 bits in length. An 8-bit word, for example, can represent an unsigned positive number in the range 0 to 11111111 binary (0 to 255 decimal) thus:

bit	7	6	5	4	3	2	1	0
bit value	2^7	2^6	2^5	2^4	2^3	2^2	2^1	2^0

In the diagram above the least significant or rightmost bit, bit 0, represents 2^0 or 1 and the most significant or leftmost bit, bit 7, represents 2^7 or 128 decimal (the convention for identifying the bits within a word is that the rightmost or least significant bit is numbered 0). The combinations of 1s and 0s of the 8-bit word thus represent an unsigned value in the range 0 to 11111111 binary (0 to 255 decimal). The general term given to an 8-bit storage word is a **byte** which is used by the majority of modern computer systems as their fundamental unit of storage. To represent values that are too large to store in 8-bits a number of bytes may be used. For example, a 16-bit number (made up from two bytes) can represent an unsigned value in the range 0 to 65535 decimal (see Appendix A for more details of the binary number system).

Many commercial and scientific calculations require the use of signed numbers and the majority of modern computer systems use *twos complement binary arithmetic* in which the most significant bit is used to store the sign (1 for a negative number and 0 for a positive number). Using twos complement binary arithmetic an 8-bit number can represent values in the range -128 to +127 and a 16-bit number values in the range -32768 to +32767 (see Appendix A). The Motorola MC68000 allows arithmetic operations to be carried out on 8-, 16- and 32-bit signed and unsigned numbers (the programmer would use the most appropriate for the application) which can represent numeric values shown in Table 1.1.

In practice it would be both difficult and error prone to enter data directly in binary form, so hexadecimal (base 16) or decimal are more commonly used. It is a relatively easy task to convert between binary and hexadecimal (Appendix A describes some conversion techniques).

Data size	unsigned range	signed range
8-bit	0 to 255	-128 to +127
16-bit	0 to 65535	-32768 to +32767
32-bit	0 to 4294967295	-2147483648 to 2147483647

Table 1.1 Numeric range of 8-, 16- and 32-bit unsigned and signed numbers

1.3.2 Representation of real numbers

Integer numbers are suitable for whole number calculations (i.e. no fractional component) and where a limited number range is acceptable. The majority of scientific and engineering applications use numbers with fractional components and which can vary in size from very small to very large values, e.g. from the size of atomic particles to intergalactic distances. For such applications programming languages provide a data type called *real numbers* which is represented internally in floating point format (see Appendix A).

Exercise 1.1 *(see Appendix G for sample answer)*

Convert the following numbers to binary and hexadecimal; do the calculation in each case and convert the result back into decimal (use signed 8-bit twos complement binary numbers):

```
   16        45        110        110
  +32       +60       - 45       + 45
```

The last calculation gives a condition called overflow; explain what has happened.

1.3.3 Character data

Character data is used within a computer to represent text (such as names and addresses) and consists of the usual printable characters, e.g. the alphabet A-Z and a-z, digits 0-9 and other characters such as +, -, *, /, !, $, %, and &.

Each character is stored in a byte of memory and represented by a particular binary pattern or character code. To enable different computers, terminals and printers to be connected together there are a number of standard character codes. The most commonly used character code is ASCII (American Standard Code for Information Interchange), which is listed in Table A.2 of Appendix A. The character A, for example, is represented by the binary pattern 01000001 (41 hexadecimal), and B by 01000010 (42 hexadecimal). The majority of computer users do not need to know or even be aware of these codes as the keyboard and display equipment converts between the characters and the internal codes automatically, i.e. if the user hits the key A on the keyboard the binary value 01000001 is sent to the computer.

Exercise 1.2 *(see Appendix G for sample answer)*

From Table A.2 of Appendix A determine the decimal, hexadecimal and binary values of the ASCII character code for the following characters:

A B C Z a b c 0 1 2 9 - ? =

Can you see anything significant about the order of the ASCII character codes for letters and digits and why the latter is useful when writing a program to read a sequence of digit characters to be converted into a numeric value ?

1.3.4 Instruction representation

A computer program is made up of a sequence of instructions which are represented by binary patterns. For example, the binary pattern 0100001001000011 (4243 hexadecimal), when executed by the Motorola MC68000 microprocessor, would set the lower 16 bits of the data register D3 to 0. Each instruction that the computer hardware can execute has a particular binary pattern, with sequences of such binary patterns in the memory of the computer forming a program. Programs in this form are in a language called *machine code*, i.e. the language the hardware of the computer understands. It is clear that if humans had to write programs in machine code, programming would be a very error prone and time consuming task. In practice, professional programmers use either an assembly language or a high-level language.

1.4 Low and high-level languages

1.4.1 Assembly languages

In assembly languages each machine instruction is represented by a meaningful mnemonic (ADD, SUB, DIV) and data specified in binary, hexadecimal, decimal and character form. For example, the MC68000 instruction which clears the lower 16 bits of data register D3 (0100001001000011 in machine code) would be written in 68000 assembly language:

```
        CLR.W       D3
```

where CLR.W is the instruction or operation-code mnemonic and D3 is the position of the data being operated upon (called the operand). The computer hardware can *only understand* machine code, so before it can be executed an assembly language program has to be converted into machine code. This is done by a program called an assembler which takes each assembly language statement and converts it on a one-to-one basis into the equivalent machine code instruction which can then be executed.

Assembly language programming is difficult because it is only one level above machine code and hence orientated to a particular computer (each type or model of central processing unit has its own machine code language). For example, a program which had been implemented in assembly language on an MC68000 microcomputer would have to be totally rewritten if transferred to an Intel 8086 based system. In addition, the programmers who had written the original software would have to learn a new assembly language for the Intel 8086 processor.

Even with the above disadvantages assembly language programming is still required in many industrial applications, in particular, in real-time control systems for the implementation of time critical modules (the majority of the system would be implemented n a high-level language).

Machine code and assembly languages are described as low-level languages in that they are orientated towards the computer hardware. High-level languages on the other hand are problem orientated and computer independent.

1.4.2 High-level problem solving languages

High-level languages are written in an English or mathematical notation which is orientated towards solving practical problems (Hunt 1982). Some examples of high-level languages are:

BASIC	Beginners All-purpose Symbolic Instruction Code: a simple language available on many home microcomputers;
FORTRAN	FORmula TRANslation: a language widely used for mathematical, scientific and engineering applications;
COBOL	COmmon Business Orientated Language: a language designed for commercial business applications;
PASCAL	a general purpose problem solving language;
C	a systems implementation language;
Modula 2	a modern general purpose and systems implementation language;
Ada	a modern systems implementation language designed for real-time applications.

After the program source code has been entered into the computer it has to be converted into machine code by a program called a compiler. Each statement in a high-level language can be converted into a number of machine code instructions. For example, consider the following Pascal statement:

```
DO:=7+2*(267-23);
```

The equivalent in 68000 assembly language is:

```
MOVE.L      #267,DO
SUB.W       #23,DO
ADD.W       DO,DO
ADD.W       #7,DO
```

and in 68000 machine code (hexadecimal byte values) is:

```
20 3C 00 00 01 0B 04 40 00 17 D0 40 06 40 00 07
```

For a more detailed example see Chapter 29 which contains a C program and a listing of the 68000 assembly language generated by the compiler.

In general the compilation process is not 100% efficient so a program written in a high-level language will take more memory and run more slowly than an equivalent assembly language program written by a **good** programmer. However, the advantages of working in a language which is orientated towards solving problems rather than towards the computer hardware means that the majority of application programs are written in high-level languages.

An additional advantage of using high-level languages is that such languages are less computer dependent than assembly languages (depending upon the quality of the international standard of the language and the particular implementation being used).

1.5 Review of information representation

ALL information within the computer, either instructions or data, is represented in binary form. To a programmer working in a high-level language this is not a problem as the compiler and run time system assign the appropriate data storage and convert information entered to and from binary. Consider the following simple Pascal program which writes a number and a character:

```
PROGRAM TEST;
CONST I=20; X='D';
BEGIN
WRITELN(' number is ',I,' letter is ',X);
END.
```

When the above program is compiled (i.e. converted into machine code), the compiler assigns the storage for any variables and sets up the values defined by CONSTant expressions, i.e. the values of I and X are converted to the equivalent 16-bit (or 32-bit) and byte binary values 0000000000010100 and 01000100 respectively (0014 and 44 hexadecimal). When the program is executed the WRITELN statement converts the internal binary representation of the integer number I (value 20 decimal) into a string of ASCII character codes (32 and 30 hexadecimal) to be transmitted to the display screen.

The display hardware then converts these into characters to be viewed, i.e. 20.

When working in a low-level language such as machine code or assembly code, the binary pattern 01000010 01000011 (42 43 hexadecimal) could represent to the computer hardware:

1 two 8-bit integer numbers: 66 and 67 decimal;
2 a 16-bit number: 16963 decimal;
3 two characters in the ASCII character code: B and C respectively;
4 the instruction to the Motorola MC68000 microprocessor to set the lower 16 bits of data register D3 to 0, i.e. CLR.W D3 in 68000 assembly language.

When working in a high-level language the compiler and run time system look after the organisation of data storage and conversion between external characters and the internal binary form. When working in machine code or assembly language the programmer is responsible for ensuring that instructions and data are separate and that the correct code is executed and data processed. It is very easy to get mixed up and try to add character data or even execute data. Careful program design and coding will avoid this problem.

Exercise 1.3 *(see Appendix G for sample answer)*

What particular problems could face an assembly language programmer when looking for a new job ?

Problems for Chapter 1

1 Why is the binary system used for information storage within modern computer systems ?

2 Convert the following numbers to binary and hexadecimal; do the calculation in each case and then convert the result back into decimal (use signed 16-bit twos complement binary numbers):

```
  15      67      189      456      1027
 +86     -86     +345     -345     +2056
```

3 Describe, in each case, the advantages/disadvantages and areas of application of:

(a) machine code programming;
(b) assembly language programming;
(c) high-level language programming.

2

Computer Hardware

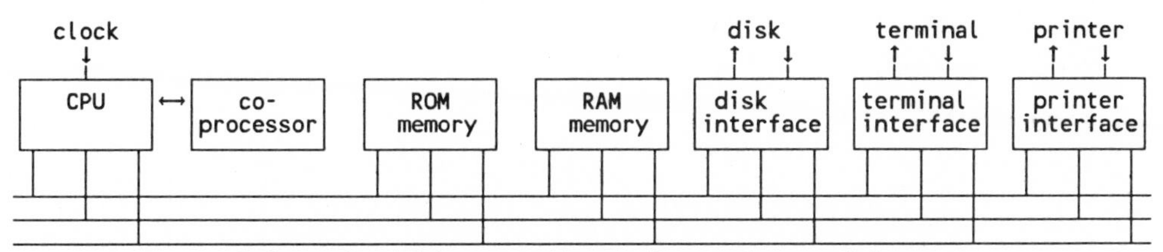

Fig. 2.1 Typical microcomputer configuration using a common bus system

Fig 2.1 is a representation of the hardware (physical components) of a simple single processor computer system comprising:

1 CPU and associated circuits, e.g. microprocessor integrated circuit chip.
2 Co-processor (if fitted), e.g. for real number calculations or graphics.
3 Primary Memory (RAM and ROM).
4 Disk interface which controls a floppy disk and/or hard disk as secondary memory for saving programs and data.
5 Terminal interface which controls the display screen and the keyboard.
6 Input/output interface devices (for connecting external devices such as printers), e.g.:
   (a) Serial I/O interface, e.g. MC6850 ACIA (Chapter 25) or MC68681 DUART (Chapter 26).
   (b) Parallel I/O Interface, e.g. MC6821 PIA (Chapters 20 and 24) or MC68230 PIT (Chapters 21, 22 and 24).
   (c) Timer controller, e.g. MC6840 PTM (Chapter 28) or MC68230 PIT (Chapter 28).

It can be seen from Fig. 2.1 that an *information highway or bus system* connects the various components of the system together:

Address Bus carries the address of the memory location or I/O device being accessed.

Data Bus which carries the data signals.

Control Bus which carries the control signals between the CPU and the other components of the system, e.g. signals to indicate when a valid address is on the address bus and if data is to be read or written.

Unless a user, usually an electronics engineer, is building components to connect directly to the bus (Clements 1987, Wilcox 1987), the physical connections and signal timing is of little interest to the majority of programmers. Even users writing assembly language programs to control external devices (e.g. motors, heaters) can do this via a parallel I/O interface such as a MC6821 PIA (Chapters 20 and 24) or MC68230 PIT (Chapters 21, 22 and 24).

2.1 Memory

There is a general rule that the faster the memory the more it costs. Although it would be desirable to have a large amount of very high speed memory for fast program execution it is not always economically possible. Within a computer system there is a hierarchy of memory:

1 high speed registers within the CPU used for the storage of temporary information and intermediate results;
2 primary memory used for storage of programs being executed and data being processed;
3 secondary memory, usually disks, used for long term storage of programs and data.

Table 2.1 shows typical sizes and access times of memory types used in modern computer systems. To simplify memory size notation the basic memory size is stated in K or M, where a K is a unit of 1024 (not 1000 as in Kilometres) and M a unit of 1048576 (because primary memory is built up as a square matrix of storage elements memory sizes are a power of 2).

	Typical size	Access time
CPU Registers	10 to 1000 byte	less than 10nSec
Primary memory	512K to 64Mbyte	less than 100nSec
Secondary memory:		
floppy disks	320K to 2Mbyte	50 to 500mSec
hard disks	20 to 2000Mbyte	less than 40mSec
magnetic tape	40 to 500Mbyte	seconds to minutes

Table 2.1 Typical microcomputer memory sizes and access times

The access time is the time between the request for information and its availability for use. This is normally stated in nSec (nanosecond = 10^{-9} or 0.000000001 of a second), μSec (microsecond = 10^{-6} or 0.000001 of a second) or mSec (millisecond = 10^{-3} or 0.001 of a second). The three orders of magnitude difference between the access times of primary and secondary memory is mainly because the former is purely electronic and the latter has mechanical moving components. In addition the technique used to access information is different in that primary memory is random access and secondary memory (disk and magnetic tape) is sequential access, i.e.:

Sequential access. To access a particular piece of data all information between the current position and the target has to be accessed, e.g. as in a magnetic tape storage system.

Random access. Any memory location may be accessed directly.

2.1.1 The organisation of the primary memory

Primary memory is used to store the machine code and data during program execution. The majority of modern computer systems use a memory store built up of bytes of storage with each byte being assigned a location address. Fig. 2.2 shows such a memory organisation with the first byte of memory having address 0, the next 1, the next 2, etc.

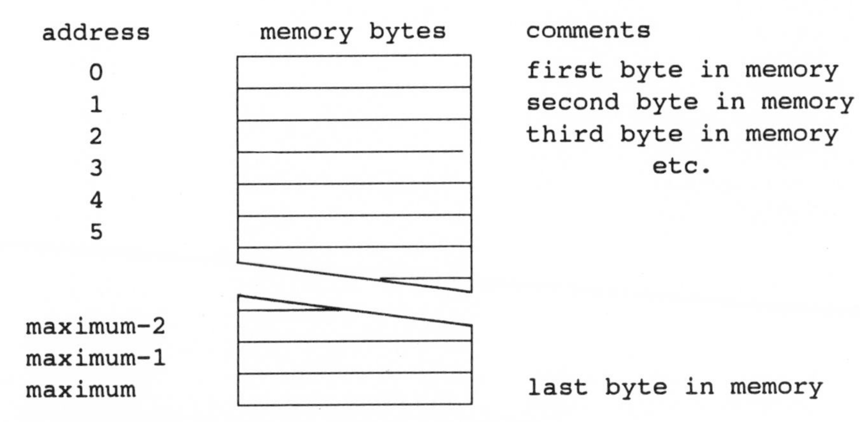

Fig. 2.2 The organisation of computer primary memory

The maximum amount of memory is limited by the number of bits used by the address bus to access memory locations. Table 2.2 lists the common microprocessors with their address and data bus sizes and the maximum amount of primary memory which can be addressed. For example:

1 The early microcomputers (e.g. Intel 8080, Zilog Z80, and Motorola 6800 series) have a 16-bit address bus which can address a maximum memory size of 65536 bytes or 64 Kbytes, i.e. 1111111111111111 in binary.
2 The Intel 8086 (used in the original IBM PC microcomputer) and Motorola MC68008 have a 20-bit address bus which can address a maximum memory size of 1048576 bytes or 1 Mbyte.
3 The Intel 80186/286 and Motorola MC68000/10 have a 24-bit address bus which can address a maximum memory size of 16777216 bytes or 16 Mbytes.
4 The Intel 80386/486 and Motorola MC68020/30/40 have a 32-bit address bus which can address a maximum memory size of 4294967296 bytes or 4 Gbytes.

Microprocessor manufacturer & type	address bus size in bits	maximum memory bytes	data bus size in bits
Intel 8080	16	64K	8
Zilog Z80	16	64K	8
Motorola 6800	16	64K	8
Intel 8088 (IBM/PC)	20	1M	8
Intel 8086 (IBM/PC XT)	20	1M	16
Motorola 68008	20	1M	8
Motorola 68000, 68010	24	16M	16
Intel 80186, 80286	24	16M	16
Motorola 68020/30/40	32	4G	32
Intel 80386, 80486	32	4G	32
Intel 80386SX	24	16M	16

Table 2.2 Common microprocessors with address and data bus sizes
Note: K = 1024 (2^{10}), M = 1048576 (2^{20}), G = 1073741824 (2^{30})

Table 2.2 shows the maximum amount of primary memory which can be addressed. In practice a computer system may be fitted with less, e.g. typically a MC68030 system has 16, 32 or 64 Mbytes. Although the primary memory is organised in bytes an instruction or data item may use several consecutive bytes of storage, e.g. using 2, 4 or 8 bytes to store 16-bit, 32-bit or 64-bit values respectively.

The size of the data bus determines the number of bits which can be transferred between system components in a single read or write operation. This has a major impact on overall system performance, i.e. a 32-bit value can be accessed with a single memory read operation on a 32-bit bus but requires two memory reads with a 16-bit bus. In practice the more powerful the processor the larger the data and address busses.

The size of the address and data busses has a major impact on the overall cost of a system, i.e. the larger the bus the more complex the interface circuits and the more 'wires' interconnecting system components. Table 2.2 shows that there are versions of some processors with a smaller data and addresses busses, e.g. the Intel 80386SX is (from a programmers viewpoint) internally identically to the 80386 but has a 20-bit address bus and a 16-bit data bus. These are used to build low cost systems which are able to run application programs written for the full processors (but with reduced performance).

2.1.1.1 RAM and ROM primary memory

The majority of computer systems contain two types of primary memory RAM and ROM:

RAM Random Access (read/write) Memory is used for storage of programs currently being executed and data being processed. Any byte may be accessed directly, i.e.random access, and read from or written to as required (when a location is written to, the previous contents are lost).

ROM Read Only (random access) Memory is used for the storage of permanent programs and data, e.g. the bootstrap loader which loads the monitor or operating system when the computer power is switched on (see Chapter 3.1). As the name implies information may be read but cannot be changed.

In addition to the primary memory there is other memory storage which is used for temporary information. These memory stores are called registers and may be found within the Control Unit, ALU and the I/O (Input/Output) device interfaces.

2.1.2 Secondary memory

The number of programs developed by a single user may be many hundreds and there must be some means for the long term storage of information. The secondary memory of a computer system (disks and magnetic tape) is used for this purpose. A floppy disk can vary in storage capacity from 320 Kbytes to 2 Mbytes and hard disks can store up to 2000 Mbytes or more. The typical access time can range from a few milliseconds for a fast hard disk, to as long as half a second for a floppy disk. This variation is due to the mechanical nature of the storage medium and the partially sequential access method (to get to a byte of information it may be necessary to read over intermediate information). The information on the disk is organised into named files and the system software provides functions for accessing these, e.g. open a file, read/write a file, close a file. Therefore even assembly language programmers rarely need to control disk I/O interfaces directly.

Exercise 2.1 (*see Appendix G for sample answer*)

1 What limits the maximum size of the primary memory of a computer system ?
2 What are RAM and ROM and what are they typically used for?
3 Why is secondary memory required in computer systems ?
4 Explain the terms sequential and random access and give examples.
5 What is the role of the bus system within a computer system ?

2.2 The CPU (Central Processing Unit) and co-processors

The CPU contains the control unit, ALU (Arithmetic Logic Unit) and associated high speed registers used for storing information during instruction processing. The processor of a microcomputer is an integrated circuit chip which contains the CPU and, in the case of a microcontroller chip, some primary memory, I/O device interfaces and other specialist facilities (Jelemensky et al 1989, Safavi 1990). Fig. 2.3 shows a (much) simplified representation of the MC68000 CPU.

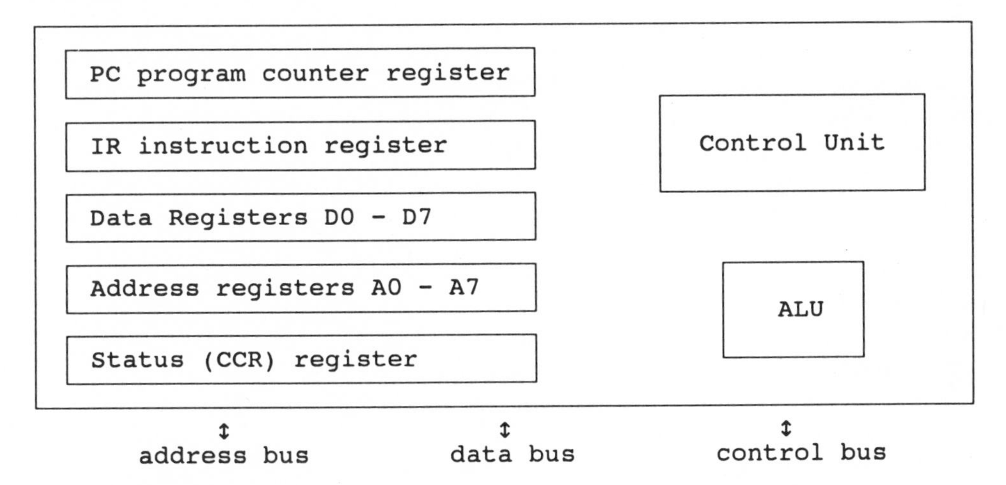

Fig. 2.3 CPU (Central Processing Unit)

2.2.1 The control unit

This component of the computer hardware has overall control of the computer system. During program execution the Control Unit fetches instructions from the primary memory, decodes them to determine the operation required, and then sets up instruction execution, e.g. to add two numbers or read a character from a keyboard. A number of registers are associated with the control unit, including:

PC or Program Counter. This register contains the address (in primary memory) of the next word of the current instruction or the first word of the next instruction. After a word has been fetched from primary memory into the instruction register the PC is automatically incremented to point at the next word.

IR or Instruction Register. After the instruction has been fetched from primary memory it is placed in the IR to be decoded.

2.2.1.1 The instruction cycle (fetch/execute cycle)

A program consists of a sequence of instructions in primary memory. Under the control of the Control Unit each instruction is processed in turn in a cyclic sequence called the fetch/execute or instruction cycle:

Fetch Cycle. A machine code instruction is fetched from primary memory (the PC points at each instruction in turn) and moved into the Instruction Register, where it is decoded (after the fetch the PC is incremented to point at the next instruction).

Execute Cycle. The instruction is executed, e.g. data is transferred from primary memory and processed by the ALU.

2.2.1.2 Instruction prefetch and pipelining

To speed up the overall operation of the CPU some microprocessors employ instruction prefetch or pipelining techniques which were first used in mainframe computers (Foster 1976). The MC68000, for example, uses a two-word (each 16-bits) prefetch mechanism comprising the IR (Instruction Register) and a one word prefetch queue. When execution of an instruction begins, the machine code operation word and the word following are fetched into the instruction register and one word prefetch queue respectively. In the case of a multi-word instruction, as each additional word of the instruction is used, a fetch is made to replace it. Thus while execution of an instruction is in progress the next instruction is in the prefetch queue and is immediately available for decoding.

In the MC68010 the prefetch queue was extended to two words (each 16-bits) in addition to the instruction register. This has particular advantages in the execution of program loops, where, if the loop instructions fit into the Instruction Register and two-word prefetch queue, a special processor mode is entered called loop mode. Instruction fetches are then suppressed (until the loop is finished) and only data is transferred between the processor and memory.

The more powerful members of the 68000 family make extensive use of pipelining techniques, e.g. the MC68040 (Edenfield et al 1990)

2.2.2 The Arithmetic/Logic Unit (ALU)

The ALU is the component of the computer system which, under the direction of the Control Unit, performs operations upon numeric and other data, e.g:

1 The integer arithmetic instructions add, subtract, multiply and divide (usually denoted by + - * / in high-level languages). Early 8-bit microprocessors could only carry out integer addition and subtraction and then only on 8-bit numbers directly. Multiplication and division and 16-bit operations had to be carried out using sequences of the available 8-bit instructions. The Motorola MC68000 can do integer addition and subtraction of 8-, 16- and 32-bit numbers, multiplication of 16-bit numbers, and divide a 32-bit number by a 16-bit number.

2 Logical instructions such as NOT, AND, OR and EOR (exclusive OR) and shift instructions.

Associated with the ALU and Control Unit of the MC68000 there are the following registers (in addition to the PC and IR - see Fig. 2.3):

Data Registers D0 to D7. These are 32-bit registers for the storage of temporary data during arithmetic and logical operations.

Address Registers A0 to A6. These registers are generally used to store addresses which point to data in memory, e.g. arrays or records.

Address Register A7. A7 serves as the Stack Pointer. The stack is a dynamic data structure used extensively by the processor during program execution.

Status/Condition Code Register. A common programming requirement is to branch to one of two possibilities depending upon the result of a calculation. For example the Pascal statement:

IF X=0 THEN Y:=20 ELSE Y:=30;

The SR (Status Register) contains a number of condition code bits which indicate the result of the last instruction (for example, if the result was zero or negative). These bits can be tested using branch instructions to control program flow (see Chapter 10).

2.2.2.1 8-bit, 16-bit and 32-bit microprocessors

The terms 8-bit, 16-bit and 32-bit, when used to refer to a microprocessor, give an indication of the power and facilities of the processor. In general, the number (8, 16 or 32) indicates the size of the data which can be processed directly by the ALU. For example, an 8-bit microprocessor can directly process 8-bit numbers with larger data types using a sequence of 8-bit instructions, e.g. using two 8-bit add instructions to add 16-bit numbers. Some extended microprocessors have instructions to process larger data (e.g. the Zilog Z80, an extended version of the Intel 8080 8-bit microprocessor, and has instructions to add and subtract 16-bit data).

2.2.3 Co-processors

Mathematical and scientific applications generally require mathematical calculations using real numbers (held in floating point form; see Appendix A). The ALU of the majority of microprocessors can only carry out calculations on integer data. Where real numbers are used there are two ways to carry out floating point calculations:

(a) by program subroutines which use the normal integer ALU,

or (b) by using a floating point co-processor chip which can be up to 100 times faster, e.g. the Motorola MC68881 (Huntsman 1983).

The modern microprocessor chips, e.g. MC68040, have a floating point co-processor on the same chip. In addition to floating point co-processors there may be other special purpose co-processors for graphics, signal processing, etc.

Exercise 2.2 *(see Appendix G for sample answer)*

1 Why is it necessary to have high speed registers within the CPU of a computer ?

2 What information does the status register of a CPU to contain and what is it used for ?

3 What applications would require a system with a floating point co-processor and why ?

2.3 Input and output

Input and output devices provide the computer user with the means to transfer information in and out of a computer system (for example, to enter a program and data, and then display the results of program execution). A typical microcomputer would have a keyboard (similar to that of a typewriter) for entry of information, and a display screen (similar to a TV set) for output information display. In addition to the display of character information it is often possible to draw diagrams on the display screen.

Many applications require information be fed directly into a computer system from the external world, e.g. readings of temperature and pressure in a washing machine controller. Parallel input/output devices such as the MC6821 PIA (Chapters 20 and 24) and MC68230 PIT (Chapters 21, 22 and 24) facilitate this.

2.3.1 I/O (input/output) interface registers

The interface circuit of an I/O device contains the circuits to control the peripheral device and status and control registers which, respectively, enable a program running in the CPU to:

(a) determine the state of the device, e.g. check if the keyboard has been hit;
and (b) control the device, e.g. to move the disk head.

2.4 Development system configurations

2.4.1 Self contained development systems

The majority of computer systems used for program development appear to the user as a self contained environment equipped with processor, primary and secondary memory, user keyboard and display. In practice, this may range from a stand alone IBM/PC compatible personal computer, through networked professional workstations to intelligent terminals attached to a mainframe computer. The user interacts with the system via an operating system which provides program development facilities, e.g. MS-DOS or UNIX (see Chapter 3.2).

The problem with using such systems for low-level program development is that the operating system environment often imposes restrictions on what user programs can do. Consider, for example, a multi-user environment where, if user programs were allowed to write data anywhere in memory, or access I/O device control registers, they could crash the whole system.

Even low-level self contained program development systems are designed with the intention that the majority of programming will be carried out using a high-level systems implementation language such as C, Modula 2 or Ada (a systems implementation language has many of the facilities of an assembly language, e.g. bit-manipulation, ability to access to I/O registers). Software systems are implemented mainly in the high-level language with the use of assembly language being restricted to specialist functions, e.g. time critical modules. The assembly language modules have to conform to restrictions imposed by the high-level language and operating system making assembly language programming *in its own right* very difficult.

2.4.2 Single board computer development systems

Single board development systems are essentially a printed circuit board with processor, primary memory and some I/O device interfaces, i.e. no secondary memory (Force 1984, Coats 1985/86). The on-board software is generally a simple monitor program (see Chapter 3.1) and no restrictions are placed on user programs. They have to be attached to a terminal and/or host computer which provides I/O facilities and secondary memory.

Fig 2.4 shows a host computer, which can vary in power from an IBM/PC compatible (Collard 1986) up to a professional workstation (Nelson & Leach 1984), attached to a single-board target system via a serial communications line (Chapter 19 will introduce serial and parallel communications techniques). A program running on the host enables a user to communicate with the target system, entering commands using the host's keyboard and displaying the results on the host's display screen. Suitable cross assemblers and compilers (see Chapter 3.3) enable programs to be developed on the host and downloaded onto the target for execution (Bramer 1990).

Fig. 2.4 also shows external experiments or devices (e.g. the Bytronic multi-application board; see Chapter 4.5) connected to the target single board computer via parallel communications lines.

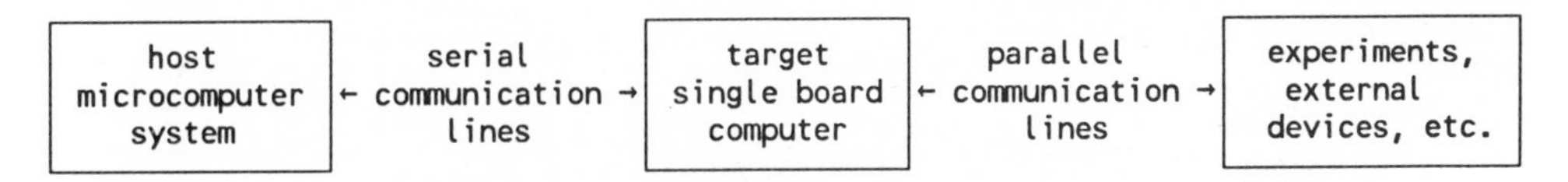

Fig. 2.4 A microcomputer acting as a host to a single board system.

Fig. 2.5 (next page) is a photograph of the Bytronic 68000 trainer. This is a typical single board 68000 based microcomputer which offers the following facilities (further details in Chapter 4):

(a) MC68000 or MC68010 CPU
(b) 64K or 128Kbytes EPROM
(c) 64K static and/or 1Mbyte of dynamic RAM
(d) MC68230 PIT (Peripheral Interface/Timer) interface chip
(e) MC68681 DUART (Dual Asynchronous Receiver/Transmitter) interface chip

Problem for Chapter 2

Examine the manuals for any microcomputer systems you have access to and in each case determine (if possible):

1 What type of microprocessor is used ?
2 Is it an 8-, 16- or 32-bit microprocessor ?
3 Does it have a co-processor or can one be fitted and if so what type ?
4 What is the size of ROM and RAM memory ?
5 What are the addresses of the ROM and RAM memory ?
6 What is the maximum amount of memory the microprocessor can address and why ?
7 What peripheral input/output devices is it equipped with ?
8 What is the secondary memory and how many bytes can it store ?
9 Does it use a proprietary or standard bus system, e.g. IBM PC/AT, VME, etc.

Fig. 2.5 Photograph of the Bytronic MC68000 single board trainer

3

Computer Software

Before a microcomputer can process information (i.e. carry out calculations or read a character from a keyboard) it requires a program. A program is a series of instructions stored in the primary memory that are executed sequentially by the processor. The programs of a computer system are called its software and include:

System software which provide aids to program development and operation of the computer system.

Application programs for solving end-user problems, e.g. word processors, spreadsheets, accounting programs, CAD design tools, etc.

3.1 Monitors and operating systems

3.1.1 System start-up or bootstrapping

When the computer is switched on it requires some instructions to initialise the hardware and start up the system software. In the past, these initial instructions had to be loaded as binary machine code into the primary memory using switches on a control panel. Today computer systems contain these initial instructions in ROM (Read Only Memory) and they are executed automatically when the computer is switched on or the reset button hit (reset is used to restart or reload the system software). This initial program in ROM may be quite complex, carrying out initial hardware tests, and then go on to provide:

(a) a **resident monitor**, which is a program that enables the user to examine and set memory contents, CPU registers, etc.;

and/or (b) a **Bootstrap Loader** which loads the system software from a disk.

The general name for ROM based resident software is **firmware**, i.e. software which is permanently fixed in ROM memory.

3.1.2 A system monitor

Programmers implementing low-level software, either in assembly language or a systems implementation language, require facilities to access to CPU registers, physical memory, I/O device registers, etc. There are a number of ways that such facilities may be provided depending upon the development system being used:

1 A **single board computer** with no disks: by a monitor program in ROM.

2 A **self contained computer system** with a disk based operating system: by a debugger program, e.g. the MS-DOS DEBUG program.

When a single board development system is switched on or reset, the resident monitor program in ROM carries out hardware tests and then prompts the user for command input. The user can enter a command to be executed by the monitor; typical facilities include:

1 Display and set the contents of the CPU registers.
2 Display and set the contents of RAM memory (values are usually displayed as hexadecimal numbers or ASCII characters and may be entered using decimal or hexadecimal numbers or ASCII characters).
3 Load instructions into memory using hexadecimal machine code.
4 Load instructions into memory using a line by line assembler (i.e. as each program statement is entered it is converted into machine code and loaded into memory).
5 Load instructions into memory from a host computer.
6 Start program execution, the user enters the start address of the program.
7 Set breakpoints within programs. The user defines breakpoints as memory addresses. If during program execution a breakpoint is reached:
    (a) program execution is suspended;
    (b) the microprocessor register contents are displayed;
  and (c) the user is prompted for a command.
  The user can then continue program execution or enter other commands.
8 Program Trace. After execution of each instruction (in the user's program) the microprocessor register contents are displayed.

The monitor provides a program development environment in which the user can load a program, set up initial values in the CPU registers and RAM memory, and then execute the program. After execution, the register contents can be displayed and the memory examined to check for correct results. Breakpoints and trace provide debugging aids to find program errors. A debugger provides similar facilities under a disk based operating system (high-level language programs may be traced line by line, variables displayed, etc).

3.1.3 Bootstrapping an operating system

Due to limitations the size of physical memory which may be fitted to a computer system ROM based software is restricted to providing facilities such as a monitor, a Bootstrap Loader or other permanent programs, e.g. a washing machine controller.

On disk based computer systems the Bootstrap Loader is a program contained in ROM which is executed when the system is switched on or reset. The Bootstrap Loader checks out the hardware and then loads the operating system from a known position on disk into primary memory. After loading the program the Bootstrap Loader transfers control to it and, after initialisation, the operating system prompts the user for command input.

A computer equipped with disks can provide a large range of system software with several languages. The operating system looks after the overall operation of the computer, the programs running in it and interaction with the user(s). The operating system will be provided on disk (secondary memory) and it must be loaded into the primary memory before it can be used. Operating systems are not normally in ROM because:

(a) Several operating systems can be supported on a single computer; the most suitable is selected depending upon the application area.
(b) Modern sophisticated operating systems require several Mbytes of memory for resident components alone.
(c) ROM based software is difficult to update, requiring an engineer to change memory chips. Disk based software is far easier to update.

The facilities provided by an operating system include:

1 Control of the disk file system, e.g. opening/closing/reading/writing, etc.
2 Editors for the creation and modification of programs and data.
3 Assemblers and compilers for programming languages.
4 A linker which links various program modules into a complete executable program.
5 Execution and debugging of systems, application and user programs.

In general, the more complex operating systems available on larger computers are used for development of programs in high-level languages. In particular, when executing programs on a multi-user machine, an assembly language program will be restricted in what it can do, e.g. not allowed to access input/output devices directly.

Exercise 3.1 *(see Appendix G for sample answer)*

1 Power up the target microcomputer displaying any systems tests.
2 After the monitor bootstraps use the help facility to display the commands available.

3.2 Software development facilities

3.2.1 Editors

An editor allows the user to:

(a) enter the program source code from a keyboard;
and (b) modify the source code to correct errors and update and extend the program.

Early editors tended to be line editors in which a single line of text was displayed and modified at a time. The majority of modern editors are of the full screen type which display a screen full of text, typically 25 lines, which is a window into the program source file. The user selects the window which shows the section of the file where the correction is to be made. A screen cursor is positioned, using keyboard keys or a mouse, at the exact place of the correction and the user can then add, change or delete characters as required. Some editors can invoke a compiler (or assembler) to compile the program source code and display an error message for each source line where an error was found (Bramer 1989b).

3.2.2 Assemblers and compilers

Assemblers convert assembly language programs (and compilers convert high-level language programs) into machine code (sometimes called object code). A program listing or error file is generated which shows any errors detected. If errors are found, these will have to be corrected and the process repeated.

Assemblers running on disk based machines can produce either absolute or relocatable code (compilers always generate relocatable code). In the case of absolute code the memory addresses that will be used to store the machine code and/or data are specified at assembly time. Relocatable code is generated from a base address of 0 and the linker then sets up the absolute addresses (see Linkers next section).

Microcomputers with ROM based assemblers generate absolute machine code directly into memory where it can be executed immediately (assuming that all modules are present - see Linkers next section). Appendix C describes the facilities of some 68000 assemblers.

3.2.3 Linkers

A complete program may be built up from a number of parts called modules which are formed using subroutines and functions (Chapter 9 introduces modular programming techniques). Each module may be in a separate file and assembled/compiled individually. When all modules are complete and error free, they are combined together using a linker to form the complete program. The linker goes through the program, assigning modules to memory, setting up links between modules, and checking for any missing or multiply defined modules. The output of the linker is a complete machine code program which can then be executed.

ROM based assemblers generate absolute machine code directly into memory (no link stage). By editing and assembling a sequence of modules a large program can be built up. In such a case care must be taken to ensure that absolute modules are separate in memory.

3.3 Cross assemblers and compilers

A native compiler or assembler produces code suitable for execution on the host computer or systems with a compatible processor. A cross assembler or cross compiler executing on one computer (the host) generates code suitable for another computer (the target) usually with a different processor. The resultant output object code is then linked (on the host) to other object code files such as:

1 Other program modules, e.g. C, Pascal or assembly language routines.
2 Libraries containing language support routines for the target, e.g. software floating point, mathematical functions, character manipulation, input/output, etc.
3 Libraries containing routines required by the target operating system or monitor, e.g. process switching and communication, interrupt handling, memory allocation, etc.

The host and target computers are usually connected via a simple asynchronous serial line with communication limited to the printable ASCII character set (e.g. as shown in Fig. 2.4). The output of the linker, which is in some binary format, must be converted into one of the standard formats used for the transfer of binary information over character oriented communications systems, e.g. Motorola S-record, Intel hex and Tektronix hex formats. A simple communications program on the host computer transmits the resultant character file to the target computer where it can then be executed (see Chapter 2.4).

Using cross software has the advantage that a single, possibly expensive, host system can be used to produce code for a range of target systems. The target only needs sufficient power and facilities to run the final application (normally much less than a full program development system) plus debugging aids (see reference Bramer 1990 for a full discussion on the importance of using modern computer tools and the advantages of using a host computer to develop common software for a variety of target systems).

In the main, the program listings in this book have been generated using a ROM based native assembler (see Appendix C). However, several examples of the use of cross assemblers will also be given.

3.4 Absolute and relocatable code

Absolute machine code programs contain instructions which refer to fixed addresses in memory where instructions and data are stored. In general, this means that the program must always be executed in the same place in memory.

Relocatable machine code contains no absolute references to particular memory locations. Instructions and data are referenced using addresses relative to the program instructions or to a base data address and thus the program can be executed anywhere in memory.

The advantage of relocatable code is that a complete program can be built up from modules which can be placed anywhere in memory, i.e. there is no need for a module to start at a particular address. When writing high-level language programs the programmer generally has no method of specifying absolute memory addresses, so the object code generated by the compiler is relocatable (there are exceptions, e.g. using pointers in C to access I/O device registers at absolute addresses in memory). Absolute addresses can usually be specified in assembly language programs thus allowing the generation of absolute code when required. Some microcomputer systems designed for program development in high-level languages do not allow absolute addresses even in assembly language programs and reference should be made to the microcomputer manuals to see if this is the case. Students attending formal courses of instruction will be given guidance on this point by the tutor.

3.5 Run-time system facilities

A modern monitor or operating system provides facilities, which can be accessed by programs, to carry out common functions, including:

1 Read a character or text string from the keyboard.
2 Write a character or text string to the display screen.
3 Open, close, read and write disk files on disk based machines.
4 Write a character or string to a printer.

In addition, a high-level language will have libraries of routines which may be accessed by application programs.

Problem for Chapter 3

For any microcomputers you have access to, determine the following:

1 What is the name of the monitor or operating system ?
2 What program debugging facilities does it provide ?
3 What other system software is available and what is its function ?
4 What applications software is available and what is its function ?

4

Introduction to the MC68000

4.1 Example MC68000 microcomputers

Low-level programming is generally used to implement software which accesses I/O device interfaces, directly manipulates information in memory, etc. In such applications an initial requirement is to obtain details of the target hardware configuration, i.e. how much ROM and RAM is fitted and what I/O devices are available. The example programs in this book were executed on three MC68000 based single board microcomputer systems:

Force 68000 board (Force 1984) equipped with 64Kbytes ROM, 128Kbytes RAM and:

(a) MC6821 PIA (Peripheral Interface Adaptor)
(b) MC6840 PTM (Programmable Timer Module)
(c) MC6850 ACIA (Asynchronous Communications Interface Adaptor).

Kaycomp 68000 board (Coats 1985/86) with 64Kbytes ROM, 64Kbytes of RAM and:

(a) MC68230 PIT (Peripheral Interface/Timer)
(b) MC68681 DUART (Dual Asynchronous Receiver/Transmitter)

Bytronic 68000 board equipped with 128Kbytes ROM, 1Mbyte of RAM and:

(a) MC68230 PIT (Peripheral Interface/Timer)
(b) MC68681 DUART (Dual Asynchronous Receiver/Transmitter)

The boards differ not only in terms of I/O devices but also in memory configuration, interrupt mechanism, etc. To test the sample programs in this book the boards were equipped with a system monitor called M68 which has a ROM based editor/assembler (implemented at Leicester Polytechnic to provide a consistent environment for a range of single board computers (see Appendix C).

4.2 The MC68000 data and address registers

4.2.1 The MC68000 data registers D0 to D7

The MC68000 CPU (Stritter & Gunter 1979, Motorola 1989) contains eight 32-bit general purpose Data Registers, D0 to D7, which are used to hold working variables used in arithmetic and logical calculations. Although the data registers are 32 bits in size they can be used for operations on 8-bit **byte**, 16-bit **word** or 32-bit **long word** data.

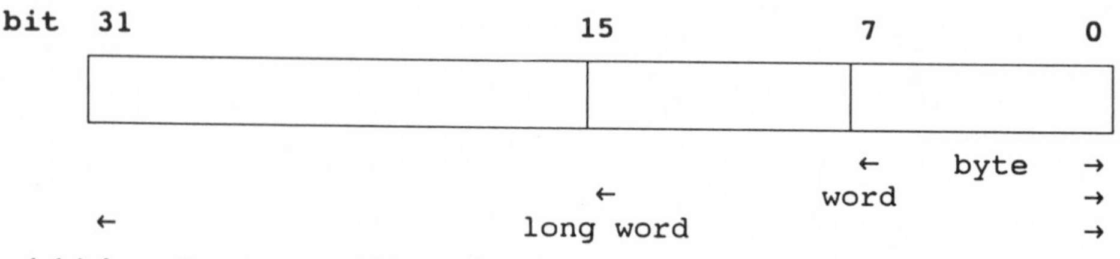

Fig. 4.1 Information storage within a MC68000 Data Register

Fig. 4.1 shows the placing of the different sizes of data within a data register. Note that operations upon:

Byte data (8-bit) will only affect bits 0 to 7, leaving bits 8 to 31 unchanged.
Word data (16-bit) will only affect bits 0 to 15, leaving bits 16 to 31 unchanged.
Long word data (32-bit) affects all the bits.

Exercise 4.1 *(see Appendix G for sample answer)*

Using the system monitor of your target MC68000 based microcomputer:

1 Clear the data registers, i.e. load 0 into them.
2 Load the following values into the specified data registers (you will need to look up how to define decimal, hexadecimal and character values):
    (a) 10 decimal into D0,
    (b) 10 hexadecimal into D1,
  and (c) the character code for 'A' into D2.
3 Display the contents of the data registers and comment on the results.

4.2.2 The address registers A0 to A7

A common programming requirement is to process lists or arrays of data. To facilitate these and similar operations the MC68000 contains eight 32-bit address registers, A0 to A7, which normally hold the addresses of data in memory (see Chapter 13 for full details). The address registers can be used to manipulate 16 or 32-bit data (not 8-bit):

1 memory addresses **must** be loaded and manipulated as 32-bit values;
2 the address registers can hold 16 or 32-bit data, e.g. 16-bit memory offsets.

For example, if a sequence of operations are to be carried out on the contents of an array, an address register is used as a *pointer* to the memory location of each element in turn:

1 the address (long word) of the first data element is loaded into an address register;
2 the operation is performed on the element using the address register as a pointer;
3 the address register is incremented by the length of a data element;
4 the operation is performed on the next data element.

Steps 3 and 4 are executed until all elements of the array have been processed.

4.3 MC68000 primary memory

The primary memory of the MC68000 is based on byte sized storage elements addressed by a 24-bit address bus which enables a memory addressing capability of 16777216 bytes or 16 Mbytes, i.e. addresses 0 to FFFFFF hexadecimal or 0 to 16777215 decimal. Although the memory is byte addressable it can be used to store 16-bit word or 32-bit long word data by using two or four bytes respectively. A restriction is that the memory address of word and long word data must be **even**, e.g. a word could start at memory addresses 1000, 1002, 1004, but not at 1001, 1003, 1005, etc. Memory content diagrams in this book will show memory organised into 16-bit words (two bytes of data).

address	contents		comments on memory contents
1000	42	40	byte values 42 in 1000 and 40 in 1001
1002	42	41	byte values 42 in 1002 and 41 in 1003
1004	42	42	byte values 42 in 1004 and 42 in 1005
1006	42	43	byte values 42 in 1006 and 43 in 1006

Fig. 4.2 Example of memory contents (values in hexadecimal)

Fig. 4.2 shows four words of memory (8 bytes) starting at address 1000 hexadecimal with the contents of each word shown as two bytes. The left-hand byte is associated with the address and the right-hand byte is the contents of the address+1. The memory contents of Fig. 4.2 could be interpreted as the byte value shown or as (values in hexadecimal):

```
16-bit word values             4240      4241      4242      4243
             at addresses      1000      1002      1004      1006
32-bit long word values        42404241  42414242  42424243
             at addresses      1000      1002      1004
eight ASCII characters         B@BABBBC  starting at address 1000
```

or a series of instructions in machine code; assembly language equivalent:

```
        CLR.W     D0        ;clear the lower 16 bits of D0
        CLR.W     D1        ;clear the lower 16 bits of D1
        CLR.W     D2        ;clear the lower 16 bits of D2
        CLR.W     D3        ;clear the lower 16 bits of D3
```

Thus the contents of a sequence of memory locations may be interpreted as byte, word or long word data (or instructions) and it is up to the programmer to know where data is and of what length. Word and long word values are stored with the most significant byte first followed by the other byte or bytes (i.e. the word value 4241 hexadecimal is stored starting at address 1002 with 42 in byte 1002 and 41 in byte 1003). The address used to access data in arithmetic and other operations is that of the most significant byte and this **must be even.** An attempt to access word or long word data at an odd address, e.g. a word value 4042 hexadecimal starting at address 1001, would cause an *address error* which would halt program execution and cause the monitor or operating system to display an error message.

Exercise 4.2 (*see Appendix G for sample answer*)

1 Using a suitable command fill part of the user RAM area of your computer with 0's, e.g. starting at address 1000 hexadecimal on the Force and Bytronic boards and 400400 on the Kaycomp board (if the addresses are not known see next section).

2 Use the monitor to load the following values into the user RAM memory area:
   (a) the byte values 01 and 23 hexadecimal;
   (b) the word value 4567 hexadecimal;
 and (c) the long word value 89ABCDEF hexadecimal.

3 Display the memory contents and check that the values are correct.

4 Using a command which operates upon word length data attempt to load the value FFFF hexadecimal into memory at an odd address and comment on what happens.

4.4 Memory and input/output device register maps

A broad specification of some single board microcomputers was presented in section 1 (of this chapter). Before attempting to implement low-level programs further information is required, e.g. how the I/O devices are interfaced, the addresses of ROM and RAM, the addresses of the monitor (or operating system) RAM areas and the addresses of the user RAM areas.

The I/O interface registers (status and control; see Chapter 2.3) appear as part of the primary memory of the MC68000 and can be accessed using instructions which access ROM and RAM memory (see Chapter 18.4). To program an I/O device the programmer must know the memory addresses assigned to its registers and the format of the data within them. Although the registers are mapped as part of the memory address space of the computer, they are not normal memory for program and data storage.

As part of the hardware specification of a low-level development system there will be a *memory map* giving ROM, RAM and I/O device addresses. Fig. 4.3 (next page) shows the memory maps for the Force, Kaycomp and Bytronic 68000 single board computer systems. There are a number of important points:

1 ROM and RAM may be of various sizes and appear at various places in the map, e.g. the Kaycomp can be equipped with 8K, 16K, 32K, 64K or 128Kbytes of ROM and 4K, 16K, 32K or 64Kbytes of RAM (64K is shown in Fig. 4.3).
2 The user RAM area for the Force and Bytronic boards is at address 1000 hexadecimal and for the Kaycomp board at address 400400.
3 The I/O device register addresses can be anywhere in the memory map not used by RAM and ROM.
4 The memory maps may in fact be more complex than Fig. 4.3 indicates, e.g. to simplify addressing the I/O device registers and memory areas are often replicated or mirrored on unused areas of the map (Coats 1985/6), i.e. the same physical memory location (or I/O register) may appear at a number of addresses in the memory map.

There are many microcomputer systems based on the MC68000 microprocessor. These will differ not only in the hardware configuration but in the monitor/operating system. If you are attending a formal course of instruction your tutor will provide information on how to use the microcomputer and any differences between the way it operates and the text of this book. Otherwise refer to Appendix C for a discussion on problems which may arise when using different MC68000 based microcomputers.

The majority of the sample program listings and runs presented in this book were generated using the M68 monitor and assembler (see Appendix C) on the Force, Kaycomp and Bytronic boards (M68 was implemented at Leicester Polytechnic to provide a consistent environment across a range of single board computers). However, the relevant programs executed without difficulty under the monitors provided with the boards, e.g. FORCEMON on the Force boards and KAYBUG on the Kaycomp.

It is worth noting that user programs executing under sophisticated disk based operating systems may not be allowed to access I/O registers and critical parts of memory (see Chapter 18.4 and Appendix C for comments on different microcomputer configurations) Such systems generally provide operating system functions to access special I/O facilities.

FORCE 68000 board

Memory	Contents	Start	End
128K RAM	exception vectors	000000	0003FF
	monitor work area and stack	000400	000FFF
	user work area	001000	01FFFF
64K ROM	system monitor	020000	027FFF
	editor/ assembler	028000	02FFFF
	unused	030000	04CF40
MC6840	PTM	04CF41	04CF4F
	unused	04CF50	05003F
MC6850	ACIA	050040	050043
	unused	050044	05CEF0
MC6821	PIA	05CEF1	05CEF7
	unused	05CEF8	FFFFFF

KAYCOMP board

Memory	Contents	Start	End
64K ROM	exception vectors	000000	0003FF
	system monitor	000400	007FFF
	editor/ assembler	008000	00FFFF
	unused	010000	3FFFFF
64K RAM	monitor work area and stack	400000	4003FF
	user work area	400400	40FFFF
	unused	410000	7FFFFF
MC68681	DUART	800000	80001F
	unused	800020	9FFFFF
MC68230	PIT	A00000	A0003F
	unused	A00040	FFFFFF

BYTRONIC board

Memory	Contents	Start	End
1M RAM	exception vectors	000000	0003FF
	monitor work area and stack	000400	000FFF
	user work area	001000	0FFFFF
64K ROM	system monitor	100000	107FFF
	editor/ assembler	108000	10FFFF
	spare	110000	11FFFF
MC68681	DUART	120000	12001F
	unused	120020	12FFFF
MC68230	PIT	130000	13003F
	unused	130040	FFFFFF

Note: memory sizes are not to scale

Fig. 4.3 Memory maps of the 68000 single board computers used in this book

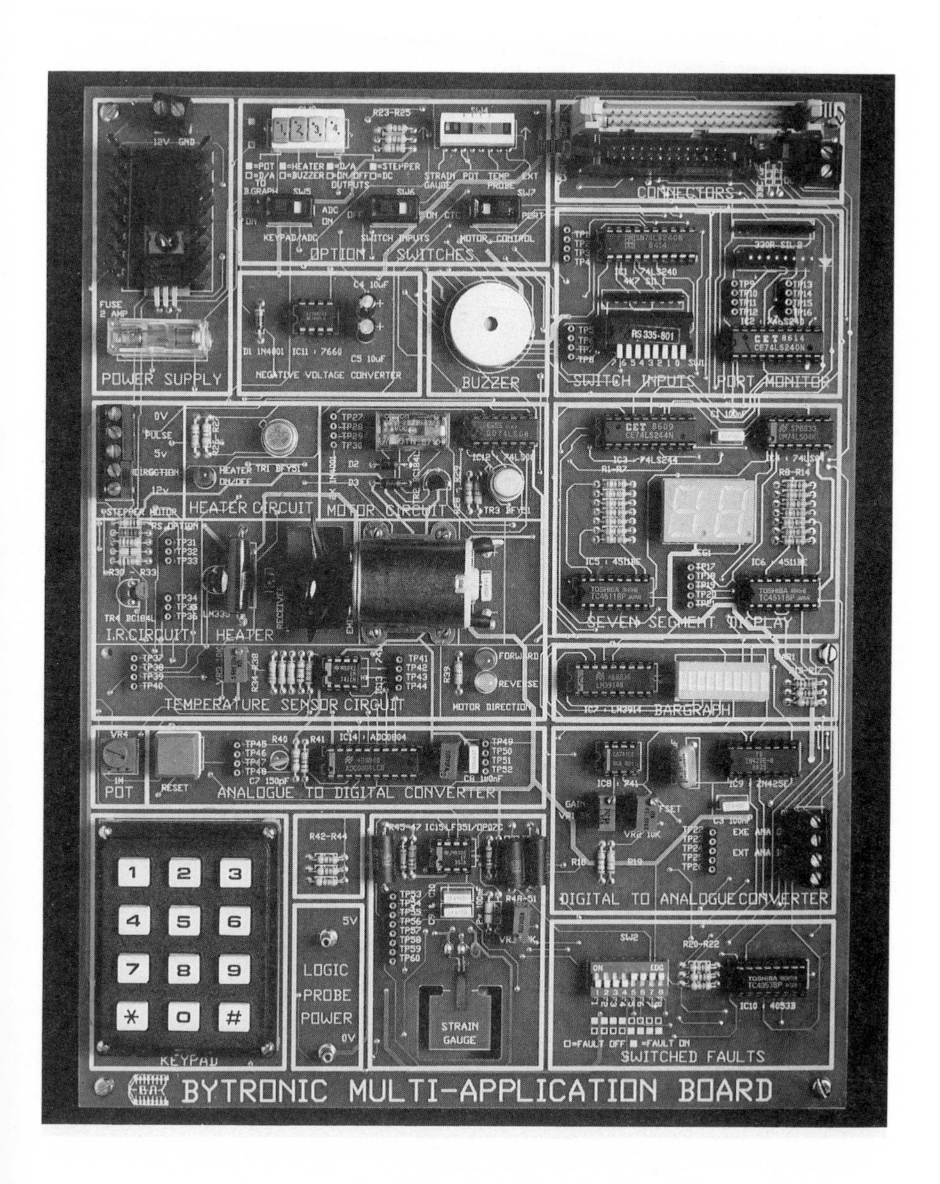

Fig. 4.4 Photograph of the Bytronic Multi-application board (Bytronic 1989)

4.5 Experiment boards for testing I/O interface programs

Programming I/O interfaces is not simple; in particular, modern general purpose interface chips (e.g. the MC68230 PIT introduced in Chapter 21) can contain twenty or thirty I/O registers, be used in a variety of different modes, and in general be very complex and difficult to operate. When faced with implementing software for a new interface device one approach is to program the device to do simple things, building up to more complex as experience is gained (this approach is taken in this book). When a thorough understanding of the device is attained it may then be linked to the application environment for testing (which may sustain damage if incorrectly operated).

For example, in the case of a parallel interface simple external circuits can be connected which give an immediate indication of program operation. Chapter 19.1 introduces parallel I/O programming techniques, Chapters 20 and 21 introduce the MC6821 PIA and MC68230 PIT respectively, and Appendix F describes some sample test circuits which can be used to test programs. Such test circuits can be built on an individual basis or a commercial experiment board purchased. There are a range of such experiment boards available containing a variety of facilities.

Fig. 4.4 (previous page) shows a photograph of the Bytronic multi-applications board (Bytronic 1989) which includes the following facilities:

1. eight LEDs (small lights) which display the state of the input signals to the board
2. eight switches which can be read by the computer
3. a buzzer for the generation of tones, e.g. as an alarm signal
4. an 8-bit analogue to digital (A to D) converter which can be read by the computer
5. a digital to analogue (D to A) converter which accepts an 8-bit value from the computer
6. a keypad (small keyboard)
7. a strain gauge; the output of which can be input to the A to D converter
8. a heater with temperature sensor; the output of which can be input to the A to D converter
9. a potentiometer; the output of which can be input to the A to D converter
10. a seven segment display (to display two digit numerical values)
11. a bargraph which can be driven from the D to A converter
12. a motor equipped with a fan (used to cool the heater); the direction of rotation is set under program control

The host computer can read the keypad, switch settings or the value from the A to D converter (attached to the potentiometer, strain gauge or temperature sensor) and control the LEDs, buzzer, heater, motor speed, the D to A converter and the seven segment display.

For example, a closed loop control experiment can be implemented to maintain the heater temperature at a particular value:

(a) by switching the heater on and off (to heat it up),
and (b) by changing the speed of the motor (using the fan to cool the heater).

The heater temperature is read using the temperature sensor via the A to D converter and the heater/motor switched on and off to set the required value (see Appendix F.6.6).

Although many of the programs in this book make use of the Bytronic multi-applications board the experiments are quite general and may be carried out using any similar devices or experimental facilities (Appendix F contains specifications and sample circuits for the experiments).

Exercise 4.3 *(see Appendix G for sample answer)*

Using the system monitor load the following machine code program into the user RAM memory area of your system, e.g. address 1000 hexadecimal on the Force or Bytronic boards or 400400 on the Kaycomp board (each value is a byte in hexadecimal):

```
20 3C 00 00 01 0B 04 40 00 17 D0 40 06 40 00 07 4E 4E
```

The program terminator in the above program is the instruction TRAP #14 (the two bytes 4E 4E hexadecimal) which returns control to the M68 monitor (used by the Force, Kaycomp & Bytronic boards). If the monitor or operating system of the target microcomputer uses some other instruction to terminate programs replace the last two bytes with the appropriate machine code, e.g.:

1 If RTS is the program terminator the last two bytes should be 4E 75, e.g.:

```
20 3C 00 00 01 0B 04 40 00 17 D0 40 06 40 00 07 4E 75
```

2 The KAYBUG monitor, which comes as standard with the Kaycomp, uses TRAP #11 followed by a word containing 0 as its program terminator, e.g.:

```
20 3C 00 00 01 0B 04 40 00 17 D0 40 06 40 00 07 4E 4B 00 00
```

Another technique to stop the program is to insert a breakpoint on the last instruction (e.g. address 1010 on the Force board) which will stop the program and display the register contents.

Start program execution (address 1000 on the Force board) and then check that the data register D0 contains the correct answer to the calculation RESULT=7+2*(267-23) (decimal values), which is equal to 495 decimal or 1EF hexadecimal.

If your system has a dissembler attempt to dissemble the program and compare the result with the listing below; comment on the results.

The above program in MC68000 assembly language is:

```
MOVE.L   #267,D0
SUB.W    #23,D0
ADD.W    D0,D0
ADD.W    #7,D0
TRAP     #14
```

You should practice using the monitor until you become proficient in its use.

5

Assembly Language Programming

The assembler syntax used in this book is that used by the majority of 68000 native assemblers. Major deviations are most likely to be found when cross assemblers are used. Cross assemblers are generally capable of generating code for a range of processors, often from different manufacturers, so the assembler syntax does not conform to any particular standard. In practice the operation-code mnemonics (instruction names) tend to be the same, with differences in operand specifications (in particular numeric and character data) and pseudo-operators (commands to the assembler). This book will indicate where variations may be expected and will give examples using the Real Time Systems XA8 (Real Time Systems 1986) and Whitesmiths AS68K (Whitesmiths 1986, 1987a, 1987b) cross assemblers (see Appendix C for an introduction to XA8 and AS68K).

5.1 Specification of numeric and other data

The majority of native MC68000 assemblers allow the specification of numeric data in (Appendix C contains details for the XA8 and AS68K cross assemblers):

Binary by prefixing the value by a %, e.g. %10101
Note: the value must contain only 0s and 1s

Hexadecimal by prefixing the value with a $, e.g. $1AF
Notes (a) The value must contain only digits and/or the letters A to F
(b) It is good practice to start with a digit, e.g. 0 thus $0FF

Decimal tends to be the default, i.e. digits with no prefix, e.g. 456

Character characters enclosed in ' marks, e.g. 'A'

Text strings characters in ' marks, e.g. 'Hello user'
Note. Each character is converted into the equivalent 8-bit binary ASCII character code to be stored in a byte (see Table A.2 of Appendix A).

Character	Decimal	Hexadecimal	Binary
'A'	65	$41	%01000001
'01'	12337	$3031	%0011000000110001
'*/+'	2764587	$2A2F2B	%001010100010111100101011

Table 5.1 Examples of data specification (68000 native assemblers)

Table 5.1 presents examples of character data, and decimal, hexadecimal and binary numeric data. The assembler converts the values into the equivalent binary patterns to be loaded into memory. The examples on each line result in the same binary value being generated, i.e. the character 'A' and the numeric values 65 and $41 are all equivalent to the binary value %01000001. When defining characters to be stored as word or long word data, the characters are left justified and padded with sufficient zeros on the right to make the data the correct size , i.e. the character 'A', when dealing with word values, would be taken as $4100, and with long word values as $41000000 (**note:** some assemblers may right justify and/or pad with space characters).

When writing assembly language programs use the most appropriate data type for the application, i.e. characters for text, binary numbers for bit patterns, hexadecimal numbers for memory addresses, and decimal numbers for mathematical calculations.

5.2 Assembly language mnemonics

Because of the requirement, even for trivial programs, to enter large amounts of numeric data in binary or hexadecimal, machine code programming is very difficult and error prone (see Exercise 4.3). In an assembly language the instructions or operations are represented by meaningful mnemonics.

5.2.1 Operation-code or instruction mnemonics

Operation-code (op-code) or instruction mnemonics are converted by the assembler into the equivalent machine code instruction and are generally assembler independent. Examples are:

CLR - to clear a value, i.e. clear or reset to 0
MOVE - to move (or copy) information, e.g. memory to registers
ADD - to add two data values
AND - to logically AND two data values
MULS - to multiply two signed data values
BSET - to set (to 1) a specified bit in a data value
ROL - to rotate the bit pattern of a data value left

Note that there may be restrictions as to upper or lower case, e.g. the Whitesmiths AS68K cross assembler will only accept lower case.

Many MC68000 instructions can operate upon byte, word or long word operands (the data to be operated upon) and the operation-code must be followed by the operand length:

.B to operate on byte (8-bit) data, e.g. CLR.B
.W to operate on word (16-bit) data, e.g. CLR.W
.L to operate on long word (32-bit) data, e.g. CLR.L

If the length definition is missing, **some** assemblers assume the operand length to be word (i.e. .W). However, it is good programming practice always to specify the length so that it is always clear from the program source code what data size being processed at any point.

5.2.2 Pseudo-operation mnemonics (assembler dependent)

Pseudo-operation mnemonics are not converted into machine code but are directives or commands to the assembler and, hence, will vary from assembler to assembler (to help in identifying possible problem areas pseudo-operators will be in **bold type** in program listings in this book). The following are examples of pseudo-operation used by native Motorola assemblers (see Appendix C for examples of XA8 and AS68K pseudo-operators):

Absolute program origin. Defines the absolute start address in memory into which the machine code is to be assembled, thus:

```
        ORG       $1000      program origin
```

Code or data following the statement is placed in memory starting at address 1000 hexadecimal. A program can contain many ORG directives to assemble code or data into different areas of memory. If no ORG statement is specified the assembler either generates relocatable object code from a base address of 0 or, in the case of a ROM based assembler, places the code at the start of the user RAM area.

Program End. This directive informs the assembler that there are no more lines of program source code.

```
        END                  end of program code
```

Define Constant. This directive is used to define storage for a variable and assign an initial value (see Chapter 8.1 for more details).

```
        DC.W      10         define a word, initial value 10
```

Exercise 5.1 *(see Appendix G for sample answer)*

What would you expect the following operation-code mnemonics to stand for:

SUB OR MULS MULU TST ROR NEG ASL BCLR JMP

5.3 Assembly language statement format

In general, assemblers use a syntax where a statement can contain up to four fields, e.g.:

```
START:  CLR.L     D0         comment field
```

where START: is the label, CLR.L is the operation-code or instruction, D0 is the operand (the data to be operated upon) and the rest is the comment field. Note that the majority of assemblers require each program statement to be on a separate line.

5.3.1 Label (or name) field

Assembly language programs require to access data in memory and branch or jump to other parts of the program depending upon control actions. In such cases it is possible to specify operand addresses (of data or instructions) directly in hexadecimal, e.g.:

```
        ORG       $1000
        ....
        ....
        CLR.L     D0         instruction at address $10FE
        ....
        ....
        JMP       $10FE      jump to instruction at $10FE
```

The JMP $10FE instruction loads the PC (Program Counter) with the address $10FE, so the next instruction to be executed is `CLR.L D0`. To write the program the programmer has to know the memory address where the CLR.L instruction is stored. A problem with specifying addresses in this way is that when the program is modified the addresses used can change, i.e. if some more instructions are inserted before the `CLR.L D0`. The way to overcome this problem, and make the code easier to understand, is to use labels to identify the address concerned. A label is a way of naming a particular address which can then be referenced by that name in other instructions, i.e. as an operand.

If a statement has a label, the assembler defines the value of the label as equivalent to the address of the first byte which contains the instruction, e.g.:

```
         ORG       $1000
         ...
         ...
LOOP:    CLR.L     D0
         ...
         ...
         JMP       LOOP      jump to instruction at label LOOP
```

By using labels the programmer does not have to know the addresses of instructions and data. The majority of assemblers read the program twice:

Pass 1 To form a symbol table of label names and their equivalent addresses.
Pass 2 To generate the machine code. When a label is referenced in an instruction its equivalent address is read from the symbol table and this is used in the machine code.

The name given to a label is selected by the programmer as meaningful in the program context. The exact rules for labels vary from assembler to assembler but the following are general guidelines for 68000 native assemblers (see the assembler manual for full details):

1 The name must start in the first column, i.e. no spaces before it, and be terminated by a colon (:).
2 It may contain up to eight characters, the first of which must be alphabetic and the others alphabetic, digits or characters such as _. The majority of assemblers allow more characters but ignore all but the first eight.
3 It is wise not to use the same name as an instruction mnemonic, register, or anything else the assembler uses, e.g. ADD, MOVE, ORG, END, D0, D7, A3, etc.
4 The name given to the label should be unique within the program, i.e. if the assembler (or linker) finds two or more identical labels it will generate an error message.

The majority of assemblers should accept the above without problems and some will allow extensions.

5.3.2 Operation-code (op-code) or mnemonic field

This field contains the mnemonic of:

(a) an operation-code or instruction, e.g. ADD.B, AND.W, CLR.L;
or (b) a pseudo-operation which is a directive to the assembler, e.g. ORG, END.

5.3.3 Operand or address field

This field contains one or more operands separated by commas, e.g. constants, registers or addresses of data to be operated upon by the instruction in the operation-code field.

5.3.4 Comment field

The remainder of the statement is the comment field which is ignored by the assembler. Some assemblers require the comment to start with a particular character (e.g. XA8 require a ; and AS68K an *). The assembler used in the examples in this book did not require this (the assembler was aware from the operation-code where the operation-code and operand fields terminated, and assumed that the remainder of the line was comment and ignored it).

Comments in the program are for documentation purposes and are ignored by the assembler. Documentation should enable other programmers to understand the purpose of the program and how it works. In addition, individual programs are soon forgotten and comments can remind the programmer about the function of sections of code when modifications are being carried out at a later date. This is particularly important in assembly language programs which the language syntax makes very difficult to read and understand. Assembly language programs in this book (apart from trivial programs) will contain a program design or description in Structured English as part of their documentation.

In the early sample programs in this book many of the comments are there to explain what the instruction itself is doing. In 'real' programs the comments should explain what the statement is doing within the overall context of the program.

5.3.5 Comment only statements

If a line starts with an * character (a ; in the case of XA8) the whole line is treated as a comment and ignored by the assembler.

5.3.6 Field separators

In general, the fields are separated by spaces and if the label field is blank it must be replaced by at least one space. To improve the appearance of the program it is wise to position the fields at particular column positions, e.g. each field taking 10 columns - many editors have a TAB key to facilitate this. Program 5.1a (next page) shows a program which calculates 7+2*(267-23) (expression evaluated left to right); in fact the same program as the machine code in Exercise 4.3. Although the statements are correct and the program will assemble without errors, the code is very difficult to read and understand. Program 5.1b shows the same program with a good layout, Program 5.1c a version for the Real Time Systems XA8 cross assembler (note the different pseudo-operators and the ; before comments) and Program 5.1d a version for the Whitesmiths AS68K cross assembler (note the difference pseudo-operators and the use of lower case characters).

Although the operation-code mnemonics have not been explained in detail yet, the sequence of instructions in the assembly language Program 5.1b is still relatively clear, i.e. MOVE to move (copy) a value, ADD to add, etc. The # prefixing a value means that the operand is a constant (otherwise it would be interpreted as a memory address). Thus an instruction mnemonic gives a good indication of the operation being performed. It is worth noting at this point that the Pascal equivalent would be:

RESULT:=7+2*(267-23);

```
-----------------------------------------------------------
* Program 5.1 - to calculate RESULT = 7+2*(267-23)
START: MOVE.L #267,D0   initialise D0 with 267
  SUBI.W          #23,D0  267-23
          ADD.W  D0,D0             2*(267-23)
     ADDI.W      #7,D0     RESULT = 7+2*(267-23)
       TRAP #14   !! STOP PROGRAM
 END
-----------------------------------------------------------
```

Program 5.1a An assembly language program in 'free format'

```
-----------------------------------------------------------
* Program 5.1 - to calculate RESULT = 7+2*(267-23)
START:     MOVE.L     #267,D0   initialise D0 with 267
           SUBI.W     #23,D0    267-23
           ADD.W      D0,D0     2*(267-23)
           ADDI.W     #7,D0     RESULT = 7+2*(267-23)
           TRAP       #14       !! STOP PROGRAM
           END
-----------------------------------------------------------
```

Program 5.1b An assembly language program with a good clear layout

```
-----------------------------------------------------------
; Program 5.1 - to calculate RESULT = 7+2*(267-23)
           .PROCESSOR M68000    ;specify processor
           .PSECT     _text     ;start code section
START:     MOVE.L     #267,D0   ;initialise D0 with 267
           SUBI.W     #23,D0    ;267-23
           ADD.W      D0,D0     ;2*(267-23)
           ADDI.W     #7,D0     ;RESULT = 7+2*(267-23)
           TRAP       #14       ;!! STOP PROGRAM
           .END
-----------------------------------------------------------
```

Program 5.1c A version for the RTS XA8 cross assembler

```
-----------------------------------------------------------
* program 5.1 - to calculate result = 7+2*(267-23)
           .text                *start code section
start:     move.l     #267,d0   *initialise d0 with 267
           subi.w     #23,d0    *267-23
           add.w      d0,d0     *2*(267-23)
           addi.w     #7,d0     *RESULT = 7+2*(267-23)
           trap       #14       *!! stop program
           .end
-----------------------------------------------------------
```

Program 5.1d A version for the Whitesmiths AS68K cross assembler

5.4 Generated machine (or object) code

The machine code generated by a program statement consists of:

(a) the operation-code which tells the processor what operation to perform,

and (b) the operand or address field which tells the processor where to find the data to be operated upon.

The labels used for operands are replaced by constants or addresses (absolute or relocatable) and the comments are ignored by the assembler. Therefore if a dissembler is used to convert the machine code back into its assembly language equivalent a great deal of the original program is missing (in fact the majority of what makes it understandable). Dissemblers are useful for checking code, i.e. if it is suspected that a program is overwriting itself.

Problem for Chapter 5

Enter Program 5.1b; modifications may have to be made to the program due to:

1. the * starting the comment line may need changing, e.g. to ; for XA8, see Program 5.1c
2. the assembler may need an ORG statement to specify the program origin in memory
3. the assembler may need a program section or segment pseudo-operator, see Program 5.1c and 5.1d (further discussion in Chapter 8.3)
4. the TRAP #14 instruction is used to stop the program under the M68 monitor and may need changing to suit other monitors, e.g. to RTS or some other TRAP instruction
5. the END pseudo-operator may need changing, e.g. .END for XA8 and .end for AS68K (see Programs 5.1c and 5.1d)
6. the AS68K assembler will require the code to be in lower case, see Program 5.1d
7. if the host system required assembly language modules to be implemented as components of a high-level language, e.g. C, the assembly language module will have to conform to any requirements imposed (e.g. calling convention, section names, data specifications, etc.)

Assemble the program, correct any errors and then execute it (if any difficult problems arise continue to Chapter 6, where assembler errors are covered, and then return to complete this problem). Students on formal courses will be given guidance on problems which may occur and on the use of the editor, assembler, etc.

Check that the result in D0 is correct, i.e. 1EF hexadecimal or 495 decimal, and that the machine code program is the same as that entered in Exercise 4.3 (this can be seen both from the assembler listing, using a dissembler or by an examination of the machine code in memory).

6

Introduction to MC68000 Instructions

This chapter introduces the addressing modes and instructions of the MC68000 processor (Motorola 1989). Although some machine code equivalents are presented, this is for information only. It is not expected that machine code would be remembered and used in practice.

6.1 Addressing modes and addressing categories

Chapter 5.4 explained that a machine code instruction consists of:

(a) the operation-code which tells the processor what operation to perform (Appendix B contains a list of these);

and (b) the operand or address field which tells the processor where to find the data to be operated upon.

In the case of the MC68000 the location of an operand (the data to be operated on) may be specified in one of three ways:

Register reference - the operand is a register (D0 to D7 or A0 to A7) specified as part of the operation-code of the instruction.

Implicit reference - the operand is implied by the instruction, i.e. no operand is specified.

Effective memory address - an *addressing mode* is specified which enables the processor to calculate where the operand is in memory (called the **effective address**). Calculating the effective address from the addressing mode may be quite simple, e.g. an address register contains a pointer to a memory location, or complex, e.g. an address in memory which is the sum of (a) an offset, (b) the contents of an address register and (c) the contents of a data register.

Addressing categories

Effective *addressing modes* can be **categorised** by the ways in which they are used, i.e:

Data References. The effective address refers to a data operand (this in fact includes everything except address registers).

Memory References. The effective address refers to an operand in memory, i.e. not in a register.

Alterable References. The effective address refers to an operand which is alterable, i.e. it can be written to.

Control References. An effective address mode which is used to refer to memory operands without associated sizes, e.g. in jump (JMP) or jump to subroutine (JSR) instructions.

The categories can be combined to define additional more restrictive classifications, e.g. a mode used to change information stored in memory would be categorised as a data alterable memory addressing mode. As an addressing mode is described it will be categorised and as each instruction is covered the addressing categories that can be used with it will be listed (see Appendix B for full details of MC68000 instructions and which addressing modes may be used).

6.2 The CLR (clear or zero) instruction

One of the first requirements in programming is to clear or reset the value of a working variable to zero. This may be done using the **CLR** (clear) instruction which has a single operand (called the destination operand). For example, to clear the byte in D0 the instruction would be:

```
        CLR.B     D0          clear bits 0 to 7 of D0 to 0
```

The machine code is:

42	00

Similarly, the word and long word CLR instructions would be:

```
        CLR.W     D1          clear bits 0 to 15 of D1 to 0
        CLR.W     D5          clear bits 0 to 15 of D5 to 0
        CLR.L     D3          clear all bits of D3 to 0
        CLR.L     D7          clear all bits of D7 to 0
```

Note that in the case of the CLR.B and CLR.W instructions the other bits of the data register are not affected. In each of the above cases the effective address of the destination operand is a data register and, accordingly, this addressing mode is called *data register direct*. The destination operand of the CLR instruction can, however, be specified using any *data alterable addressing mode*, e.g. to clear memory locations (see Appendix B for full details).

Unless a data register contains some useful information it is a good idea always to use the CLR.L instruction (or similar) to clear or initialise all the bits before it is used. This will save confusion when using the monitor to examine the data registers for byte and word results.

To simplify the following sections, word length instructions will mainly be used (i.e. .W). Unless otherwise specified, the instructions can also be carried out on byte (.B) and long word (.L) operands.

Data register direct addressing mode

When using *data register direct addressing mode* the operand is in a data register. This addressing mode can be used for data source operands (i.e. where the data is coming from) and data alterable destination operands (i.e. where the data is going to).

6.3 The MOVE instruction

It is often necessary to move data between registers or registers and memory. The **MOVE** instruction *copies* the value of the source operand into the destination operand **without** affecting the contents of the source. For example, the instruction to move the word value from data register D1 to data register D0 would be (bits 16 to 31 of D0 are not affected):

```
        MOVE.W      D1,D0        copy word from D1 to D0
```

The machine code is:

30	01

Byte and long word operands may be moved using the .B and .L operand lengths. In the case of the MOVE.B and MOVE.W instructions the higher (or more significant) bits in the destination register are not affected. In the above example both the source and destination operands are specified using *data register direct addressing mode.* The source operand for MOVE may be specified using *any addressing mode* and the destination operand by a *data alterable addressing mode* (see Appendix B for full details).

The MC68000 assembler expects (in the operand field of an instruction with two operands) the source operand, followed by a comma, then the destination operand. The specification of source followed by destination is the opposite of some other assemblers, e.g. the Intel 8086.

6.3.1 Using MOVE to move constants into data registers

Constants are values that do not change throughout the program execution. To move a constant into a data register the MOVE instruction is used with the constant as the source operand and the target data register as the destination operand, e.g.:

```
        MOVE.W      #10,D0       move 10 decimal into D0
```

The machine code is:

30	3C
00	0A

The symbol # tells the assembler that the value following is a constant (sometimes called a literal) and the MOVE instruction moves (copies) the value following into the destination operand. The machine code occupies two words, the operation-code or op-code ($303C), which is immediately followed by the operand ($000A, i.e. decimal 10). After the op-code $303C has been fetched and decoded, the operand $000A is read from the memory word immediately following the op-code and the PC is adjusted to point to the next instruction. This addressing mode is categorised as memory data and is called *immediate addressing*, i.e. the operand immediately follows the instruction in memory.

Immediate byte (.B) and long word (.L) operands can be moved using the same instruction format (with .B and .W the other bits of the destination data register are not affected). In all cases the constants must be able to fit within the operand length specified, for example, the following statements would be rejected by the assembler:

```
        MOVE.W      #123456,D0
        MOVE.W      #'ABC',D1
```

Both the value 123456 and the three characters 'ABC' require more than the 16 bits available for word length operands.

When dealing with word or long word values it should be remembered that characters are left justified and padded with zeros, e.g.:

```
MOVE.W    #'A',D1
MOVE.L    #'ABC',D2
```

In the first example $4100 is moved into D1 (i.e. zero into bits 0 to 7 and the character A into bits 8 to 15) and in the second $41424300 is moved into D2.

6.3.2 The move quick (MOVEQ) instruction

In many cases the value of a constant is small and it will fit into a single byte, i.e. in the range -128 to +127. For constants in this range there is a special **quick form** of MOVE using immediate addressing for the source operand and a data register for the destination operand. When executed the byte value is sign extended to long word size and all 32 bits are copied to the destination register, e.g.:

```
MOVEQ     #10,DO     move 10 into DO
```

The machine code is:

70	0A

In this case, the instruction occupies one word in memory (with a byte for the op-code $70 and a byte for the immediate operand $0A) and results in the value $0000000A being moved into D0. Consider:

```
MOVEQ     #$80,DO    move $80 into DO
```

Due to the sign extension to 32 bits, this would result in the value $FFFFFF80 being moved into D0, i.e. the sign bit (bit 7) of the source operand ($80 or 10000000 binary) has been copied into bits 8 to 31 of the destination operand. When using MOVEQ care must be taken to see that the operand will fit into a byte:

```
MOVE.B    #'A',DO    move character A into DO
MOVEQ     #'A',DO    move character A into DO
```

In the first example the operand length is defined as part of the op-code (.B) so the operand is taken correctly as the byte value $41 (ASCII character A). In the second example there is no length specified and, depending upon the assembler, the operand may be assumed to be of length word (the default) and taken as $4100 (character A padded with 0). The assembler then gives an error message saying that this word length operand will not fit into the byte length operand required for MOVEQ. A way to overcome this problem is to use EQU pseudo-operator to define byte length character operands for MOVEQ (see Chapter 7.2).

The advantage of the quick form of MOVE is that the machine code occupies only one word of memory as against two or three words for the standard word and long word move immediate, and therefore executes faster, i.e. fewer memory accesses are required during instruction execution.

Exercise 6.1 *(see Appendix G for sample answer)*

Program 6.1 (below) clears registers, moves constants and registers, etc. Note:

1 The CLR instructions are used to clear data registers which will be used for byte and word size data (there is no point clearing registers used for long word data).
2 The comments in Program 6.1 describe the instructions. **Comments in a normal program should say why the operation is being carried out.**

After making any modifications required for the computer/assembler configuration being used, type in Program 6.1 and assemble it (your tutor, or your microcomputer reference manual, will provide information on the use of the system editor, assembler, and linker, and how to execute programs - see Appendix C for a discussion on different systems and assemblers). Obtain an assembler listing (this will be discussed in the next section).

```
----------------------------------------------------------------
* Program 6.1 - clears registers and loads values
START:  CLR.L    D0          clear D0
        CLR.L    D1          clear D1
        CLR.L    D4          clear D4
        CLR.L    D5          clear D5
        CLR.L    D6          clear D6
        MOVE.W   #56,D0      move word 56 decimal to D0
        MOVE.W   D0,D1       move word from D0 to D1
        MOVE.L   #$12AB,D2   move long word 12AB hex to D2
        MOVE.L   D2,D3       move long word D2 to D3
        MOVE.W   #'AB',D4    move word ASCII codes A & B to D4
        MOVE.B   #'A'-1,D5   move byte ASCII code A - 1 to D5
        MOVE.W   #'A',D6     move word ASCII code A to D6
        MOVE.L   #'A',D7     move long word ASCII code A to D7
        TRAP     #14         !! STOP PROGRAM !!
        END                  end of program
----------------------------------------------------------------
```

Program 6.1 A simple program using CLR and MOVE

```
--------------------------------------------------------------------------------
 1                          * Program 6.1 - clears registers and loads values
 2 001000 4280              START:  CLR.L    D0          clear D0
 3 001002 4281                      CLR.L    D1          clear D1
 4 001004 4284                      CLR.L    D4          clear D4
 5 001006 4285                      CLR.L    D5          clear D5
 6 001008 4286                      CLR.L    D6          clear D6
 7 00100A 303C     0038             MOVE.W   #56,D0      move word 56 decimal to D0
 8 00100E 3200                      MOVE.W   D0,D1       move word from D0 to D1
 9 001010 243C 000012AB             MOVE.L   #$12AB,D2   move long word 12AB hex to D2
10 001016 2602                      MOVE.L   D2,D3       move long word D2 to D3
11 001018 383C     4142             MOVE.W   #'AB',D4    move word ASCII codes A & B to D4
12 00101C 1A3C     0040             MOVE.B   #'A'-1,D5   move byte ASCII code A - 1 to D5
13 001020 3C3C     4100             MOVE.W   #'A',D6     move word ASCII code A to D6
14 001024 2E3C 41000000             MOVE.L   #'A',D7     move long word ASCII code A to D7
15 00102A 4E4E                      TRAP     #14         !! STOP PROGRAM !!
16                                  END                  end of program
START      00001000 L
--------------------------------------------------------------------------------
```

Fig.6.1 Assembler listing of Program 6.1 using a ROM based native 68000 assembler

6.4 The assembler listing and error reports

Fig. 6.1 (last page) is an assembler listing of Program 6.1 (the precise format of the listing depends upon the assembler - Appendix C presents some more examples). The listing is divided into a number of fields:

Line number: the line number in the program source code.

Memory address: the address (in hexadecimal) where the first byte of the associated machine code will be placed. The address may be absolute (i.e. the actual memory address that will be used), or relocatable from address 0. In the latter case the linker will place the code at an absolute memory address.

Generated operation-code: a hexadecimal word showing the instruction op-code value.

Operands: if the source or destination operands are immediate values, offsets or addresses these are shown in hexadecimal (operands such as registers are specified within the op-code and hence do not appear here). If there is insufficient space for the operands, e.g. two long word values, they may continue onto the next line.

Source Program code: the remainder of the line contains the assembly language statement.

Consider the statement at line 13 of Fig. 6.1, `MOVE.W #'A',D6`, where the address is $1020, the op-code is $3C3C and the immediate source operand is $4100 (character A followed by a zero byte); there is no destination operand value (the D6 destination operand being specified within the op-code).

The M68 ROM based assembler used to generate Fig. 6.1 automatically starts assembling programs at the start of the user RAM area ($1000 for the Force and Bytronic boards, as shown, and $400400 for the Kaycomp board - see Fig. 4.3 for memory maps). This may be changed by using a program origin ORG statement. When a disk based assembler is used (see Appendix C) the program will start from a base address of 0 (unless an absolute origin is specified) and will be allocated actual addresses at link time.

Following the listing of the code a symbol table is displayed which lists labels and their equivalent values and other symbols, e.g. constants defined using the EQU pseudo-operator (see Chapter 7.2). Disk based assemblers will also indicate if the symbol was absolute or relocatable and what program section it is in.

The assembler listing is useful for finding errors and obtaining instruction addresses which can then be used when debugging the program, i.e. for displaying memory contents or using breakpoints. The sample programs in this book are generally in the form of assembler listings (with symbol tables if relevant) generated using the M68 ROM based assembler (targeted at the Force, Bytronic or Kaycomp boards). Listings generated by the XA8 and AS68K cross assemblers will also be presented.

6.4.1 Correction of assembly errors

Assembly languages are oriented towards processor architecture rather than solving end-user problems and as a result program implementation is difficult and prone to errors. At the simplest level errors in programs tend to be either semantic or syntactic:

syntactic: the program constructs or statements do not conform to the syntax (rules) of the programming language being used;

semantic: the program does not do what the specification states.

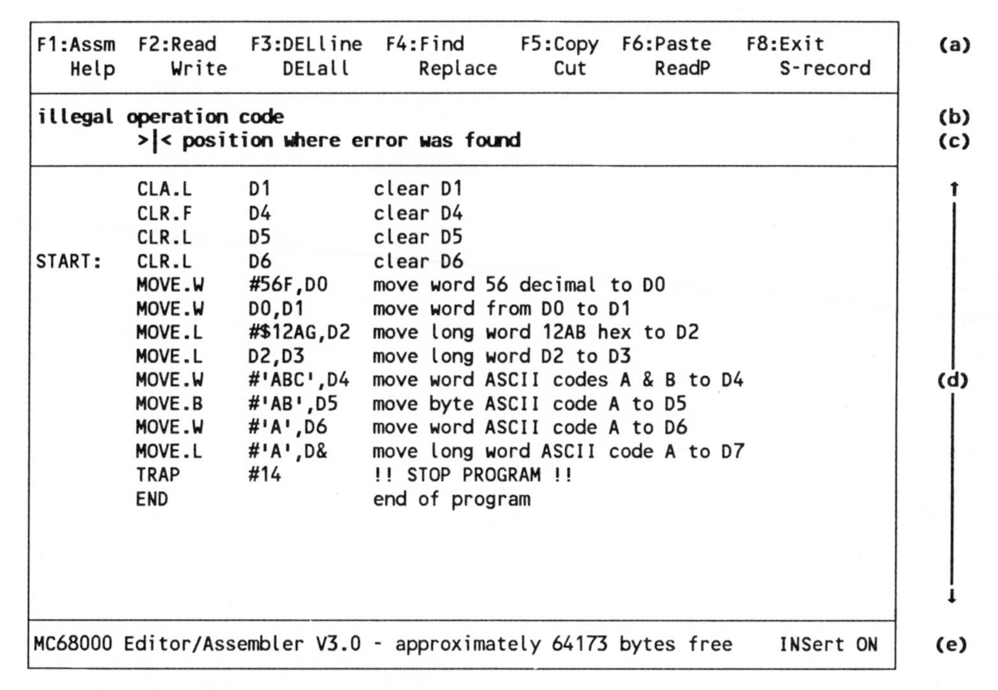

Fig. 6.2 M68 ROM based assembler error message (see text for discussion)

Syntax errors are found by the assembler and reported on the display screen, in the assembler listing file or in an error file. Fig. 6.2 shows a screen dump from the M68 ROM based editor/assembler in which an assembly error message is reported:

(a) menu of commands invoked by function keys, e.g. `F1: Assm` assemble program
(b) error found by the assembler, i.e. `illegal operation code`
(c) position on the line (following) where the error was found
(d) program code with the line where the error was found at the top, i.e. `CLA.L D1`
(e) status line indicating version of editor/assembler, free space, etc.

sample statement		typical assembler error message
CLA.L	DO	undefined op-code, i.e. CLA
CLR.F	D1	invalid operand size, i.e. .F
CLR.W	DO	undefined symbol (DO instead of D0)
MOVE.W	#56F,D0	illegal character (F in a decimal)
MOVE.L	#$12AG,D2	illegal character (G in a hex)
MOVE.W	#'ABC',D4	value too large ('ABC' needs 3 bytes)

Table 6.1 Sample statements and typical assembler error messages

The exact form of the error reports varies from assembler (Appendix C presents some more assembler listings with error messages and explanations of the causes). Table 6.1 presents sample program statements with errors and typical assembler error messages.

Common error messages associated with symbols are:

Redefined symbol - two or more labels of the same name have been found
Undefined symbol - a label specified in an operand could not be found

A common cause of the latter is mistyping DO for data register D0; the assembler would consider DO is a label and act accordingly.

Semantic errors can be very difficult to find. The use of modern Software Engineering techniques (Steward 1987) which aid in program specification and design should ensure that the design is correct before starting to write code. Before attempting to write assembly language programs one should be reasonably proficient in using such techniques and implementing systems in high-level languages.

High-level languages contain program and data constructs which aid in program implementation from a structured design. Assembly languages, on the other hand, have no high level program and data constructs and the processor oriented instructions and addressing modes make syntax errors common. Even after removing the syntax errors, and assuming that the design is correct, one can still get execution or run-time errors.

6.5 Debugging execution time errors

The assembler can only find errors in the syntax of the assembly language statements. Even if the assembler finds no errors the program may still contain run-time errors which cause it to fail or give faulty results when executed.

When coding a program the programmer works from a program design which was derived from a specification. It is possible that the resultant program code does not do what the design states it should (it is assumed that the specification and design were correct). For example, consider a program statement which moves an initial value into Data Register D0 (where it will then be used as a basis for further calculations):

```
        MOVE.L      #56,D0          move an initial value into D0
```

The programmer, however, when typing in the statement hits the 1 key instead of the 0, so:

```
        MOVE.L      #56,D1          move an initial value into D0
```

To the assembler the statement is valid and would assemble without error. At execution time, however, the result of any calculations which used the initial value in D0 would be incorrect (the value in D0 could be any 32-bit number). This error is fairly obvious and not difficult to find. If the cause of a run-time error is not obvious from looking at program source code the program has to be debugged at run-time:

Generating more test data where, once an error is found, more test data is constructed to provide further information about program flow, data values, etc. This can be use to determine if the error effects particular test values or is more general.

Brute force approach in which write statements are placed at strategic points in the program to display the values of intermediate variables and narrow down the cause of errors. Low-level system monitors provide facilities which enable programs to write information to the user screen (see Chapter 9.4).

Using debugging tools such as breakpoints or program trace:

Using breakpoints. Using addresses from the assembler listing breakpoints are placed at strategic points in the code. When a breakpoint is reached the program is halted and the monitor displays the register contents. From examination of intermediate results (register and memory contents) it can be determined whether processing up to that point is correct:

if not an error has occurred and register/memory contents may give indication of where it occurred (otherwise work backwards);

if OK continue to the next breakpoint.

Program Trace is similar to breakpoints but as each instruction is executed the register contents are displayed. Monitors usually allow the programmer to set the number of instructions to trace, then the program is halted. A combination of breakpoints and trace can track down many errors, i.e. setting a breakpoint at a point before an error occurs and then continuing with trace to find it.

Unless used with care the above approaches can be very hit and miss, producing vast amounts of irrelevant output which has to be searched through. In addition, such techniques can introduce further errors, e.g. when debugging time critical code or interrupt driven I/O systems (see Chapter 18.3) where debugging upsets the timing of the program. In practice the above techniques have to be used with more formal debugging techniques:

Debugging by Induction in which 'clues' to the cause the error are extracted from test data. Such clues should provide information about the program flow, modification of data, etc., providing a basis to derive a hypothesis for the cause of the error. The program is modified to test the hypothesis, which if correct, fixes the error.

Debugging by Deduction in which all conceivable causes of the error are listed. Test data is analysed to eliminated causes until one remains. The program is modified to test the hypothesis, which if correct, fixes the error.

In practice a combination of techniques is used. Proficiency in program implementation and testing is achieved by experience and there is no short-cut or magic formula.

Problem for Chapter 6

1. Correct any errors in your version of the Exercise 6.1 program and assemble it.
2. If required, link the program and generate the executable code.
3. If required, download the program into the target 68000 system.
4. Execute the program.
5. Display the contents of the data registers.
6. Are they what you would expect ?
7. Could the program be simplified, e.g. by using MOVEQ instructions ?
8. Use the monitor TRACE facility to trace the program execution.
9. Put some breakpoints into the program and rerun it.
10. Introduce some syntax and run-time errors into your program (or working with a partner introduce errors into each other's programs). Track them down, i.e. by using assembly time error messages and run-time debugging facilities.

Review of Chapters 4, 5 and 6

Chapter 4 presented an overview of the basic architecture of the MC68000 processor and Chapters 5 and 6 introduced assembly language programming.

4.1 Specifications of the 68000 single board computers used in this book were outlined.

4.2 The eight Data Registers D0 to D7 can hold byte (8-bit), word (16-bit) and long word (32-bit) data. Byte and word operations do not affect the other bits in the register.

The eight Address Registers A0 to A7 can hold word (16-bit) and long word (32-bit) addresses and related data (byte operations are not allowed).

4.3 MC68000 primary memory is based on byte (8-bit) storage elements addressed by a 24-bit address bus allowing an address range of 0 to FFFFFF hexadecimal (0 to 16777215 decimal) or 16 Mbytes. Word (16-bit) and long word (32-bit) information can be stored using two or four byte storage elements (which **must** start on an even word boundary) with the address of the first or most significant byte being used in operand references.

4.4 Introduced memory and I/O register maps and presented (Fig. 4.3) memory maps of the Force, Kaycomp and Bytronic single board computers. I/O registers appear as part of the primary memory and can be accessed using normal memory manipulation instructions.

4.5 Described in outline the facilities of the Bytronic multi-application board.

5.1 Numeric data may be specified in binary (prefix %), hexadecimal (prefix $) or decimal (no prefix). Character information is enclosed in quote marks '. Character strings when stored in words or long words are left justified and padded with zeros.

5.2 In assembly language the machine code instructions are represented by meaningful mnemonics (e.g. CLR to clear or zero a value). The mnemonics can be of two types:

(a) op-code mnemonics which are converted into the equivalent machine code instruction;
(b) pseudo-operation mnemonics which are directives to the assembler, e.g. ORG.

When an MC68000 instruction can operate upon byte, word or long word data the op-code must be followed by an operand length specification (.B byte, .W word, .L long word).

5.3 Assembly language statements are composed of up to four fields:

```
START:    CLR.L      D0           comment field
```

Label field: is used to label or name the instruction. The label is equivalent to the address of the first byte used to store the instruction.
Operation-code field: contains the operation or pseudo-operation mnemonic.
Operand field: contains the operands (where the information to be operated upon by the instruction is to be found).
Comment field: contains comments for documentation purposes (if a line starts with an * the whole line is treated as a comment).

5.4 The machine code generated by a program statement consists of:

(a) the operation-code which tells the processor what operation to perform,
and (b) the operand or address field which tells the processor where to find the data to be operated upon.

6.1 The location of an operand (the data to be operated on) may be specified by:

Register reference - the operand is a register (D0 to D7 or A0 to A7)

Implicit reference - the operand is implied by the instruction.

Effective memory address - an *addressing mode* is specified which enables the processor to calculate where the operand is in memory (called the **effective address**).

Effective *addressing modes* can be categorised by the ways in which they are used, i.e: to reference data, memory, alterable and control.

6.2 The **CLR** instruction is used to clear or reset (zero) a destination operand specified using a *data alterable addressing mode*, e.g.:

```
CLR.W     D0        clear word in D0
CLR.B     D1        clear byte in D1
```

Byte and word operations on data registers do not affect the other bits.

When using *data register direct addressing mode* the operand is in a data register. This mode can be used for data source operands and data alterable destination operands.

6.3 The **MOVE** instruction copies the source operand (specified using *any addressing mode*) into a destination operand (specified using a *data alterable addressing mode*). The source operand is specified, followed by a comma and the destination operand, e.g.:

```
MOVE.W    D0,D1     move word from D0 to D1
MOVE.L    D2,D3     move long word from D2 to D3
```

If the source operand is a constant the *immediate addressing mode* can be used (the source operand is stored immediately following the operation-code in memory), e.g.:

```
MOVE.W    #10,D0    move 10 decimal into D0
MOVE.L    #'ABC',D1 move characters ABC into D1
```

The second example moves the value $41424300 into D1 (the ASCII character codes for ABC padded by a zero byte).

The quick form may be used with byte sized source operands and a destination operand in a data register (the immediate operand is sign extended to 32 bits), e.g.:

```
MOVEQ     #10,D0    move 10 decimal into D0
```

6.4 The assembler listing displays the source code with the generated machine code and associated addresses. In addition, errors in the syntax of the language are reported (e.g. incorrect operation-code mnemonics or operand specifications, missing labels, etc.).

6.5 Even if the program assembles correctly the program can still contain run-time errors which cause the program to give incorrect results. The program then has to be debugged at run-time using monitor facilities such as breakpoints, trace, etc.

7

MC68000 Arithmetic Instructions

7.1 Binary arithmetic instructions

7.1.1 Binary addition (ADD) and subtraction (SUB) instructions

There are a number of binary addition and subtraction instructions using various addressing modes which operate on byte, word and long word operands. For example, the **ADD** addition and **SUB** subtraction instructions allow the following operand combinations:

Source in a data register and a destination specified using *data register direct* or *any memory alterable addressing mode*.

or Source specified using *any addressing mode* and the **destination in a data register**.

```
ADD.W    D0,D1       add word in D0 to contents of D1
SUB.L    D2,D3       subtract long word D2 from D3
ADD.W    #10,D0      add the constant 10 to word in D0
SUB.L    D3,LABEL    subtract D3 from long word in memory
ADD.B    LABEL,D1    add contents of byte in memory to D1
```

Depending upon the result bits are cleared or set in the CCR (the Condition Code Register which will be covered in detail in Chapter 10.2), i.e. N if negative, Z if zero, V if arithmetic overflow (result too large to represent in operand size) and C if carry (see Appendix B for details of which CCR bits are affected by particular instructions). Alternative program paths dependent upon the result of the last calculation may then be taken by using conditional branch instructions (see Chapter 10.2) which test the CCR bits.

If the source operand is immediate (i.e. a constant), **ADDI** and **SUBI** can be used with the destination operand specified using any data *alterable addressing mode* (not just a data register as in the case of ADD and SUB):

```
ADDI.W    #50,D0        add the word constant 50 to D0
SUBI.L    #$1F,D3       subtract long word $1F from D3
ADDI.W    #50,LABEL     add word to LABEL in memory
```

In Program 7.1 (which is identical to Program 5.1) SUBI, ADD and ADDI are used in a simple arithmetic calculation. Note the use of add in line 4 to perform a multiply by two.

```
---------------------------------------------------------------------------
1                          * Program 7.1 - to calculate RESULT = 7+2*(267-23)
2 001000 203C 0000010B START:    MOVE.L     #267,D0    initialise D0 with 267
3 001006 0440      0017          SUBI.W     #23,D0     267-23
4 00100A D040                    ADD.W      D0,D0      2*(267-23)
5 00100C 0640      0007          ADDI.W     #7,D0      RESULT = 7+2*(267-23)
6 001010 4E4E                    TRAP       #14        !! STOP PROGRAM
7                                END
---------------------------------------------------------------------------
```

Program 7.1 Sample Program (identical to Program 5.1b)
Program start address is $1000 for the Force/Bytronic board (see memory map)

For immediate source operands in the range **1 to 8**, the quick form of add and subtract **ADDQ** and **SUBQ**, can be used to save memory and speed up processing:

```
ADDQ.W     #2,D0          add word 2 to D0
SUBQ.L     #6,D1          subtract long word 6 from D1
ADDQ.W     #5,LABEL       add word 5 to memory
```

Although the size of the constant is more limited than MOVEQ (1 to 8 rather than -128 to 127), the operand size may be specified as byte, word or long word and the destination operand may be specified using *any alterable addressing mode.* Program 7.2 (a modified version of Program 7.1) shows the use of ADDQ (note in line 5 the op-code is $5E40 while it was $06400007 in line 5 of Program 7.1 making the program one word shorter).

```
-------------------------------------------------------------------------------
1                             * Program 7.2 - to calculate RESULT = 7+2*(267-23)
2 400400 203C 0000010B START:      MOVE.L      #267,D0     initialise D0 with 267
3 400406 0440      0017            SUBI.W      #23,D0      267-23
4 40040A D040                      ADD.W       D0,D0       2*(267-23)
5 40040C 5E40                      ADDQ.W      #7,D0       RESULT = 7+2*(267-23)
6 40040E 4E4E                      TRAP        #14         !! STOP PROGRAM
7                                  END
-------------------------------------------------------------------------------
```

Program 7.2 Modification of Program 7.1 to use ADDQ (in line 3)
Program start address is $400400 for the KAYCOMP board (see memory map)

The following instructions add 5 to the word contents of D0:

```
ADD.W      #5,D0          add the word constant 5 to D0
ADDI.W     #5,D0          add the word constant 5 to D0
ADDQ.W     #5,D0          add the word constant 5 to D0
```

ADD, ADDI and ADDQ can be used to add an immediate operand to a data register. If the immediate or quick immediate form of an instruction exists it is good practice to use it. Many assemblers given ADD and SUB and will generate the most efficient version. It is recommended practice, however, for the programmer to use the correct mnemonic so that when reading the program it is always clear which operands are used.

7.1.2 The exchange, negate and sign extend instructions

The **EXG** instruction exchanges the contents of one 32-bit register with those of another. The source and destination registers can be either data or address registers, e.g. (the length is not specified; all 32 bits are exchanged):

```
EXG        D0,D1      exchange D0 and D1
EXG        A1,D3      exchange A1 and D3
```

The **NEG** instruction negates an operand (i.e. changes its sign) by subtracting the destination operand, specified using any *data alterable addressing mode*, from 0 using twos complement binary arithmetic. When using byte (.B) and word (.W) operands in a data register the other bits are not affected. For example:

```
NEG.W      D0         negate word value in D0
NEG.L      LABEL      negate long word in memory
```

The sign extend (**EXT**) instruction extends the sign bit of a **data register** from a byte to a word (i.e. bit 7 is copied to bits 8 to 15) or from a word to a long word (i.e. bit 15 is copied to bits 16 to 31). To sign extend a byte to a long word use two instructions, e.g.:

```
EXT.W     D0        extend byte sign to word in D0
EXT.L     D0        extend word sign to long word in D0
```

Exercise 7.1 (*see Appendix G for sample answer*)

Design, code and test a program to calculate (using word length operands):

```
RESULT = 2*(355+4*(89*2-7)-13*8)
```

The RESULT (1870 decimal or $74E) should be in a data register, e.g. D0. Use repeated addition to carry out the multiply instructions (or read the next section to use multiply).

7.1.3 Binary multiply instructions (MULS and MULU)

The majority of the instructions covered so far could operate upon byte, word and long word operands. The multiply instructions of the MC68000 are limited to multiplying a pair of word (16-bit) operands to form a long word (32-bit) result.

There are two multiply instructions, namely **MULS**, for signed numbers, and **MULU**, for unsigned numbers. In both cases, a word (16-bit) source operand, specified by a *data addressing mode*, is multiplied by a word (16-bit) destination operand, which **must** be in a data register. The long word (32-bit) result is placed back into the destination data register (CCR bits N and Z are set according to the result V and C are cleared):

```
MULS.W    #30,D0      D0 = D0*30
MULS.W    VALUE,D1    D1 = D1*VALUE (VALUE is in memory)
MULU.W    #$4F,D5     unsigned multiply D5=D5*$4F
```

7.1.4 Binary divide instructions (DIVS and DIVU)

Binary divide instructions divide a word (16-bit) source operand, specified by a *data addressing mode*, into a long word (32-bit) destination operand which **must** be in a data register. A long word (32-bit) result (quotient and remainder) is left in the data register. There are two divide instructions, **DIVS** and **DIVU** for signed unsigned numbers respectively.

```
DIVS.W    #10,D0      divide long word in D0 by 10
DIVU.W    VALUE,D1    divide D1 by contents of VALUE
```

Note that division by 0 causes a TRAP (see Chapter 23.2), CCR bits N and Z are set according to the result, V is set if division overflow occurs and C is cleared. The 32-bit result in the destination data register consists of:

(a) the quotient in the lower 16 bits (bits 0 to 15);
and (b) the remainder in the upper 16 bits (bits 16 to 31).

If, in the first example above, D0 contained the long word value 23 decimal, the result of D0/10 would be $00030002 in D0 (quotient 2 and remainder 3). The remainder can be accessed using **SWAP** (which swaps the lower and upper 16 bits of a data register):

```
SWAP      D0          swap lower and upper 16 bits of D0
```

7.2 Specifying constants and using expressions

Constants are used extensively within programs to specify non varying data, including:

(a) characters and text for screen display,
(b) numeric values for use within calculations,
and (c) addresses of I/O (input/output) registers.

In practice the majority of constants are used on a *one off* basis (e.g. to initialise a value to 56, add 5 to a value, multiply a value by 10, etc.), being of importance only at a particular point in a program. In such cases the numeric/character constant should be specified in the instruction as an immediate operand. In other cases a constant may have a wider content:

1 a constant has significance throughout the program and may be used in many places;
2 a program or module may by used in a range of applications where the constants, although fixed for each application, will vary from application to application.

Consider a program designed to run on a range of 68000 systems which have a MC68230 PIT interface. Depending upon the exact hardware configuration the memory addresses assigned to the MC68230 PIT I/O registers will vary. When implementing the first version of the program the I/O register addresses could be specified as numeric constants in the operands of every instruction where they are used. The problem then arises when the program is to be ported to another system, and, every occurrence of an I/O register addresses in the program has to be modified. Not only is this a cumbersome task but it is always possible to make errors either by missing some of the modifications or typing the new values incorrectly. Using the equate (EQU) pseudo-operator overcomes this problem.

```
--------------------------------------------------------------------------------
1        00000002     CONST_1: EQU     2          define CONST_1 = 2
2        000000FF     DATA1:   EQU     $FF        define DATA1 = $FF
3        00000015     DATA2:   EQU     %10101     define binary bit pattern
4        00004142     CHARS:   EQU     'AB'       define CHARS = 'AB'
5        00400000     PIT_1:   EQU     $400000    define first I/O register
6        00000041     CHARB:   EQU.B   'A'        define character A byte length
7        00004100     CHARW:   EQU.W   'A'        define character A word length
8        41000000     CHARL:   EQU.L   'A'        define character A long word
9                              END

CONST_1   00000002 E   DATA1    000000FF E   DATA2    00000015 E
CHARS     00004142 E   PIT_1    00400000 E   CHARB    00000041 E
CHARW     00004100 E   CHARL    41000000 E
--------------------------------------------------------------------------------
```

Fig. 7.1 Examples of the use of the EQU pseudo-operator

Fig. 7.1 shows how the equate pseudo-operator **EQU** is used to equate a name to a value (in a similar way to a label name being equivalent to an address). The names and equated values are stored in the symbol table along with labels and their equivalent addresses. When the name is used in the program its equivalent value will then be substituted in a similar way to the situation when a label is used and its equivalent address is substituted.

The equate pseudo-operator can have a length appended, e.g. in Fig. 7.1 the definitions of CHARB (byte value $41, the character code for A) CHARW (word value $4100, the character code for A followed by a 0 byte) and CHARL (long word value $41000000, the character code for A followed by three zero bytes).

Using EQU can overcome the problem of some assemblers defaulting to word length when specifying single character immediate operands to MOVEQ (see Chapter 6.3):

```
CHARA:    EQU.B     'A'           define a byte size character
          MOVEQ     #CHARA,D0     move character A into D0
```

CHARA is defined as length byte and this then works correctly when used in MOVEQ.

Constants are usually specified and documented at the start of the relevant program or subroutine source code. To modify the program for another application where the constant is different, e.g. a different I/O register address, an edit of the specification and then reassembly is required.

Note that *every day* constants used in the program, e.g. multiply a value by 2, should be specified as a numeric or character value in the operand. Constants specified using EQU should have a wider content.

7.2.1 Expressions

Assemblers generally allow expressions when specifying operands, initialising variables, etc. Fig. 7.2 (next page) shows expressions used in operands of the EQU pseudo-operator and MOVE instruction. The operators vary from assembler to assembler but can include:

+	addition
-	subtraction
*	multiplication
/	division - producing truncated integer result
-	unary minus
>>	shift right
<<	shift left
&	logical AND
!	logical OR
()	to change the order of evaluation

Expressions are evaluated left to right using the following operator precedence:

1 parenthetical expression (innermost first)
2 unary minus
3 shift
4 and, or
5 multiplication, division
6 addition, subtraction

It must be emphasised that these expressions are calculated using integer arithmetic at assembly (or link) time and are not the same as expressions in a high-level language which are executed at run-time. Symbols which may be included in expressions are:

1 labels and their associated values; assemblers generally place limitations on the way labels may be used in expressions, in particular on relocatable values (which may have to be calculated at link time);
2 names defined using the EQU (or similar) pseudo-operator;
3 numeric or character constants.

```
------------------------------------------------------------------------------
1           00400000        PIT_1:   EQU       $400000          define PIT I/O register 1
2           00400001        PIT_2:   EQU       PIT_1+1          define PIT I/O register 2
3           00000040        CHAR_B:  EQU.B     'A'-1            byte value 'A'-1, i.e. $40
4           00000064        TEST:    EQU       $64              define a test value
5 400400 103C      002F              MOVE.B    #'0'-1,D0        move value '0'-1, i.e. $2F
6 400404 223C 000000C8               MOVE.L    #TEST*2,D1       move value TEST*2 into D1
7 40040A 343C      00C7              MOVE.W    #TEST*2-1,D2     move value TEST*2-1 into D2
8                                    END

PIT_1       00400000 E    PIT_2       00400001 E     CHAR_B      00000040 E
TEST        00000064 E
------------------------------------------------------------------------------
```

Fig. 7.2 Examples of expressions in operands

7.3 Using data registers to hold variables

A variable is a data element which will vary in value during the execution of a program (as opposed to a constant whose value does not change). When using a 68000 variables may be stored using a number of techniques, e.g.:

Variables in registers. A data or address register holds the value of the variable during calculations, i.e. the result of the sample programs 7.1, 7.2 and Exercise 7.1 was a variable held in D0.

Named variables in memory. A variable is assigned a memory area which is labelled (with the name of the variable). The label may then be used in instruction operands to access the variable, with the *value of the variable* being the *value of the contents* of the memory location(s). High-level languages often use this technique to hold global variables (see Chapter 29 for an example of a C program).

Variables in memory accessed via a base register. An address register is used as a pointer to the base or start of a data area in memory which is allocated on program startup. Offsets are used to access particular elements of the data area (see Chapters 13, 14, 15 and 29).

Variables on the stack. The stack is a dynamic data structure used extensively by the processor which can also be used by programs to hold variables (see Chapter 14). High level languages tend to use the stack to hold local variables in subroutines.

Exercise 7.2 *(see Appendix G for sample answer)*

In the programs so far the initial values used in calculations were constants (specified using immediate operands), i.e. in Exercise 7.1 the values 2, 355, 4, 89, 2,7, 13 and 8 were constants and the variable RESULT was returned in a data register. Design, code and test a program to calculate (using word length variables in data registers):

```
RESULT= A*(355+4*(89*2-B)-13*8)
```

Assign data registers to hold A and B which will have to be loaded with test values (of A and B) prior to executing the program. For example, assuming that A and B are held in D3 and D4 respectively, the monitor would be used to load the initial values into those registers before program execution. The program will have to be executed a number of times using different values of A and B to ensure that it is working correctly.

7.4 Multiple-word and BCD arithmetic

If greater precision is required than can be obtained using the normal arithmetic instructions, multiple-word arithmetic can be carried out (e.g. a 64-bit addition could be carried out as a sequence of eight 8-bit, four 16-bit or two 32-bit add operations):

ADDX	add extended	destination = source + destination + X
SUBX	subtract with extend	destination = source - destination - X
NEGX	negate with extend	destination = 0 - destination - X

The C (carry) and X (extend) bits of the Condition Code Register (see Chapter 10.2) transfer information between successive operations. The C bit is a normal carry bit set by arithmetic instructions. The extend bit X is set to the same value as C by arithmetic instructions, but is not affected by instructions such as MOVE which may alter the C bit.

```
        ADD.L     D3,D1          add least significant 32 bits
        ADDX.L    D2,D0          add most significant 32 bits + X
```

The ADD adds the least significant 32-bit halves in D3 and D1 and places the 32-bit result in D1 (the carry from this addition would set the carry bit C and the extend bit X). The ADDX adds the most significant 32-bit halves plus the value of the extend bit X and places the 32-bit result in D0 (i.e. D0 = D0 + D2 + X). The resultant 64-bit result is now in D0 (most significant half) and D1 (least significant half). The carry from ADDX sets the C and X bits which may then be used for further ADDX instructions (e.g. 96-bit add).

7.4.1 BCD (Binary Coded Decimal) arithmetic

BCD (binary coded decimal) arithmetic stores each decimal digit in 4 bits. For example, the number 16 decimal would be stored as $10 in binary and $16 in BCD (i.e. 8 bits can be used to store unsigned binary numbers in the range 0 to 255 decimal and BCD numbers in the range 0 to 99 decimal). The range of values that can be represented using BCD in a given word size is less than could be represented using binary. The advantage of BCD is that it is easier to write input/output routines. The MC68000 provides three instructions for BCD arithmetic, ABCD (add BCD), NBCD (negate BCD) and SBCD (subtract BCD) (for full details see reference Motorola 1989). Some high-level languages allow the programmer to select the use of binary or BCD operations when performing calculations.

Problem for Chapter 7

Implement programs to calculate (A and B being word length variables in data registers):

```
    RESULT = (A*30+B*20)/10
    RESULT = (A/(10*B))+(C*B/A)
```

1. DIVS and DIVU expect a 32-bit long word dividend so use EXT.L if previous calculations have generated a word length value.
2. Integer arithmetic is used and any remainders should be ignored. Comment on how this would affect the accuracy of the results.
3. Comment on what happens if A and B become too small or too large, e.g. if zero is entered for A or B in the second problem.

8

Variables in memory

The simple programs implemented as exercises or problems in Chapters 6 and 7 used variables held in the data registers. There are only eight data registers and a program of any size or complexity must be able to store variables in memory (loading variables from memory into the data registers for processing and then saving the results to memory). Chapter 7.3 briefly summarised the techniques used to store variables on the MC68000 and this Chapter will describe the second of these, i.e:

Named variables in memory. A variable is assigned a memory area which is labelled (with the name of the variable). The label may then be used in instruction operands to access the variable, with the *value of the variable being the contents* of the memory location(s).

8.1 The DC and DS pseudo-operators

8.1.1 The DC (Define Constant) pseudo-operator

The DC (Define Constant) pseudo-operator is used to define a storage area for byte, word or long word data and assign an initial value to that data.

```
--------------------------------------------------------------------------------
1        00004000                       ORG     $4000
2 004000 19                   VALUE1:   DC.B    25          byte with contents 25 decimal
3 004001 5A                   CHAR:     DC.B    2+'X'       byte with contents ASCII 'X'+2
4 004002 041D                 VALUE2:   DC.W    1053        word with contents 1053
5 004004 0006F8C0             VALUE3:   DC.L    456896      long word contents 456896
6 004008 31 32 33 00          CHARS:    DC.W    '123'       three characters in words
7                                       END
VALUE1     00004000 L    CHAR      00004001 L     VALUE2     00004002 L
VALUE3     00004004 L    CHARS     00004008 L
--------------------------------------------------------------------------------
```

Fig. 8.1 Specification of a data area using the DC pseudo-operator

Fig. 8.1 is an assembler listing which shows the specification of data using the DC pseudo-operator. The resultant label values and object code is:

```
              address         contents

VALUE1  →  004000          | 19 | 5A |    ← CHAR is at 4001
VALUE2  →  004002          | 04 | 1D |
VALUE3  →  004004          | 00 | 06 |
           004006          | F8 | C0 |
CHARS   →  004008          | 31 | 32 |
           00400A          | 33 | 00 |
```

The variable name or label is given the value of the address of the first byte used to store the variable. The *value of the variable is the contents* of the address(es), i.e. the value stored in the memory location(s):

VALUE1	=	$4000 holding the byte value 25 decimal ($19)
CHAR	=	$4001 holding the ASCII character 'X'+2 in a byte ($5A)
VALUE2	=	$4002 (plus $4003) holding the word value 1053 ($41D)
VALUE3	=	$4004 (plus $4005, $4006 and $4007) holding the long word value 456896 ($6F8C0)
CHARS	=	$4008 (plus $4009, $400A and $400B) holding the character codes for '123' followed by 0 ($31323300)

When generating word and long word data the most significant byte is stored in the first memory location (i.e. in the address associated with the label) with the less significant byte(s) in the following location(s). When assigning characters to word and long word storage, the characters are stored one per byte, left justified and padded with sufficient zeros to make the length correct (i.e. zero on the end of '123' in Fig. 8.1 line 6).

Although DC stands for Define Constant it only sets up an initial value in memory which may be modified at run time, i.e. by moving a new value into the variable.

```
--------------------------------------------------------------------------------
1          00004000                ORG      $4000
2 004000 05 0A 14        DATA:     DC.B     5,10,20      define three bytes
3 004003 48 45 4C 4C 4F  TEXT:     DC.B     'HELLO'      define five bytes (characters)
4 004008 0038 0420       DATA1:    DC.W     56,1056      define two words
5 00400C 00004002        D_ADD:    DC.L     DATA+2       address of third element of DATA
6 004010 31 32 33 00     CHAR:     DC.L     '123'        define three characters
7                                  END
DATA      00004000 L    TEXT      00004003 L    DATA1     00004008 L
D_ADD     0000400C L    CHAR      00004010 L
--------------------------------------------------------------------------------
```

Fig. 8.2 Using the DC pseudo-operator to initialise a series of variables

Fig. 8.2 shows how a series of initial values (e.g. to initialise an array) may be specified using the DC pseudo-operator. The label values will be DATA=$4000, TEXT=$4003, DATA1=$4008, D_ADD=$400C (holds the address of the third element of DATA) and CHAR=$4010 (CHAR is padded with a zero), thus:

	address	contents		
DATA →	004000	05	0A	
	004002	14	48	← TEXT at 4003
	004004	45	4C	
	004006	4C	4F	
DATA1 →	004008	00	38	
	00400A	04	20	
D_ADD →	00400C	00	00	
	00400E	40	02	
CHAR →	004010	31	32	
	004012	33	00	

Note that DC (and DS below) for word and long word data must start on an even word boundary. Care must therefore be taken to ensure that the definition of byte data does not result in an odd address, e.g.:

```
          ORG       $4000
VALUE1:   DC.B      67          define a byte
VALUE2:   DC.W      1567        define a word
```

The assembler will either put the word VALUE2 on the next even word boundary (at $4002) or would generate an error message due to the attempt to start the word value at address $4001 (i.e. following the byte VALUE1 in address $4000).

Some assemblers have a pseudo-operator which will force the next byte of object code to be generated at an even address (i.e. to ensure that any following instructions or data start at an even address). For example, XA8 uses .EVEN (see also Program 13.2):

```
VALUE1:   .BYTE     67          ;define a byte
          .EVEN
VALUE2:   .WORD     1567        ;define a word
```

8.1.2 The DS (Define Storage) pseudo-operator

In many cases, data storage will have to be defined for intermediate results of calculations where initial values are not relevant. The intermediate storage could be set up using DC pseudo-operators with an initial value of 0, or the DS pseudo-operator may be used.

Fig 8.3 shows the use of the DS (Define Storage) pseudo-operator to reserve blocks of memory which will store byte, word and long word data. In each of the DS specifications of Fig. 8.3 twenty bytes are reserved for the associated data area (i.e. 20 bytes = 10 words = 5 long words) and the labels are equivalent to the first byte of the associated data area, i.e. TABLE = $4000, WTABLE = $4014 and LARRAY = $4028. Note that the memory allocated by DS is **not** initialised in any way.

```
1            00004000                  ORG     $4000
2 004000     00000014     TABLE:       DS.B    20          reserve 20 bytes
3 004014     00000014     WTABLE:      DS.W    10          reserve 10 words
4 004028     00000014     LARRAY:      DS.L    5           reserve 5 long words
5                                      END
TABLE        00004000 L   WTABLE       00004014 L   LARRAY     00004028 L
```

Fig. 8.3 Using DC to reserve storage for a sequence of variables

Do not confuse the use of the DS and DC directives:

```
        ORG       $4000
X:      DC.B      10       define a byte with initial value 10
Y:      DS.B      10       reserve 10 bytes of storage
Z:      DC.B      $0A      define a byte with initial value $A
```

The first DC pseudo-operator reserves a byte of storage at label X (address $4000) and sets the initial contents to 10. The DS pseudo-operator reserves ten bytes of storage from label Y (address to $4001 to $400A). The second DC pseudo-operator reserves a byte of storage at label Z (address $400B) with the initial value 0A hexadecimal (10 decimal).

8.2 Accessing operands in memory

The DC and DS pseudo-operators reserve memory areas for data storage with each data definition having a label associated with it which can then be used for instruction operands.

```
1                              * Program 8.1: SUM=A+B using word data in memory
2 001000 3039 00001014 START:     MOVE.W    A,D0       get value of A into D0
3 001006 D079 00001016            ADD.W     B,D0       add B to it
4 00100C 33C0 00001018            MOVE.W    D0,SUM     put result into SUM
5 001012 4E4E                     TRAP      #14        !! STOP PROGRAM
6                              * data area
7 001014 01C8          A:         DC.W      456        A initial value 456
8 001016 FFB2          B:         DC.W      -78        B initial value -78
9 001018 0000          SUM:       DC.W      0          SUM is initialised to 0
10                                END
START      00001000 L    A         00001014 L    B          00001016 L
SUM        00001018 L
```

Program 8.1 Program using operands in memory

In Program 8.1 there are three variables A, B and SUM which are referenced as operands in the program instructions. Using the DC.W pseudo-operator, the contents of A, B and SUM are assigned the initial values 456, -78 and 0 respectively (lines 7,8 and 9). Instructions refer to the variables by using the label name in the operand field (A and B in lines 2 and 3 as source operands and SUM in line 4 as a destination operand). After A+B is calculated the result is saved into the variable SUM (address $1018), the contents of which would be checked using the system monitor. DS.W could have been used to reserve a word of storage for SUM, however, the contents would not have been initialised and could therefore contain anything. Using DC.W the value of SUM is initialised to 0 and any changes can be detected (e.g. in the case of the program not working correctly). It is always good programming practice to initialise variables, otherwise, when the contents are examined it will not be known if the values are those left by the last program or are the results of calculations in the current program.

The addressing mode used to access the variables A, B and SUM in Program 8.1 is called *absolute or direct memory addressing* (the operand address is stored as an absolute value following the op-code in memory, $1014 in line 2, $1016 in line 3 and $1018 in line 4). Absolute addressing can be used for all types of operand reference: data, memory, alterable and control. There are two forms of absolute addressing, short and long.

Short Absolute. The low-order half of the effective address is stored in the word following the operation-code in memory, for example (with the generated machine code):

```
ADD.W      $4000,D0
```

D0	78
40	00

The first word contains the op-code ($D078) followed by a word containing the operand address ($4000). When executed the high-order half of the effective address is obtained by extending the sign bit (bit 15) of the low-order address. This allows quick accessing of address locations 0 to $007FFF and $FFFF8000 to $FFFFFFFF (first and last 32Kbytes of memory) which are often used for common subroutines, data and I/O device registers.

Long Absolute. The effective address is stored in two words following the operation-code in memory, for example:

```
ADD.W      $400000,D0
```

D0	79
00	40
00	00

Which form of absolute address is generated at assembly time depends upon the assembler being used (in particular when the effective address will fit into the short form). In general assemblers resolve all forward references into the long form.

When writing relocatable code absolute addressing should be avoided except for references to fixed addresses (e.g. I/O registers and monitor routines), and to external routines where the references will be satisfied by the linker (see Chapter 15 for further discussion).

The executable code of Program 8.1 is terminated by the TRAP #14 instruction (line 5), which returns control to the M68 monitor when the program is finished. If the TRAP #14 was omitted the executable code would not be terminated, and at execution time the CPU would attempt to execute the data as though it was instructions. The binary patterns that make up the data may, in fact, form legal instructions, or may give an illegal instruction error, but in either case the program would not work correctly. It is essential that executable code and data are separate (see next section for further discussion).

When writing programs the operand format for absolute and immediate addressing should not be confused (it is easy to forget the #), i.e.:

```
ADD.W   #$1000,D0    add the constant $1000 to D0
ADD.W   $1000,D0     add contents of address $1000 to D0
```

In the first case *immediate addressing mode* is used to add the **constant 1000 hexadecimal** to D0. In the second case, *absolute memory addressing mode* is used to add the **contents of memory address 1000 hexadecimal** (which may be any word value) to D0. The symbol # tells the assembler that immediate addressing is being used, otherwise absolute memory addressing is used. Consider the following outline program:

```
         ORG        $1000
         ADD.W      SUM,D0     add contents of SUM to D0
         ADD.W      #SUM,D0    add value of SUM to D0
         .....
         .....
         .....
         ORG        $4000
SUM:     DC.W       25
```

The first instruction adds **the contents** of SUM (25 decimal) to D0, while the second instruction adds **the value** of SUM to D0, i.e. the address $4000. It must be remembered that the assembler defines the label as the address of the first memory byte storing the instruction or data and this should not be confused with the **contents** of memory.

Exercise 8.1 *(see Appendix G for sample answer)*

Rewrite the program of Exercise 7.2 using variables (words) in memory (some systems may require separate program and data sections to be defined - see next section), i.e:

```
RESULT = A*(355+4*(89*2-B)-13*8)
```

Before executing the program the memory locations assigned to A and B will have to be loaded with suitable test data and after execution RESULT examined. The memory addresses assigned to A, B and RESULT can be obtained from the assembler listing (or linker memory map) and the system monitor will provide facilities for the setting and examination of memory contents. Execute the program a number of times using different values of A and B to ensure that it is working correctly.

8.3 Program and data sections

The MC68000 separates memory references (reads from and writes to memory as the program is executing) into two classes:

(a) **Program references** which refer to the section of memory which contains the program being executed;

and (b) **Data references** which refer to the section of memory which contain data.

The areas of memory which contains the program and its data are called, respectively, the *program space* and the *data space*. When a program is executing:

1 the instructions are **always** read from the program space;

2 (a) operand reads (data) are **generally** from the data space

and (b) operand writes are **always** to the data space.

Operand reads are not always from the data space because there is generally some constant data within the program space, e.g. immediate operands and constant data defined such as text messages to be displayed on the screen during program execution.

In Program 8.1 the data space was separated from the program space purely by program comments, i.e. the target environment was a single board microcomputer and both would be loaded into the same physical memory area. However, in other environments there is a definite difference between program and data space:

An embedded system (e.g. a washing machine controller) would have the program space in ROM and the data space in RAM.

A multiprocessing environment allows a number of separate processes (programs) to run on the same computer concurrently. Not only are processes protected from each other (so one program is unable to overwrite another program) but the program space of each process would be in a RAM memory area which was write protected by a memory management unit (i.e. it cannot be written to). The data space would be in a separate RAM memory area which could be read from or written to.

Any attempt to write to the program space would result in a run-time error which would terminate program execution and display a message, e.g. bus error (see Chapter 23.2).

8.3.1 The XA8 .PSECT (program section) pseudo-operator

The .PSECT pseudo-operator is used by the XA8 cross assembler to start or reopen a program section (Motorola assemblers use the pseudo-operator SECTION):

```
        .PSECT    psect_name        ;open/reopen a program section
```

XA8 uses the name `_text` for the executable program code section and `_data` for the data section. Within the program source code file the program and data sections (and there may be many) start with a .PSECT directive. When the linker links the file (or files):

1. the `_text` sections are joined together and placed in memory starting at a specified address (which may be *blown* into EPROM).
2. the `_data` sections are joined together and placed in memory starting at a specified address (which will be in RAM).

Program 8.1a and 8.1b (next page) are versions of Program 8.1 for the XA8 and AS68K cross assemblers. Note the different pseudo-operators, e.g. XA8 uses .PROCESSOR to define the processor, .PUBLIC to make the label names public for a program memory map (see below), .WORD instead of DC.W and .END instead of END. The difference in memory addresses between the Program 8.1, 8.1a and 8.1b (i.e. SUM is at address $1018 in Program 8.1 and $0004 in 8.1a and 8.1b) is that the M68 assembler generates absolute code (Program 8.1) while XA8 and AS68K generate relocatable code. The addresses shown are offsets from the start of the program or data section.

When the relocatable code is linked the linker will generate absolute code to be downloaded into the target 68000 board. For example, the commands to assemble, link and generate the S-record file of Program 8.1a (for the Force/Bytronic board) are:

```
x68000 +o p8_1a.s
linkx -a -db 0x4000 -tb 0x1000 p8_1a.o
relx -v -o map
hexx -s -o p8_1a.mx
```

x68000 (assembler) assembles the source file p8_1a.s and generates the object file p8_1a.o

linkx (the linker) links the object code file p8_1a.o (no other object code files are specified in this case) generating absolute code (option -a) with the `_data` section at $4000 (option -db 0x4000) and the `_text` section (the program code) at $1000 (option -tb 0x1000) producing an output file called xeq

relx is used to examine object code files; in this case it produces a program memory map of the executable program file xeq into a file called map

hexx reads file xeq and generates a file called p8_1a.mx in Motorola S-record format for downloading onto the Force/Bytronic board

When debugging programs the absolute addresses used by variables may be obtained from the memory map. The following is the map generated by relx for Program 8.1a which shows START at absolute address $1000, A at $4000, B at $4002 and SUM at $4004:

```
0x00001000A START
0x00004000A A
0x00004002A B
0x00004004A SUM
```

Note:

1 only labels named using the .PUBLIC pseudo-operator appear in the map;
2 the capital A on the end of the numbers indicates an absolute address.

Problem for Chapter 8

Rewrite and test the programs of the Problem for Chapter 7 using variables in memory (word length operands), i.e:

```
RESULT = (A*30+B*20)/10
RESULT = (A/(10*B))+(C*B/A)
```

```
--------------------------------------------------------------------------------
 1                        ; Program 8_1a: SUM=A+B using word data in memory
 2                        ;
 3                                  .PROCESSOR M68000    ;specify processor
 4                                  .PUBLIC    START,A,B,SUM
 5                                  .PSECT     _text     ;start program section
 6 000000  3039           START:    MOVE.W     A,D0      ;get value of A into D0
           00000014
 7 000006  D079                     ADD.W      B,D0      ;add B to it
           00000016
 8 00000C  33C0                     MOVE.W     D0,SUM    ;put result into SUM
           00000018
 9 000012  4E4E                     TRAP       #14       ;!! STOP PROGRAM
10                        ;
11                                  .PSECT     _data     ;start data section
12                        ; data area
13 000000  01C8           A:        .WORD      456       ;A initial value 456
14 000002  FFB2           B:        .WORD      -78       ;B initial value -78
15 000004  0000           SUM:      .WORD      0         ;SUM initialised to 0
16                                  .END
--------------------------------------------------------------------------------
```

Program 8.1a Program using operands in memory (XA8 assembler)

```
--------------------------------------------------------------------------------
 1                             * program 8_1: sum=a+b using word data in memory
 2                                       .text              *start code section
 3 00000 3039 00000000 D start:          move.w   a,d0      *get value of a into d0
 4 00006 d079 00000002 D                 add.w    b,d0      *add b to it
 5 0000c 33c0 00000004 D                 move.w   d0,sum    *put result into sum
 6 00012 4e4e                            trap     #14       *!! stop program
 7                             * data area
 8                                       .data              * start data section
 9 00000 01c8                  a:        .word    456       *a initial value 456
10 00002 ffb2                  b:        .word    -78       *b initial value -78
11 00004 0000                  sum:      .word    0         *sum initialised to 0
12                                       .end
                    code segment size = 20
                    data segment size = 6
--------------------------------------------------------------------------------
```

Program 8.1b Program using operands in memory (AS68K assembler)

Review of Chapters 7 and 8

Chapter 7 introduced arithmetic instructions and their use with variables in data registers and Chapter 8 then went on to describe the use of named variables in memory.

7.1 The binary addition and subtraction instructions allow operands as follows:

Source in a data register and a destination specified using *data register direct* or any *memory alterable addressing mode.*

or Source specified using *any addressing mode* and the **destination in a data register.**

```
ADD.W   D0,D1      add word in D0 to contents of D1
SUB.L   D2,D3      subtract long word D2 from D3
ADD.W   #10,D0     add the constant 10 to word in D0
SUB.L   D3,LABEL   subtract D3 from long word in memory
```

If the source operand is a constant the immediate ADDI and SUBI instructions can be used (destination specified using a *data alterable addressing mode*):

```
ADDI.W    #50,D0        add the word constant 50 to D0
ADDI.L    #50,LABEL     add word to LABEL in memory
```

ADDQ and SUBQ can be used for constant source operands in the **range 1 to 8** (length byte, word or long word and the destination specified using an *alterable addressing mode*):

```
ADDQ.W    #2,D0         add word 2 to D0
SUBQ.L    #6,D1         subtract long word 6 from D1
ADDQ.W    #5,LABEL      add word 5 to memory
```

The **EXG** instruction exchanges the contents of one 32-bit register with another:

```
EXG         A1,D3     exchange A1 and D3
```

The **NEG** instruction negates the destination operand (length is byte, word or long word and specified using a *data alterable addressing mode*), e.g.:

```
NEG.W     D0         negate word value in D0
NEG.L     LABEL      negate long word in memory
```

The **EXT** instruction extends the sign of a data register from byte to word or word to long word, e.g.:

```
EXT.W     D0        extend byte sign to word in D0
EXT.L     D0        extend word sign to long word in D0
```

MULS (signed) and **MULU** (unsigned) binary multiply instructions multiply a word (16-bit) source operand (specified using a *data addressing mode*) with a word destination operand in a data register, and form a long word (32-bit) result in the data register, e.g.:

```
MULS.W    #30,D0     D0 = D0*30
MULS.W    VALUE,D1   D1 = D1*VALUE (VALUE is in memory)
MULU.W    #$4F,D5    unsigned multiply D5=D5*$4F
```

The **DIVS** (signed) and **DIVU** (unsigned) binary divide instructions divide a word (16-bit) source operand (specified using a *data addressing mode*) into a long word (32-bit) destination operand in a data register. The result is in a data register (quotient in bits 0-15 and remainder in bits 16-31) , e.g.:

```
        DIVS.W     #10,D0      divide long word in D0 by 10
        DIVU.W     VALUE,D1    divide D1 by contents of VALUE
```

The **SWAP** instruction swaps the lower and upper 16 bits of a data register, e.g.:

```
        SWAP       D0          swap lower and upper 16 bits of D0
```

7.2 The **EQU** pseudo-operator equates a name with a value, e.g.:

```
CHARA:      EQU.B      'A'        define a byte character A
```

Expressions are generally allowed when specifying operands, initialising variables:

```
            MOVE.B     #'0'-1,D0       move value '0'-1, i.e. $2F
            MOVE.W     #TEST*2-1,D2    move value of TEST*2-1 to D2
```

7.3 A variable is a data element which will vary in value during the execution of a program (as opposed to a constant whose value does not change). Variables may be stored as:

Variables in registers. A data or address register holds the value of the variable.

Named variables in memory. A variable is assigned a memory location which is labelled and may then be referred to as an operand in instructions. The *value of the variable* is the *value of the contents* of the memory location(s).

Variables in memory accessed via a base register. An address register is used as a pointer to the base of a data area in memory (offsets are used to access the data).

Variables on the stack. The stack is a dynamic data structure used extensively by the processor which may also be used by programs to hold variables.

7.4 introduced multiple-word and BCD arithmetic.

8.1 The **DC** (define constant) pseudo-operator is used to define data areas and their initial values, e.g.:

```
VALUE1:     DC.B       10,'A'       define two bytes
VALUE2:     DC.W       5678         define a word value
```

The **DS** (define storage) pseudo-operator is used to define areas of storage (the areas reserved are not initialised in any way), e.g.:

```
ARRAY:      DS.W       100           array of 100 word elements
```

8.2 The *absolute addressing mode* can be used to specify all types of operand reference (data, memory, alterable and control), e.g.

```
            MOVE.W     $4000,D0       move contents of $4000 to D0
            ADD.W      LABEL,D5       add contents of LABEL to D5
```

The absolute address is stored in memory following the operation-code either in **short** form (it can fit into 16 bits) or **long** form (32 bits).

8.3 The areas of memory which contains the program and its data are called, respectively, the *program space* and the *data space*. When a program is executing:

1 the instructions are **always** read from the program space;
2 (a) operand reads (data) are **generally** from the data space
and (b) operand writes are **always** to the data space.

Subsection 8.3.1 described the .PSECT pseudo-operator used by the XA8 cross assembler to open or reopen program sections which are used to define program and data space.

9

Modular Programming with the MC68000

9.1 Modular programming

A large software system can consist of many individual programs, each of which may consist of 200000 lines or more of program statements. Attempting to design, code and test a large program as a single entity is virtually impossible, so it is broken down into logical components called modules (Steward 1987). When writing programs modules are implemented as subroutines (also called procedures or functions) with a program consisting of a main routine (usually called the main program) which makes calls (transfers control) to subroutines to carry out specific tasks. When a task is complete the subroutine returns control to the main program which calls the next subroutine, etc. (many main programs are a sequence of subroutine calls). Subroutines called by the main program may call other subroutines and some languages allow subroutines to call themselves (called recursion).

During the program specification and design stages logical tasks or sequences of tasks, which will become modules, are identified and specified (see Chapter 17 for further discussion). Once a specification of the logical steps performed by each module has been drawn up (modules can contain other modules and call other modules as required) the individual modules can then be designed, coded and tested. Modules are kept as independent as possible by controlling the passing of information between modules and only allowing an individual module access to data that it actually uses. Errors then tend to be localised to a module or a group of modules making them much easier to find than if all modules had access to all the data (passing information between assembly language modules is covered in Chapter 16). Once working satisfactorily, modules can be integrated to build up to the complete package. As modules are integrated with others, errors can occur due to faulty interaction of modules. Modular programming does tend to limit such errors and isolate them to particular sets of modules within the package. In addition, if errors are found whilst the package is being used in its application environment or when upgrades are required, it is easy to identify the relevant modules and modify just these.

A program may require the same sequence of operations at many different points during execution (e.g. to read a decimal number from a keyboard). It is possible at each point to repeat the identical sequence of instructions but this would be wasteful and error prone. Subroutines enable a sequence of instructions to be written once and *called* from elsewhere in the program as required. If a subroutine has a wider context it may be placed in a library to be accessed by other programs (possibly written by other programmers).

The majority of programming languages provide libraries of modules to carry out common tasks. For example, a high-level language could provide the following facilities:

1 text input from the users keyboard and output to the display screen;
2 file I/O, i.e. open, close, read write, etc.;
3 manipulating data structures, e.g. text string processing;
4 mathematical routines, e.g. sin, cos, tan, log, etc.;
5 exception and interrupt handling (as in Modula 2 and Ada).

Today, the main application of assembly languages is in the implementation of critical low-level facilities within programs written mainly in a high-level language. Assembly languages do not have libraries of support routines provided as standard (tasks such as those listed above being carried out from modules written in the high-level language). However, the monitor or operating system of a low-level development system does usually provide basic facilities which can be accessed from assembly language programs. This Chapter:

(a) introduces the use of subroutines for the MC68000 (section 3);

(b) describes a library of assembly language subroutines (section 3 and Appendix D; and (c) describes the run-time facilities offered by a typical system monitor (section 4).

9.2 Writing subroutines

A subroutine is an independent section of program code which can be accessed from other parts of the program. For example, subroutine RDDECW in the library of Appendix D reads a decimal number from the keyboard (reads digits until a non-digit is hit) and places the resultant binary word value in data register D0. When a program wishes to read a number, the subroutine is called (i.e. control is transferred to it) using a JSR (jump to subroutine) instruction. After the subroutine has performed its task the last instruction is RTS (return from subroutine) which returns control to the instruction immediately following the original JSR in memory. The program is resumed and can use the number in D0, etc.

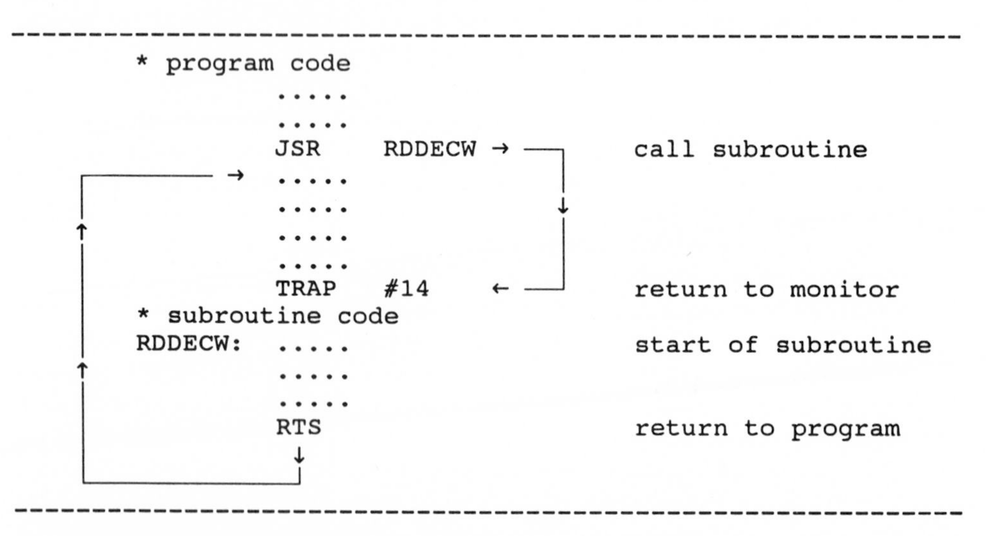

Fig. 9.1 Flow of program execution when a subroutine is called

Fig 9.1 shows an *outline* of a program calling a subroutine. The main program instructions are executed sequentially until the `JSR RDDECW` instruction transfers control to the subroutine. The subroutine instructions are then executed until the `RTS` returns control to the calling program, which then continues. The actions performed by the MC68000 are:

1 When a JSR (and BSR) instruction is executed (for full details see Chapter 14.2):
   (a) the current value of the PC (which points to the next instruction) is pushed or moved as a long word value onto a data area called the stack.
   (b) The subroutine start address specified by the operand is moved into the PC; the next instruction to be executed will be the first instruction of the subroutine code.
2 The subroutine will then be executed until terminated by an RTS instruction, which pops or moves the long word value off the top of the stack into the PC.
3 The instruction executed after the RTS will be that following the original JSR.

Subroutines may be called as many times as required and control will always, assuming the stack is intact, return to the instruction following the corresponding JSR. The structure of the stack allows subroutines to call other subroutines or even themselves, i.e. recursion. For example, Fig. 9.2 shows an outline main program which calls two subroutines (see Appendix D for the actual subroutine code):

1 RDDECW which reads a decimal number from the keyboard into D0 by:
   (a) calling subroutine RDCHAR to read a character sequence from the keyboard
   and (b) converting the character sequence into a decimal number.
2 WRDECW which displays in decimal the number in D0 by:
   (a) converting the numeric value into a character sequence
   and (b) calling subroutine WRCHAR which writes a character to the display screen.

```
-----------------------------------------------------------------
* main program
START:     JSR        RDDECW     read first number
           MOVE.W     D0,D1      and save it into D1
           JSR        RDDECW     read second number
           ADD.W      D1,D0      add the numbers
           JSR        WRDECW     and display the result
           TRAP       #14        !! STOP PROGRAM
*
* subroutine to read a decimal number into D0
RDDECW:    ......                subroutine code
           ......
           JSR       RDCHAR      read character from the keyboard
           ......
           ......
           RTS                   return to calling program
*
* subroutine to write the numeric decimal value of D0
WRDECW:    ......
           ......
           JSR       WRCHAR      write a character to the screen
           ......
           ......
           RTS                   return to calling program
           END
-----------------------------------------------------------------
```

Fig. 9.2 Program (outline) calling two subroutines

9.2.1 Subroutine parameters

Subroutine parameters are:

(a) the data or address values passed into the subroutine by the calling program for processing;

and (b) values returned to the calling program on exit from the subroutine.

Consider a subroutine which calculates the factorial of an integer number:

1. the program places the integer number in D0;
2. the program calls the factorial subroutine by using JSR;
3. the subroutine calculates the factorial of the number in D0, and then places the result in D0;
4. the subroutine returns control to the calling program by using RTS;
5. the program uses the factorial value in D0.

In the operations described above, parameters were passed into and out of the subroutine via the data register D0 (a factorial subroutine will be presented in the Problems for Chapter 11 and a recursive version in the Problem for Chapter 14). There may be a number of input and output parameters, input parameters only (e.g. write a character to the screen), output parameters only (e.g. read a character from the keyboard), or none at all (the subroutine performs some task not needing or returning data, e.g. clear the display screen). When working in 68000 assembly language parameters can be passed in and out of subroutines in a number of ways. The majority of examples presented in this book use registers for parameter passing but other techniques are also described.

Exercise 9.1 *(see Appendix G for sample answer)*

Give three reasons for the use of subroutines in programming.

9.3 Use of the MC68000 assembly code library in Appendix D

Appendix D contains the assembly code for a general purpose library of subroutines which can be used by MC68000 assembly language programs. The library contains routines to read characters and decimal numbers from the keyboard, and write characters, text messages and decimal and hexadecimal numbers to the display screen. In the previous chapters, after a program had been executed, the results had to be obtained by using the system monitor to examine the contents of data registers and/or memory addresses. In general the monitor is clumsy to use and presents data in hexadecimal form, leaving the user to convert to decimal if necessary. The subroutines overcome this problem by allowing the program to output results directly to the display in hexadecimal or decimal. In addition, text can be displayed which indicates the progress and results of the program. For example, Program 5.1 calculated 7+2*(267-23) and placed the result in D0 where the user could read it as $1EF. It would be far better to present the result on the display screen:

```
The result of 7+2*(267-23) = 495
```

If a program is not working correctly the subroutines can be used to write out intermediate results to determine the problem areas (e.g. use WRREGS, described below, to display the contents of all the registers). When the program is finally working these intermediate displays of information can be removed.

In the case of a formal course the tutor will have already entered, assembled and tested the subroutines and will provide further documentation on their use with the target microcomputer (or describe similar facilities provided by the monitor, see section 4 below). Otherwise, the code will have to be entered and tested as described in Appendix D, remembering the idiosyncrasies of the assembler and monitor (see Appendix C).

9.3.1 Calling the library subroutines

The subroutines are called (control transferred to them) by the use of the JSR (jump to subroutine) instruction. In all cases the contents of working registers D0-D7 and A0-A7 are not affected unless explicitly stated (e.g. RDDECW uses data register D0 to return the number read from the keyboard). Thus the user does not have to worry about the routines affecting important values stored in registers.

The following lists the subroutines giving the name, the function and an example of use where appropriate.

WRCHAR. Writes the ASCII character (byte) in D0 to the display (bits 8 to 31 are ignored):

```
MOVE.B    #'X',D0   display an X on the screen
JSR       WRCHAR
```

WRLINE. Writes a new-line to the screen (this generally means sending the ASCII characters CR (Carriage Return) and LF (Line Feed) to the display).

WRSPAC. Writes a space character to the display.

WRTEXT. Write a text string on the screen; the text following the JSR in memory:

```
JSR     WRTEXT
DC.W    'hello user$'
```

(a) The text is defined, using the DC.W pseudo-operator, immediately after the JSR instruction in memory.
(b) The text **must** be terminated by a $ (otherwise WRTEXT will be unable to find the end of text and the program will not work).
(c) This method of passing parameters to subroutines by mixing instructions and data requires great care, and no attempt should be made to write subroutines using this technique until one is experienced in assembly language programming. In fact many tutors regard the technique as suspect because of the mixing of code and data within the same program area and prefer the techniques of WRTXTA and WRTXTS below. Using WRTEXT for messages does have the advantage that when reading the program the particular message being displayed is immediately apparent (rather than being in a separate data area elsewhere in the program) thus making the program source code more legible.

WRTXTA and WRTXTS are similar to WRTEXT except that the address of the start of the text is in address register A0 for WRTXTA and on the stack for WRTXTS (the use of address registers will be covered in detail in Chapter 13 and the stack in Chapter 14), e.g.:

```
            MOVEA.L  #MESS1,A0  load address of message into A0
            JSR      WRTXTA     and message
            .......
            PEA.L    MESS2      push address onto the stack
            JSR      WRTXTS     and message
            .......
            TRAP     #14        !! STOP PROGRAM
*
* data area
MESS1:      DC.W     'hello user$'
MESS2:      DC.W     'goodbye user$'
            END
```

(a) WRTXTA: the address of the message is placed in the address register A0.
    WRTXTS: the address of the message is pushed onto the stack.
(b) The text is defined in a data area (at the end of the program code or in a different program section or segment) using the DC pseudo-operator or similar.
(c) The text must be terminated by the $ character.

WRHEXD. Writes, in hexadecimal, the digit in bits 0 to 3 of D0.

WRHEXB. Writes, in hexadecimal, the byte in bits 0 to 7 of D0.

WRHEXW. Writes, in hexadecimal, the word in bits 0 to 15 of D0.

WRHEXL. Writes, in hexadecimal, the long word in bits 0 to 31 of D0.

WRDECW. Writes the signed decimal value of the word (16 bits) in D0 (a space is output before the number and leading zeros are suppressed), for example:

```
            MOVE.W   #4567,D0   display a decimal value
            JSR      WRDECW
```

RDCHAR. Reads a character from the keyboard and places the byte ASCII character in D0 (bits 8 to 31 are cleared). Note that the character is not automatically echoed onto the screen (if character echo is required call WRCHAR after RDCHAR as follows):

```
            JSR      RDCHAR     read keyboard character into D0
            JSR      WRCHAR     and echo it onto the screen
```

RDDECW. Reads a signed decimal value from the keyboard into D0 (16-bit word value with bits 16 to 31 sign extended).

```
            JSR      RDDECW   read decimal number into D0
```

The digits are echoed as they are read and the number is terminated by a non-digit character. The minus '-' may be hit as the first character to enter a negative value.

RDCHECK is used to check if a keyboard key has been pressed; if so the Z bit in the CCR (Condition Code Register) is cleared (see Chapter 10.2 for details of the CCR).

```
JSR      RDCHECK    keyboard been hit ?
BEQ      NO_CHAR    if not go to NO_CHAR
JSR      RDCHAR     yes, read the character
```

WRREGS. Writes the contents of all the registers. This is very useful for debugging programs at run time by using it to display suspect values.

DELAY. Delays program execution for a number of milliseconds; the delay value (word) follows the JSR in memory:

```
JSR      DELAY    delay program execution
DC.W     1000     for one second
```

Examples of subroutine use will be given below and in the sample programs in this book.

9.3.2 Linking the library routines to a program

The previous section described the function of each subroutine and how to call it. However, the addresses of the subroutines (i.e. the values of WRDECW, RDDECW, etc.) have not yet been defined. How this is done depends upon the assembler, linker and operating system or monitor. On a formal course the tutor will provide documentation on the techniques to use, otherwise refer to the microcomputer manuals. The following subsection describes a technique common to many native MC68000 assemblers.

9.3.2.1 Using the XREF (external reference) pseudo-operator

The XREF external reference pseudo-operator (which is used by many native MC68000 assemblers; more details in Chapter 16.1) informs the assembler that the names specified are not defined within the current program source file but are external (i.e. in another object code file or a library file to be linked by the linker).

In Program 9.1 (next page) the XREF statement (line 5) informs the assembler that the names WRLINE, WRDECW and WRTEXT are external to the current program (e.g. in another object code file which will be added at link time). Once specified the subroutines can be called by the program using the JSR instruction. If the associated XREF was missing the subroutine calls to WRLINE, WRTEXT and WRDECW would generate an assembly error (undefined symbol). When executed Program 9.1 would display:

```
The result of 7+2*(267-23) = 495
```

The M68 ROM based assembler used to assemble Program 9.1 contained the library routines within itself and no explicit link stage was required. When assembling the library names referenced were replaced with the corresponding absolute addresses in ROM, e.g. it can be seen that the subroutine WRLINE is at address $2802C on the FORCE board:

(a) from the symbol table at the end of the listing, i.e. `WRLINE 0002802C`

and (b) from the object code in line 11 where it is referenced, i.e. at address 001010 is the object code is 4EB9 0002802C where 4EB9 is the op-code of JSR and 0002802C the absolute address of the operand (subroutine WRLINE in ROM).

```
--------------------------------------------------------------------------------
 1                          * Program 9.1 - to calculate RESULT = 7+2*(267-23)
 2                          *
 3                          * Library subroutines defined using the XREF
 4                          *   external reference pseudo-operator.
 5                                   XREF       WRLINE,WRDECW,WRTEXT
 6                          *
 7 001000 203C 0000010B START:       MOVE.L     #267,D0    initialise D0 with 267
 8 001006 0440      0017             SUBI.W     #23,D0     267-23
 9 00100A D040                       ADD.W      D0,D0      2*(267-23)
10 00100C 0640      0007             ADDI.W     #7,D0      RESULT = 7+2*(267-23)
11 001010 4EB9 0002802C              JSR        WRLINE
12 001016 4EB9 00028038              JSR        WRTEXT     display message
13 00101C 20 54 68 65 20             DC.W       'The result of 7+2*(267-23) = $'
14 00103C 4EB9 0002806E              JSR        WRDECW
15 001042 4E4E                       TRAP       #14        !! STOP PROGRAM
16                                   END
WRLINE     0002802C X    WRDECW    0002806E X    WRTEXT    00028038 X
START      00001000 L
--------------------------------------------------------------------------------
```

Program 9.1 Using XREF to reference external routines (Force board)

```
--------------------------------------------------------------------------------
 1                         ; Program 9.2 - to calculate RESULT = 7+2*(267-23)
 2                         ;
 3                         ; Library subroutines defined using the EXTERNAL
 4                         ;   external reference pseudo-operator.
 5                                  .EXTERNAL  WRLINE,WRDECW,WRTXTA
 6                         ;
 7                                  .PROCESSOR M68000     ;define processor
 8                                  .PSECT     _text      ;open code section
 9 000000  203C            START:   MOVE.L     #267,D0    ;initialise D0 with  267
           0000010B
10 000006  04400017                 SUBI.W     #23,D0     ;267-23
11 00000A  D040                     ADD.W      D0,D0      ;2*(267-23)
12 00000C  06400007                 ADDI.W     #7,D0      ;RESULT = 7+2*(267-23)
13 000010  4EB9                     JSR        WRLINE
           00000000
14 000016  207C                     MOVEA.L    #MESS,A0   ;load address of message
           0000002A
15 00001C  4EB9                     JSR        WRTXTA     ;display message
           00000000
16 000022  4EB9                     JSR        WRDECW
           00000000
17 000028  4E4E                     TRAP       #14        ;!! STOP PROGRAM
18                         ; message data area
19 00002A  20546865 MESS:           .TEXT      " The result of 7+2*(267-23) = $"
20                                  .END
--------------------------------------------------------------------------------
```

Program 9.2 Using the XA8 .EXTERNAL pseudo-op to reference external routines

Program 9.2 is an amended version of Program 9.1 for the XA8 cross assembler. It uses WRTXTA to display the message (the message text is in a separate data area at the end of the program, not within the code as with WRTEXT). Note that because the message is constant data it is within the program section `_text`, not in the data section `_data` (see Chapter 8.3). The addresses of the library routines are set to 0 (lines 13, 15 & 16) to be replaced with the actual addresses when the program is linked to the library.

9.4 Using monitor facilities via the TRAP instruction

In general a system monitor will provide run-time facilities for a program executing under it. The MC68000 has an instruction, TRAP, which is used by programs to communicate with the monitor or operating system (TRAP will be discussed in Chapter 16.7 and Chapter 23). TRAP transfers control to the monitor, which, after carrying out the task returns control to the program (in a similar manner to a subroutine call). TRAP has a single operand in the range 0 to 15, so:

```
        TRAP        #15         execute trap instruction 15
```

In practice monitors and operating systems use traps for a variety of functions (see microcomputer manual for details). The following describes some of the facilities offered by the M68 monitor which are similar to the KAYBUG monitor (Coats 1985/86).

9.4.1 M68 and KAYBUG monitor calls

M68 and KAYBUG provide facilities via TRAP #11. Monitor functions are specified by a word following the TRAP instruction and include:

Function 0: Stop the program and return control to the monitor:

```
        TRAP        #11
        DC.W        0           stop program execution
```

Function 1: Output character in D0 to the display screen

```
        MOVE.B      #'A',D0     move character into D0
        TRAP        #11
        DC.W        1           display the character
```

Function 5: Output a text string (terminated by a 0 byte); address is in A6:

```
        MOVEA.L     #MESS,A6    move message address into A6
        TRAP        #11
        DC.W        5           display the message
```

Function 10: Output a word in hexadecimal; value is in D0:

```
        TRAP        #11
        DC.W        $0A         display word value in D0 in hex
```

Function 12: Output a newline:

```
        TRAP        #11
        DC.W        $0C         output a newline to the screen
```

Function 27: Check if keyboard has been hit, set Z if NOT, clear if character available

```
        TRAP        #11
        DC.W        $1B         check if keyboard has been hit
        BNE.S       READ        if so Z is cleared
```

Program 9.3 (a modified version of Program 9.2) shows the use of the monitor facilities via TRAP #11 calls. When executed under the M68 monitor Program 9.3 would display:

```
The result of 7+2*(267-23)  = 01EF
Program terminated with M68 system call (TRAP #11)
```

```
1                           * Program 9.3 - to calculate RESULT = 7+2*(267-23)
2                           *
3                           * Use M68 monitor facilities to display information
4                           *
5 001000 203C 0000010B START:     MOVE.L      #267,D0      initialise with  267
6 001006 0440       0017          SUBI.W      #23,D0       267-23
7 00100A D040                     ADD.W       D0,D0        2*(267-23)
8 00100C 0640       0007          ADDI.W      #7,D0        RESULT = 7+2*(267-23)
9 001010 4E4B                     TRAP        #11          output a newline
10 001012 000C                    DC.W        $0C
11 001014 2C7C 00001026           MOVEA.L     #MESS,A6     load message address
12 00101A 4E4B                    TRAP        #11          display message
13 00101C 0005                    DC.W        $05
14 00101E 4E4B                    TRAP        #11          display value in HEX
15 001020 000A                    DC.W        $0A
16 001022 4E4B                    TRAP        #11          !! STOP PROGRAM
17 001024 0000                    DC.W        0
18                          * message data area
19 001026 20 54 68 65 20MESS:     DC.B        ' The result of 7+2*(267-23) = '
20 001044 00                      DC.B        0             terminate with 0
21                                END
START       00001000 L      MESS         00001026 L
```

Program 9.3 Using M68 monitor facilities via TRAP #11

Exercise 9.2 (*see Appendix G for sample answer*)

What are the advantages of using the library subroutines (or similar) for input and output over using the system monitor to load and display register and memory contents directly ?

Making use of library or monitor facilities rewrite the program of Exercise 8.1 to:

(a) read the values of the variables A and B from the keyboard (display a suitable message to prompt the user for input);

(b) display the result of the calculation on the screen with a suitable message.

Users of a disk based system will have to link the library routines to the program using the system linker.

Problem for Chapter 9

Rewrite the programs of the Problems for Chapter 8 to read the input variable values from the keyboard and to display the result onto the screen.

10

Program Control Structures

Although program flow is normally sequential, at particular points in a program, control structures can be used for:

1 a choice of alternate paths through the program, i.e. an IF 'condition' structure;
2 executing a sequence of instructions a number of times, i.e. loop or iterative structure.

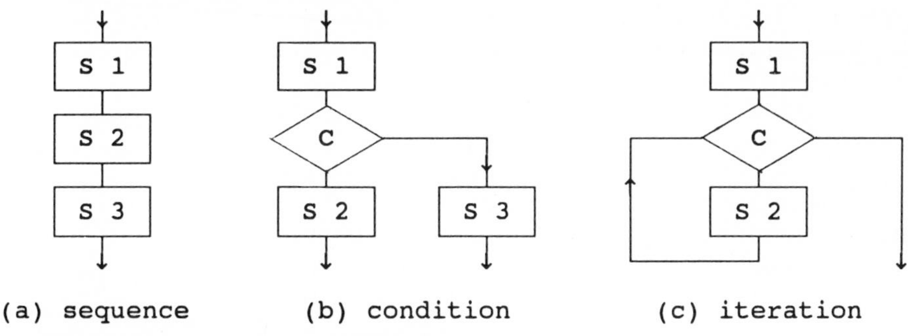

Fig. 10.1 Examples of program structures

Fig. 10.1 shows some program structures:

(a) **program sequence**: sequential execution of a number of statements (S1, S2 & S3);
(b) **IF structure**: alternate paths (S2 or S3) are taken depending upon condition C;
and (c) **LOOP structure**: a program loop (executing S2) depending upon condition C.

The coding these structures in MC68000 assembly language will be introduced below.

10.1 The branch (BRA) and jump (JMP) instructions

The branch (**BRA**) and jump (**JMP**) instructions, which can be considered as equivalent to the GOTO statement available in many high-level languages, transfer control to the address in the operand field. Branch and jump do, however, have a much broader use in assembly languages in that they are used to create the IF and LOOP structures provided by complex high-level control statements in high-level languages.

10.1.1 The branch (BRA) instruction

The branch instruction takes the following forms:

```
BRA.S     TEST
BRA.L     TEST
```

The BRA instructions would cause the processor to branch (i.e. transfer control) to the instruction labelled TEST. The branch address is stored as an offset relative to the current instruction in **short** (.S) or **long** (.L) form.

.S short form. The instruction is stored in one memory word made up of a one byte operation-code followed by a one byte signed displacement or offset, e.g:.

```
BRA.S      TEST
```

60	0A

where $60 is the operation-code and the offset is $0A. When using short form the branch is limited to -126 to +129 memory locations relative to the current instruction. The offset is added to the PC (which holds the address of the current instruction+2) to form the effective address of the operand, i.e. the new value of the PC.

.L long form. The one byte op-code is identical to the above but the byte following is zero, indicating that the offset is stored in the next word as a 16-bit signed number:

```
BRA.L      TEST
```

60	00
00	0A

where the operation-code is $6000 and the offset is $000A. Using this long form, the branch is limited to -32766 to +32769 memory locations relative to the current address.

The short form should be used where possible to save memory and speed up loop execution. If the offset will not fit into a byte, the assembler will display an error message; simply alter the .S to .L to overcome the problem.

The addressing mode described above is called *program counter relative with displacement* (i.e. the effective address of the operand is PC+offset), and is the only addressing mode which can be used with branch instructions. As the operand address is relative to the PC the instructions generated will be relocatable (PC relative addressing will be discussed further in Chapter 15).

Although the branch range is limited, by using modular programming techniques no single module should be so large that the limit on the size of the .L offset becomes a problem.

10.1.2 The jump (JMP) instruction

In the case of the **JMP** instruction the destination operand is specified using a *control addressing mode* (see Appendix B), which allows jumps to anywhere in memory, e.g.:

```
JMP        TEST        jump to memory address TEST
```

In this example, the destination is specified using *absolute addressing mode*, and control is transferred to the address equivalent to the label TEST, i.e. the value (address) of TEST is moved into the PC.

In practice JMP instructions should only be used when a branch is not possible (the branch instructions have the advantage of being relocatable and the short form only takes one word of memory).

10.2 The CCR and conditional branch instructions

10.2.1 The Condition Code Register (CCR)

The BRA and JMP instructions are unconditional transfers of control (i.e. control is always transferred). A common programming requirement is to branch to one of two possibilities depending upon the result of a calculation (an IF structure). Consider the Pascal statement:

```
IF X=0 THEN Y:=20 ELSE Y:=30;
```

If the value of X is zero then Y is assigned the value 20, otherwise, (if X is not 0) Y is assigned the value 30. In assembly language programming there is a similar requirement to test the results of calculations or values of data.

Within the MC68000 processor there is a word sized register called the **Status Register (SR)**, which contains information on the state of the processor (full details are in Chapter 23.1). The low-order byte of the Status Register is the **Condition Code Register (CCR)** which contains a number of bits which depend upon the result of certain instructions. These bits are given single character names that indicate their function.

- Z **Zero** bit that is set to 1 if the previous instruction gave a zero result otherwise it is reset or cleared to 0.
- N **Negative** bit takes on the value of the most significant bit of a result. Thus if twos complement signed arithmetic is being used, this will be set to 1 if the result is negative otherwise it is reset. If unsigned arithmetic is being used, this bit is ignored.
- C **Carry** bit holds the carry from the most significant bit produced by arithmetic instructions or shifts.
- X **Extend** bit is always set to the same value as the carry bit whenever it is affected by an arithmetic instruction. This bit is used in multi-precision arithmetic operations.
- V **Overflow** bit is set if an arithmetic operation resulted in an overflow (i.e. the result was too large to be represented by the operand size and is therefore incorrect). For example, if an addition of two signed byte operands resulted in a positive value larger than 127, the overflow bit would be set.

Within the Condition Code Register (CCR) the bits are positioned as follows:

bit	7	6	5	4	3	2	1	0
				X	N	Z	V	C

Bits 5, 6 and 7 are not used. There are special forms of the MOVE, ANDI, ORI and EORI instructions (see Chapter 12) which can be used to manipulate bits within the CCR. For example, the **MOVE to CCR** instruction:

```
        MOVE.W    #0,CCR       clear all the condition codes
        MOVE.W    #5,CCR       set Z and C and clear the others
```

The source operand can be specified using *any data addressing mode*, and although the length must be specified as .W, only the low-order byte is moved into the CCR.

To read the contents of the CCR of the MC68000, the whole of the Status Register must be read using the **MOVE from SR** instruction:

```
        MOVE.W    SR,D0         move the status register to D0
```

The low-order byte of D0 will contain the CCR condition code bits (the rest of the SR will be covered in Chapter 23.1). The destination operand is specified using a *data alterable addressing mode*. Do not try to move values into the SR until its operation is fully understood.

In the case of the **MC68010/20/30/40 processors** MOVE from SR is a privileged instruction which may not be used by user programs executing in User Mode (see Chapter 23.1). User programs that wish to read the CCR should use the **MOVE from CCR** instruction:

```
        MOVE.W    CCR,D0        move the condition codes to D0
```

The destination is specified using a *data alterable addressing mode* and, although the operand length is word, the CCR is moved into bits 0 to 7 and bits 8 to 15 are cleared (this instruction does not exist on the MC68000).

Many instructions alter the values of the condition codes. Refer to Appendix B to determine which condition code bits are affected by particular instructions.

After an instruction, or a sequence of calculations, the value of particular bits in the condition code register can be tested using the conditional branch instructions described in the next section (e.g. branch to an address if the result of an addition was zero, else execute the next instruction in sequence). **There are no conditional JMP instructions.**

10.2.2 Conditional branch instructions

Conditional branches, similar in structure to **BRA**, branch according to the values of the condition codes. The structure of the conditional branch is:

```
          Bcc.S     TEST
or        Bcc.L     TEST
```

If the condition cc is TRUE, control is transferred to the address TEST (*program counter relative with displacement addressing mode* is always used). If the condition is FALSE, execution continues with the next instruction in sequence. The values of condition cc are shown in Table 10.1 where NOT indicates logical inversion, e.g. consider branch if positive (BPL) in which the N (Negative) bit must NOT be set (note that 0 is treated as a positive number).

cc	Condition tested	Bit tested	Branch **IF**
BEQ	equal (zero)	Z	zero bit is set
BNE	not equal (not zero)	NOT Z	Z bit is NOT set
BMI	minus (negative)	N	N bit is set
BPL	plus (positive)	NOT N	N bit is NOT set
BCS	carry set	C	carry is set
BCC	carry clear	NOT C	carry is NOT set
BVS	overflow set	V	overflow set
BVC	overflow clear	NOT V	overflow NOT set

Table 10.1 Single bit condition code tests

The instructions in Table 10.1 test single bits in the condition code register. Table 10.2 shows more complicated tests using combinations of bits for unsigned and signed numbers. Note that some assemblers do not allow BHS or BLO, so the equivalent BCC or BCS must be used. Table 10.3 shows the combination of CCR bits tested for the various tests.

Condition tested	Signed numbers	Unsigned numbers
equal	BEQ	BEQ
not equal	BNE	BNE
greater than	BGT	BHI
greater than or equal	BGE	BHS or BCC
less than	BLT	BLO or BCS
less than or equal	BLE	BLS

Table 10.2 Combination bit condition code tests

When writing program code the same instructions are used for signed and unsigned addition and subtraction, but the comparison instructions are different (see Table 10.2). The tests for unsigned numbers are useful when testing and comparing address values, e.g. to check if the end of an array has been reached.

Mnemonic	Condition	Encoding	Test
T	true	0000	1
F	false	0001	0
HI	high	0010	$\overline{C}.\overline{Z}$
LS	low or same	0011	C+Z
CC(HS)	carry clear	0100	$\overline{C}$
CS(LO)	carry set	0101	C
NE	not equal	0110	$\overline{Z}$
EQ	equal	0111	Z
VC	overflow clear	1000	$\overline{V}$
VS	overflow set	1001	V
PL	plus	1010	$\overline{N}$
MI	minus	1011	N
GE	greater or equal	1100	$N.V+\overline{N}.\overline{V}$
LT	less than	1101	$N.\overline{V}+\overline{N}.V$
GT	greater than	1110	$N.V.\overline{Z}+\overline{N}.\overline{V}.\overline{Z}$
LE	less or equal	1111	$Z+N.\overline{V}+\overline{N}.V$

Table 10.3 CCR bits tested (‾ is logical NOT, . is logical AND, + is logical OR)
The encoding is used within the operation-code, see Appendix B Table B.3 & B.4

10.3 The test (TST) and compare (CMP) instructions

The **TST** instruction compares the operand with 0 and sets the condition code bits Z and N accordingly (X is not affected, V and C are cleared). The operand can be of length byte, word or long word and specified using *any data alterable addressing mode*, e.g.:

```
        TST.W     D0          test word value in D0
        BPL.S     POS         and branch to label POS if >= 0
```

The **CMP** instruction compares two byte, word, or long word operands and sets the condition code bits without affecting either of the operands, i.e. the instruction subtracts the source operand from the destination, sets the CCR bits and then throws away the result. The source operand of CMP may be specified using *any addressing mode* and the destination operand **must** be a data register, e.g.:

```
CMP.W      D0,D1      compare word D0 and D1
BEQ.S      LABEL      branch to LABEL if D1 = D0
.....                 execution continues if D1 <> D0
```

It is easy to become confused with the **sense of the compare instruction** when using tests such as less than (BLT, BLE, BLO or BLS) or greater than (BGT, BHI, BGE, BHS), e.g.:

```
CMP.W      D1,D0      compare D1 with D0
BLT.S      LABEL      branch if D0 < D1 ???
```

Is the branch taken if D1 < D0 or D0 < D1 ? Consider the two instructions separately:

(a) the CMP is a subtract D0-D1 (destination-source);
then (b) the branch compares this result against 0 (as in the TST instruction above).

In the above example the BLT branch is taken if the result of (D0-D1) < 0, i.e. if D0 < D1. Another way is to consider the two instructions together, where the relationship between the compare and branch is:

	DESTINATION operand of **CMP**	branch test condition	**SOURCE** operand of **CMP**
i.e: branch if:	D0	LT	D1

The branch is taken if D0 < D1 (destination less than source).

If the source operand is a constant **CMPI** should be used (length byte, word or long word and the destination operand specified using *any data addressing mode*), e.g.:

```
CMPI.W     #25,D0     compare D0 with 25
BEQ.S      EQUAL      if D0 = 25 goto label EQUAL
BLT.S      LESS       if D0 < 25 goto label LESS
.....                 execution continues if D0 > 25
```

When comparing characters use ' marks, e.g.:

```
CMPI.B     #'X',D0    is character in D0 'X' ?
BEQ.S      X          if so goto label X
.....                 execution continues if not 'X'
```

Other versions of the compare instruction are:

CMPA where the destination operand is an address register and the source operand may use *any addressing mode* (see Chapter 13).

CMPM to compare memory locations using *address register indirect with postincrement addressing mode* (see Chapter 13 Exercise 13.1).

If CMP is used with the above addressing modes many assemblers will generate the correct instruction (other assemblers will generate an error message). It is recommended practice, however, to use the correct version to make the program code more legible.

10.4 IF control structures

10.4.1 The IF condition .... structure

The IF condition .... construct allows a branch to be taken if a condition is true, i.e.:

If the condition C is true the branch is taken and statements S are executed. In an assembly language program the IF structure may be coded using a branch instruction, for example:

```
* program statements in Structured English
* IF D0 = 25
*     D0 = D0 + 1
* ENDIF
          CMPI.W    #25,D0
          BNE.S     ENDIF1     if D0 <> 25 goto end of IF
          ADDQ.W    #1,D0      if D0 = 25 increment D0
ENDIF1:   .....                next instruction
```

This example shows that an IF construct is coded to branch around the conditional statements. In other words, the branch instruction tests the **opposite** condition from the Structured English construct.

Labels as above can be used to show the parts of the structure and the end of the IF. It is good programming practice, however, to use labels relevant to the meaning of the application program. Remember not to use the same label twice.

10.4.2 The IF condition .... ELSE .... structure

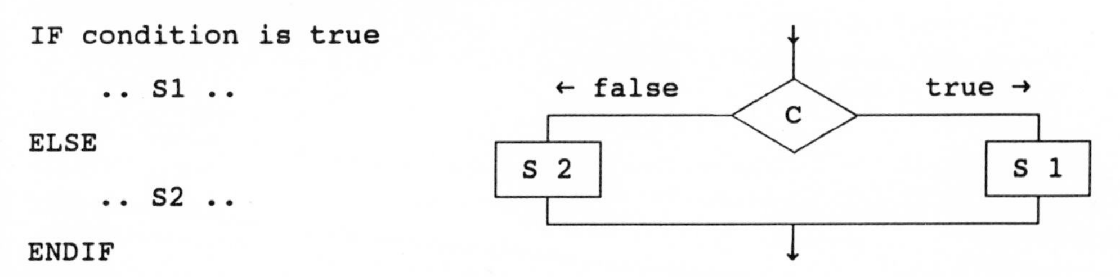

If the condition C is true, statements S1 are executed, else, if the condition is false, statements S2 are executed, e.g.:

```
* program statements in Structured English
* IF D0 = 25
*       D0 = D0 + 1
* ELSE
*       D0 = D0 + 10
* ENDIF
          CMPI.W    #25,D0
          BNE.S     ELSE1
          ADDQ.W    #1,D0        if D0 = 25 then D0=D0+1
          BRA.S     ENDIF1
ELSE1:    ADD.W     #10,D0       if D0 <> 25 then D0=D0+10
ENDIF1:   .....                  next instruction
```

10.5 The set according to condition (Scc) instruction

Many high-level languages have logical variables which are stored in a byte, with TRUE represented by 11111111 binary and FALSE by 00000000 binary. A single IF type operation which tests the condition codes and sets a logical variable is **Scc** (cc as in Tables 10.1 and 10.2). If the condition cc is true the destination operand, of length byte and specified using a *data alterable addressing mode*, is set TRUE otherwise FALSE.

```
          SEQ       LOG1      if Z=0 set LOG1 to TRUE
```

Exercise 10.1 (*see Appendix G for sample answer*)

Design, code and test a program which reads a number from the keyboard and displays a message indicating whether the value was zero, positive or negative.

Problem for Chapter 10

Design, code and test a program which reads characters and displays a message indicating the category of each character

(a) a digit 0..9;
(b) an upper case letter A..Z;
(c) a lower case letter a..z;
(d) some other printable ASCII character, i.e. !"£$%^&* etc.;
or (e) a control character.

Remember that the ASCII codes for 0 to 9, A to Z and a to z are in ascending order, see Appendix A Table A.2. If you are in real difficulty look at Program 11.1 on the next page.

Extend the program:

(f) to count the number of characters in each category;
(g) when the <ESC> character is pressed display the total of each category and exit.

11

Program Loop Control

11.1 WHILE and UNTIL loops

A common program requirement is to execute a sequence of instructions *WHILE* a condition is satisfied or *UNTIL* a condition is satisfied.

11.1.1 The WHILE loop

The WHILE loop can be represented as follows:

```
LOOP WHILE condition C is satisfied

    ..S..

ENDLOOP
```

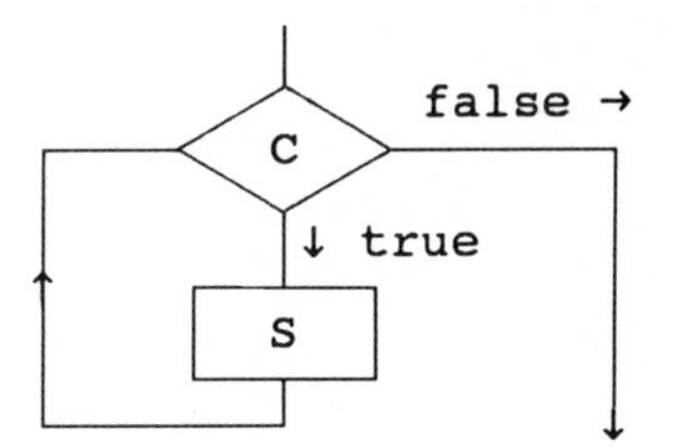

While the condition C is satisfied, the statements S within the loop are executed. Once C is **not** satisfied the loop terminates and program execution continues to the statement following the end of the loop.

```
--------------------------------------------------------------------------------
 1                           * Program 11.1 - to read and echo characters
 2                           *   while the characters are capital letters
 3                           *
 4                           * READ(character)
 5                           * LOOP WHILE character is in the range 'A' to 'Z'
 6                           *     WRITE(character)
 7                           *     READ(character)
 8                           * END LOOP
 9                           * STOP
10                           *
11                           * !! Reference external library subroutines
12                                     XREF      RDCHAR,WRCHAR
13                           *
14 001000 4EB9 0002807E LOOP:          JSR       RDCHAR      read a character
15 001006 0C00     0041                CMPI.B    #'A',D0
16 00100A 6D0E                         BLT.S     STOP        if value < 'A' stop
17 00100C 0C00     005A                CMPI.B    #'Z',D0
18 001010 6E08                         BGT.S     STOP        if value > 'Z' stop
19 001012 4EB9 00028028                JSR       WRCHAR      in range 'A'..'Z' echo
20 001018 60E6                         BRA.S     LOOP
21 00101A 4E4E          STOP:          TRAP      #14         !! STOP PROGRAM
22                                     END
--------------------------------------------------------------------------------
```

Program 11.1 Program using a WHILE loop (Force board version)

Program 11.1 contains a program loop reading and echoing characters **while** capital letters (A..Z) are entered; any other character stops the program. The program begins with a *Structured English* description of the algorithm used, with the comments in the program code explaining how the algorithm has been implemented. There is not an exact one-to-one relationship between the Structured English statements and sequences of assembly language statements. To try and give a one-to-one relationship would require either:

(a) Writing convoluted Structured English to make mapping into assembly language easy (the Structured English is there to document the algorithm not to replicate the assembly code).

or (b) Writing convoluted assembly code to map the Structured English exactly (assembly code is written to gain efficiency by using the most appropriate instructions of the processor, and such a mapping is inconsistent with this aim).

11.1.2 The UNTIL loop

The UNTIL loop can be represented as follows:

```
LOOP

    ..S..

UNTIL a condition is satisfied
```

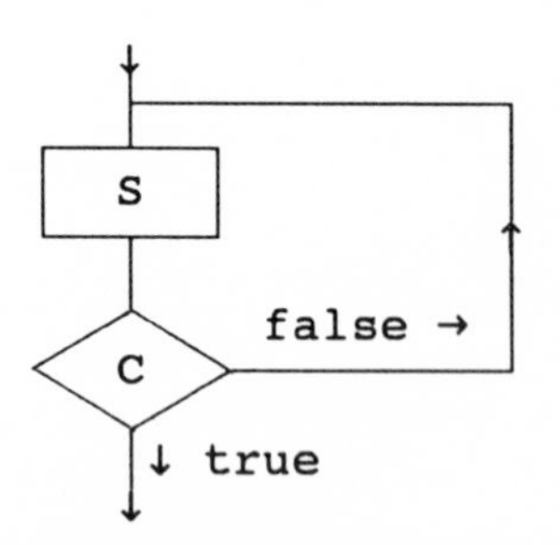

In the **while loop** the condition is tested at the start of the loop, thus if it was satisfied immediately the loop is never executed. In the **until loop** the condition is tested at the end of the loop, so even if the condition is satisfied immediately the statements within the loop are executed at least once. Program 11.2 reads and echoes characters **until** an X is hit (note that the X will be echoed onto the screen).

```
------------------------------------------------------------------------------
 1                          * Program 11.2 -
 2                          *   read and echo characters UNTIL an X is hit
 3                          *
 4                          * LOOP
 5                          *     READ(character)
 6                          *     WRITE(character)
 7                          * UNTIL character='X'
 8                          * STOP
 9                          *
10                                   XREF      RDCHAR,WRCHAR
11 001000 4EB9 00028084 LOOP:        JSR       RDCHAR     read a character
12 001006 4EB9 00028028              JSR       WRCHAR     echo the character
13 00100C 0C00      0058             CMPI.B    #'X',D0    if X exit from loop
14 001010 66EE                       BNE.S     LOOP
15 001012 4E4E                       TRAP      #14        !! STOP PROGRAM
16                                   END
------------------------------------------------------------------------------
```

Program 11.2 Program using an UNTIL loop (Force board version)

Exercise 11.1 *(see Appendix G for sample answer)*

Design and code a subroutine RDBINW which reads a 16-bit number in binary from the keyboard (return the value in D0). Test it with a suitable program. A Structured English description of RDBINW is:

```
SUBROUTINE RDBINW(VAR BINARY_VALUE:INTEGER)
BINARY_VALUE=0
READ(character)
LOOP WHILE (character = '0') OR (character = '1')
    WRITE(character)
    BIT_VALUE = character - ASCII '0'
    BINARY_VALUE = (2 * BINARY_VALUE) + BIT_VALUE
    READ(character)
END LOOP
RETURN
```

11.2 The FOR loop

Many loops are simple in the sense that a number of statements are executed for a number of times, e.g.:

```
counter=0
loop
    ..S..
    counter = counter + 1
until counter > 9
```

In this case the loop is executed ten times with the variable 'counter' taking the values 0 to 9 (e.g. for use as an array index). A simpler way of representing this loop construct is:

```
loop for counter = 0 to 9
    ..S..
end loop
```

Program 11.3 uses a FOR loop to read and echo 10 characters. On each iteration 1 is added to the loop counter in D1 (line 15) and if the value is 10 (lines 16 and 17) the loop terminates (line 18), otherwise it repeats again (goto line 13).

If the value of the loop counter is not used for anything else it is simpler to implement a LOOP FOR n DOWNTO m construct using the decrement and branch instruction DBRA (some assemblers use the mnemonic DBF for this instruction - check the manuals), e.g.

```
          DBRA      D1,LOOP     loop while D1 <> -1
```

DBRA has a data register as the source operand and a label as the destination operand. The data register (always of length word) is decremented and if the result is -1 the loop terminates, otherwise a branch is made to the label. In Program 11.4 a loop using DBRA is executed 10 times (line 15), reading characters and displaying them. The initial value of the loop counter (line 12) is specified as one less than the required loop count, due to the loop terminating when the count equals -1. This is different from many other processors which terminate loops when the count equals 0. Remember that DBRA operates on word length values in the data registers, and terminates the loop when the loop count is -1.

```
--------------------------------------------------------------------------------
 1                        * Program 11.3 - to read and echo ten characters
 2                        *
 3                        * LOOP FOR ten times
 4                        *     READ(character)
 5                        *     WRITE(character)
 6                        * END LOOP
 7                        * STOP
 8                        *
 9                        * the loop counter is held in D1
10                        **
11                                  XREF      RDCHAR,WRCHAR
12 001000 4281            START:    CLR.L     D1          zero loop counter
13 001002 4EB9 00028084   LOOP:     JSR       RDCHAR      read a character
14 001008 4EB9 00028028             JSR       WRCHAR      and echo it
15 00100E 5241                      ADDQ.W    #1,D1       increment loop counter
16 001010 0C41      000A            CMPI.W    #10,D1
17 001014 66EC                      BNE.S     LOOP        if D1<>10 loop again
18 001016 4E4E                      TRAP      #14         !! STOP PROGRAM
19                                  END
--------------------------------------------------------------------------------
```

Program 11.3 FOR loop using separate increment, compare and branch (Force version)

```
--------------------------------------------------------------------------------
 1                        * Program 11.4 - to read and echo ten characters
 2                        *
 3                        * LOOP FOR ten times
 4                        *     READ(character)
 5                        *     WRITE(character)
 6                        * END LOOP
 7                        * STOP
 8                        *
 9                        * the loop counter is held in D1
10                        *
11                                  XREF      WRCHAR,RDCHAR
12 001000 7209            START:    MOVEQ     #9,D1       set loop counter -1
13 001002 4EB9 00028084   LOOP:     JSR       RDCHAR      read a character
14 001008 4EB9 00028028             JSR       WRCHAR      and display it
15 00100E 51C9      FFF2            DBRA      D1,LOOP     loop while D1 <> -1
16 001012 4E4E                      TRAP      #14         !! STOP PROGRAM
17                                  END
--------------------------------------------------------------------------------
```

Program 11.4 FOR loop using a single DBRA instruction (Force board version)

Sometimes it is necessary to terminate a loop prematurely, e.g. if an error occurs. This can be done either by using a separate CMP and branch or by decrement and branch instructions which also test the condition codes (DBcc: cc being the condition to be tested, see Tables 10.1 and 10.2. When a DBcc instruction is executed the sequence is as follows:

(a) if the condition cc is TRUE the loop terminates,

else (b) the data register is decremented then:

(i) if the result is -1 the loop terminates,

else (ii) a branch is made to the label.

```
--------------------------------------------------------------------------------
 1                          * Program 11.5 - read and echo ten characters
 2                          *   if a space character is hit exit the loop
 3                          *
 4                          * LOOP FOR ten times
 5                          *     READ(character)
 6                          *     WRITE(character)
 7                          *     IF character = ' ' EXIT LOOP
 8                          * END LOOP
 9                          * STOP
10                          *
11                          * the loop counter is held in D1
12                          *
13                                    XREF      WRCHAR,RDCHAR
14                          *
15 400400 7209              START:    MOVEQ     #9,D1       set loop counter -1
16 400402 4EB9 0000807E     LOOP:     JSR       RDCHAR      read a character
17 400408 4EB9 00008028               JSR       WRCHAR      and display it
18 40040E 0C00      0020              CMPI.B    #' ',D0     was it a space
19 400412 6704                        BEQ.S     ENDLOOP     if a space EXIT loop
20 400414 51C9      FFEC              DBRA      D1,LOOP     branch if D1 <> -1
21 400418 4E4E              ENDLOOP:  TRAP      #14         !! STOP PROGRAM
22                                    END
--------------------------------------------------------------------------------
```

Program 11.5 Using separate CMP for loop EXIT (Kaycomp board version)

```
--------------------------------------------------------------------------------
 1                          * Program 11.6 - read and echo ten characters
 2                          *   if a space character is hit exit the loop
 3                          *
 4                          * LOOP FOR ten times
 5                          *     READ(character)
 6                          *     WRITE(character)
 7                          *     IF character = ' ' EXIT LOOP
 8                          * END LOOP
 9                          * STOP
10                          *
11                          * the loop counter is held in D1
12                          *
13                                    XREF      WRCHAR,RDCHAR
14                          *
15 400400 7209              START:    MOVEQ     #9,D1       set loop counter -1
16 400402 4EB9 0000807E     LOOP:     JSR       RDCHAR      read a character
17 400408 4EB9 00008028               JSR       WRCHAR      and display it
18 40040E 0C00      0020              CMPI.B    #' ',D0     was it a space
19 400412 57C9      FFEE              DBEQ      D1,LOOP     branch if D1 <> -1
20                          *                                and not a space
21 400416 4E4E                        TRAP      #14         !! STOP PROGRAM
22                                    END
--------------------------------------------------------------------------------
```

Program 11.6 Using DBEQ for loop EXIT (Kaycomp board version)

In Program 11.5 separate branch and DBRA instructions are used to read and echo ten characters; if a space character is entered the loop EXITs (lines 18 and 19) otherwise DBRA is executed (line 20). Program 11.6 is similar but uses a single DBEQ instruction (lines 18 and 19).

The MC68010 (Motorola 1989) has a special **loop mode** of operation for use with DBcc instructions which is used to speed up simple loops such as memory copy or search.

11.2.1 Nested loops

A loop structure can also contain other loops. In the following example the inner loop would be executed 200 times, i.e. 20 times for each execution of the outer loop. When coding such structures use different registers for each loop control variable.

```
loop for 10 times
    loop for 20 times
        ....
        ....
    end loop
end loop
```

Exercise 11.2 (*see Appendix G for sample answer*)

Using loops and subroutines design, code and test a program to display the pattern:

```
**********??????????**********
**********??????????**********
**********??????????**********
**********??????????**********
**********??????????**********
```

Problems for Chapter 11

1 Design and code a subroutine to calculate the factorial of a word length integer N (due to the 16-bit word length calculations the maximum value of N will be 7, i.e. 8! = 40320). On entry D0 should contain N and on exit should contain N!. Test the subroutine using a suitable program and test data.

Algorithm and Structured English:

N! = N * (N-1) * (N-2) 1
for N >= 1

```
FACTOR=1
LOOP WHILE N>1
    FACTOR = FACTOR * N
    N = N - 1
END LOOP
```

2 Using modular programming techniques design, code and test a program which displays a Christmas tree pattern. Use loops to display this pattern (i.e. make a single call to WRCHAR from within a loop to display the * characters, not fourteen calls to WRTEXT).

Extend the program so that the height of the tree can be entered at run time.

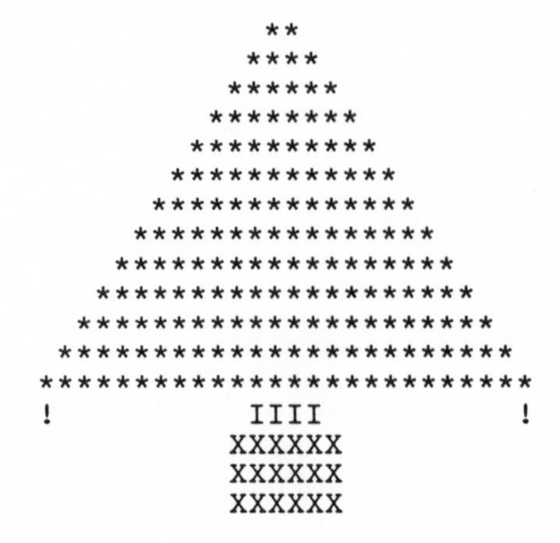

12

Logical, Shift, Rotate and Single Bit Instructions

This group of instructions allows operations on single bits (or groups of bits) within bytes, words or long words. The instructions can be used, for example, to test if certain bits are set in an input/output device interface thereby indicating whether the device is ready for input or output commands.

12.1 NOT, AND, OR and EOR (Exclusive OR) logical instructions

The **NOT** instruction inverts the value of a bit (i.e. 0 becomes 1 and 1 becomes 0). The other instructions operate on two bits (for example A and B). A truth table can show the combinations of the bits A and B and the resultant values:

A	B	AND	OR	EOR
0	0	0	0	0
0	1	0	1	1
1	0	0	1	1
1	1	1	1	0

In summary:

AND if A=1 **AND** B=1 the result is 1, otherwise the result is 0
OR if A=1 **OR** B=1 the result is 1, otherwise the result is 0
EOR if (A **OR** B is 1) **AND** (A **AND** B is **NOT** 1) the result is 1, otherwise result is 0

Logical instructions allow bits within data to be manipulated. For example, the MC68000 **NOT** instruction inverts or complements every bit in a destination operand of length byte, word or long word, which can be specified using a *data alterable addressing mode*, e.g.:

```
NOT.B      D0          invert bits 0 to 7 of D0
NOT.L      LABEL       invert long word in memory
```

If in the first example, if bits 0 to 7 of D0 were 00001010 before the operation, then the result (of NOT.B D0) would be 11110101.

The MC68000 **AND** and **OR** instructions perform their logical operations between each bit of the source operand and the corresponding bits in the destination operand. The result is placed into the destination operand, e.g.:

```
AND.B      D0,D1        AND byte values in D0 and D1
AND.B      #$0F,D0      AND byte in D0 with $0F
AND.W      D4,LABEL     AND D4 with value in memory
OR.W       #$0FF00,D0  logical OR with D0
OR.L       LABEL,D2     logical OR memory with D0
```

The operand length of **AND** and **OR** may be byte, word or long word and specified:

(a) **source in a data register** and destination in a data register or specified by a *alterable memory addressing mode*,

or (b) source specified by a *data addressing mode* and **destination in a data register.**

The **EOR** (Exclusive OR) instruction operates upon byte, word or long word data, and performs an exclusive OR, between a source operand which **must be in a data register**, and a destination operand specified using a *data alterable addressing mode*, e.g.:

```
EOR.B      D0,D1        exclusive OR D0 and D1
EOR.W      D2,LABEL     exclusive OR D2 and memory
```

If the operations were carried out on the byte values 00001010 and 01001100, the results would be AND = 00001000, OR = 01001110 and EOR = 01000110.

The immediate form of the logical instructions (**ANDI, ORI and EORI**) use an immediate operand of length byte, word or long word as the source, and a destination operand that may be specified using a *data alterable addressing mode*, e.g.:

```
ANDI.B     #$0F,D0        AND byte $0F with in D0
ORI.L      #$1F,LABEL     OR long word $1F with memory
EORI.W     #$0F,D2        exclusive OR word $0F and D2
```

It is good programming practice to use the immediate forms ANDI, ORI or EORI even when using a destination operand in a data register.

The instructions ANDI, ORI and EORI, with an immediate byte (.B) length source operand, can be used to alter bits in the CCR (Condition Code Register), e.g.:

```
ANDI.B     #$F7,CCR     clear the N bit
ORI.B      #4,CCR       set the Z bit
EORI.B     #4,CCR       invert the Z bit
```

Without affecting the state of the other CCR bits the first example clears the N (Negative) bit, the second sets the Z (Zero) bit, and the third inverts the Z bit.

When the processor is in supervisor mode (see Chapter 23.1), the logical instructions ANDI, ORI and EORI with a word (.W) length source operand, can be used to alter individual bits in the SR (Status Register).

12.2 Shift and rotate instructions

Shift and rotate instructions allow bits to be moved around within a register or a memory location. The source and destination operands of the shift and rotate instructions are:

1. Destination operand is a data register. All operand lengths are allowed and the number of bits to shift or rotate are set by:

   (a) an immediate source operand in the range 1 to 8,

   or (b) a source operand in another data register, having a value in the range 0 to 63.

2. Destination operand in memory (specified using a *memory alterable addressing mode*). The operand length is word, and the shift or rotate is by 1 bit position only.

Many of the shift and rotate instructions make use of the X (Extend) bit in the CCR.

12.2.1 Arithmetic shifts

Arithmetic shifts assume that the values being shifted are twos complement binary values (i.e. signed binary numbers).

ASL (Arithmetic Shift Left) shifts the destination operand left by the number of bits specified:

(a) shifting in 0s from the right,
and (b) shifting the most significant bit out into C and X.

For example, ASL.B:

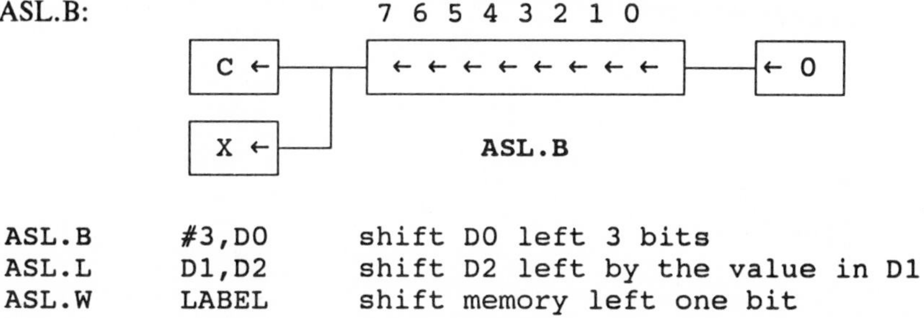


```
ASL.B     #3,D0      shift D0 left 3 bits
ASL.L     D1,D2      shift D2 left by the value in D1
ASL.W     LABEL      shift memory left one bit
```

If in the first example D0 held the binary value 00001010 (decimal 10) before the operation, the result would be 01010000 (decimal 80), and C and X would both equal 0. Therefore using ASL to shift a binary number left one bit is equivalent to a multiply by 2, i.e. the first example above is shifted left three bits to give the value ((10*2)*2)*2 = 80.

The C and X condition code bits are set to the value of the last bit shifted out. V (overflow) indicates if a sign change occurred during the shift, and Z (Zero) and N (Negative) are set according to the result.

ASR (Arithmetic Shift Right) shifts the destination operand right by the number of bits specified:

(a) shifting or extending the sign bit from the left,
and (b) shifting the least significant bit out into C and X.

For example ASR.B:

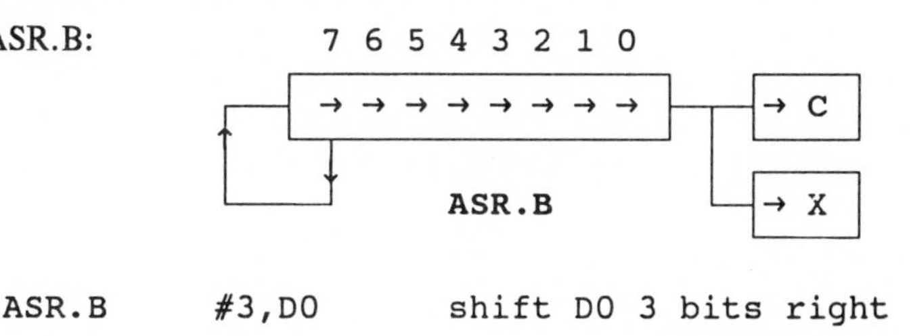


```
ASR.B     #3,D0      shift D0 3 bits right
```

If D0 contained 01010000 (decimal 80) before the operation it would be 00001010 (decimal 10) after, or if it was 10110000 (decimal -80) before the operation it would be 11110110 (-10 decimal) after. Thus a division by two is performed for each bit position shifted.

The Z and N bits are set according to the result and V is cleared. If the shift count is zero, both ASL and ASR clear the C bit and leave the X bit unaffected.

Using arithmetic shifts is a very efficient way of carrying out multiply and divide of 2, 4, 8, etc., in time critical code.

12.2.2 Logical shifts

Logical shifts are similar in form to arithmetic shifts except that the operands are assumed to be simple bit patterns (i.e. not signed numbers).

LSL (Logical Shift Left) is identical to ASL except that the V (overflow) bit is always set to 0.

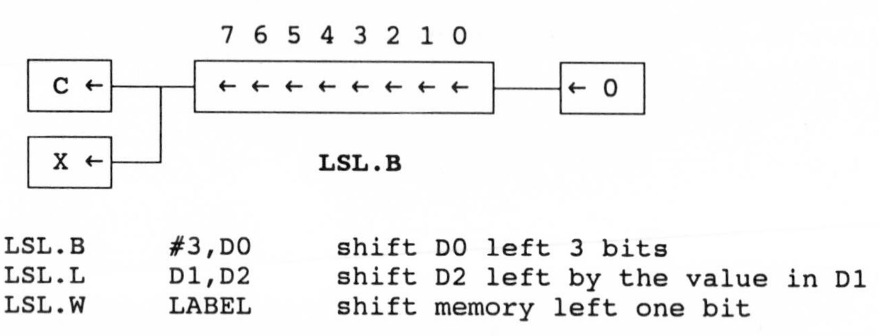


```
LSL.B      #3,D0      shift D0 left 3 bits
LSL.L      D1,D2      shift D2 left by the value in D1
LSL.W      LABEL      shift memory left one bit
```

LSR (Logical Shift Right) is identical to ASR except that the bits shifted in from the left are 0s, not copies of the sign bit.

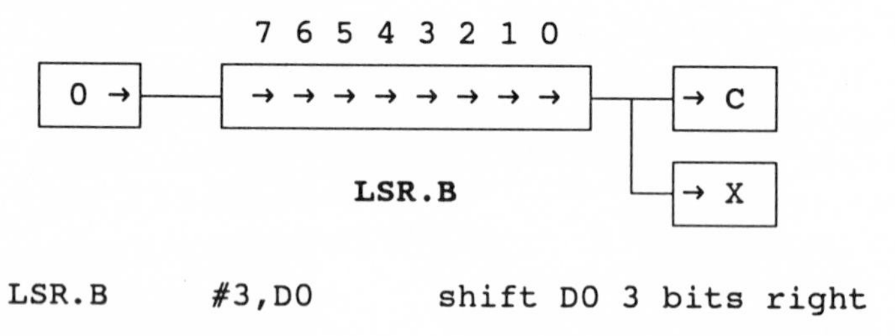


```
LSR.B      #3,D0       shift D0 3 bits right
```

12.2.3 Rotate instructions

With the shift instructions the bits shifted out were moved into C and X and then lost, and the bits shifted in were 0s or a copy of the sign bit. In the case of rotate instructions, bits rotated out at one end are rotated into the other end.

ROL (Rotate Left) and **ROR (Rotate Right)** rotate the operand by the specified number of bits, and leave in C a copy of the last bit moved. N is set if the most significant bit of the result is set, Z is set if the result is zero, V is cleared and X is not affected. For example ROL.B and ROR.B:

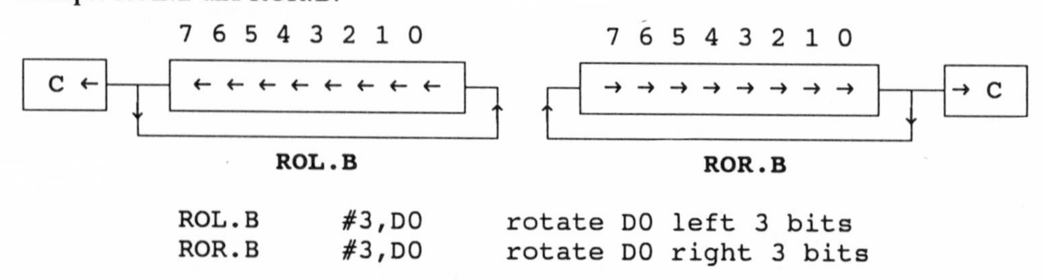


```
ROL.B      #3,D0      rotate D0 left 3 bits
ROR.B      #3,D0      rotate D0 right 3 bits
```

If D0 contained 00001010 before the operation it would contain 01000001 after the ROR.B instruction and 01010000 after the ROL.B instruction.

The rotate with extend instructions, **ROXL** and **ROXR**, are similar to ROL and ROR, except that the rotate goes via the X bit in the CCR (Condition Code Register), i.e. each bit is rotated out into X (and C) as X is rotated in at the other end. N is set if the most significant bit of the result is set, Z is set if the result is zero, and V is cleared. For example ROXL.B and ROXR.B:

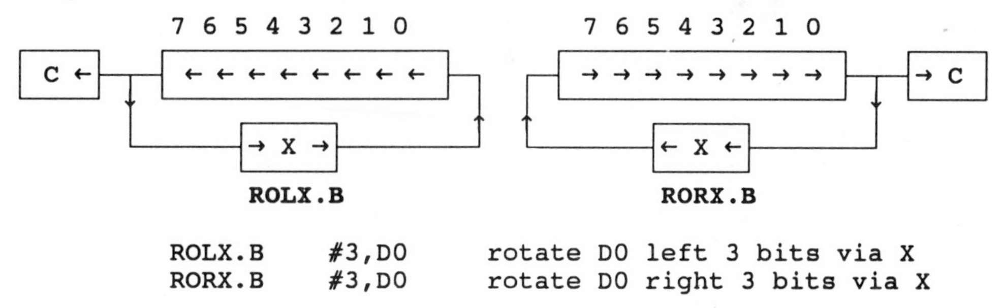


```
ROLX.B     #3,D0      rotate D0 left 3 bits via X
RORX.B     #3,D0      rotate D0 right 3 bits via X
```

Exercise 12.1 *(see Appendix G for sample answer)*

Extend the program written in Exercise 11.1 (to test the RDBINW subroutine) by adding a subroutine WRBINW which displays in binary the word value stored in D0. Use shift and/or rotate instructions to get the individual bits and test them.

12.3 Operations on single bits

The instructions **BCLR** (bit clear), **BSET** (bit set), **BCHG** (bit change) and **BTST** (bit test) operate upon a single bit in the destination operand. The bit position is specified by a source operand which is either immediate or in a data register (the least significant bit is bit number 0). The destination operand may be:

(a) in a data register, in which case the operand length is always long word,

or (b) in memory (specified using a *data alterable addressing mode*), in which case the operand length is always byte.

```
BCLR.L   #0,D0      test & clear bit 0 of D0
BSET.B   #3,LABEL   test & set bit 3 of memory byte
BSET.L   D0,D1      test & set bit in D1 specified by D0
BCHG.L   #23,D0     test & change bit 23 of D0
```

If the operand length is not specified, the assembler will assume long word in a data register and a byte in memory, i.e. the **only** possible values. The sequence of events in these operations is:

(a) test the value of the bit specified and set the Z bit to this value (the other CCR bits are not affected);

then (b) set the specified bit according to the operation.

The **BTST** instruction tests the value of a bit in a destination operand (specified using a *data addressing mode*) and sets or clears the Z bit accordingly (the other CCR bits are not affected). BTST can be considered as the functional equivalent of the first part of the above operations, e.g.:

```
BTST.L    #3,D0       test bit 3 of D0
BTST.B    #1,LABEL    test bit 1 of byte LABEL
BTST.L    D0,D1       test bit in D1 specified by D0
```

Exercise 12.2 *(see Appendix G for sample answer)*

Modify the program of Exercise 12.1 by implementing another version of the subroutine WRBINW which displays in binary the word value stored in D0. In this case use single bit instructions to get the individual bits and test them.

12.4 Indivisible test and set an operand (TAS) instruction

The **TAS** instruction tests a **byte** operand, specified using a *data alterable addressing mode*, and sets the N and Z bits accordingly. The high order bit (bit 7) of the operand is then set to 1. For example,

```
TAS.B    D0           test & set D0
TAS.B    LABEL        test & set LABEL
```

This instruction is indivisible (i.e. will not be interrupted) and can be used to allow the synchronisation of several processors, e.g. the operand could be an inter-processor flag.

Problem for Chapter 12

When serial data (see Chapter 19.2) is transmitted over a noisy communications system (e.g. a telephone line) a parity bit can be added to check for transmission errors. To generate **even parity** the data bits are summed and the parity bit is added to make the total an even number (or odd for odd parity). From Table A.2 it can be seen that in the ASCII character code bit 7 (the most significant bit) is always 0. When transmitting ASCII characters two parity check techniques can be used:

(a) The parity is added on the end of the data bits (i.e. eight data bits plus one parity bit are transmitted, bit 0 first). For example, the character A would be 001000001 (parity bit is 0) and C would be 101000011 (parity bit is 1).

or (b) Bit 7 of the ASCII code is replaced by the parity bit (i.e. seven data bits plus one parity bit are transmitted). For example, the character A would be 01000001 (parity bit is 0) and C would be 11000011 (parity bit is 1).

At the receiver, the parity is generated again and compared with the value of the parity bit received. If it is different an error has occurred which will be indicated by an error bit in the interface status register. The parity bit is removed when the data is passed to the computer (to give normal ASCII characters). The parity check can only detect certain error conditions (e.g. if two bits are in error the data could pass the parity error check). More complex error detection (and correction) systems are available if more secure data communications are required.

Design and code a subroutine which, when called with an ASCII character in D0, sets (or clears) bit 7 to give even parity (i.e. seven data bits plus one parity bit). Design and code a test program (use the WRBINW subroutine written in Exercises 12.1 or 12.2 to display the binary values) and test it with suitable test data.

Review of Chapters 10, 11 and 12

Chapters 10 and 11 described the implementation of basic program control structures (IF, IF .. ELSE, WHILE, LOOP .. UNTIL, FOR, etc.) in MC68000 assembly language. Chapter 12 covered logical and shift instructions.

10.1 The **BRA** (branch) instruction causes an unconditional transfer of control to the specified address. The address is stored as an offset from the current address (called program counter relative addressing) in either (a) short form (as a byte) allowing a branch of -126 to +129 locations relative to the current address, or (b) in long form (as a word) allowing a branch of -32766 to +32769 locations relative to the current address. By using offsets relative to the PC the branch instruction are relocatable.

```
        BRA.S       LABEL           short branch
        BRA.L       LABEL           long branch
```

The JMP (Jump) instruction causes an unconditional transfer of control to the specified address (specified using a *control addressing mode*) which can be anywhere in memory.

10.2 The **CCR (Condition Code Register)** is the lower byte of the **SR (Status Register)** and contains bits which depend upon the result of certain instructions. The bits are **Z** (Zero), **N** (Negative), **C** (Carry), **X** (Extend) and **V** (Overflow). The values of the CCR bits can be altered using special forms of MOVE, ANDI, ORI or EORI operations and read using MOVE from SR (MOVE from CCR on the MC68010)

Conditional branch instructions (similar in structure to the unconditional branch) branch according to the values of the condition codes. The structure is (short and long form):

```
        Bcc.S       LABEL
        Bcc.L       LABEL
```

The values of cc are EQ, NE, MI, PL, CS, CC, VS, VC, GT, HI, GE, HS, LT, LO, LE and LS (for full details see Tables 10.1 and 10.2).

10.3 The test (TST) instruction tests an operand (specified by a *data alterable addressing mode*) and sets the Z and N bits accordingly (C and V are cleared and X is not affected):

```
        TST.W       D0          test word value in D0
        BPL.S       POS         branch to label POS if >= 0
```

The compare (CMP) instruction compares a source operand (specified using *any addressing mode*) with the contents of a destination data register and sets the CCR bits accordingly. The compare immediate (CMPI) instruction compares an immediate source operand with a destination operand (specified using a *data alterable addressing mode*) and sets the CCR bits accordingly. Both forms of compare can compare byte, word or long word operands. The sense of the compare operation is:

DESTINATION operand of CMP   branch test condition   SOURCE operand of CMP

For example:

```
        CMPI.W      #25,D0          compare D0 with 25
        BLE.S       LABEL           branch to LABEL if D0 <= 25
```

10.4 described the IF conditional ...., and IF conditional .... ELSE structures and their implementation in assembly code:

```
IF condition is true                IF condition is true
    ..S..                               ..S1..
ENDIF                               ELSE
                                        ..S2..
                                    ENDIF
```

11.1 described the WHILE and UNTIL loop structures:

```
LOOP WHILE a condition is satisfied     LOOP
    ..S..                                   ..S..
END LOOP                                UNTIL a condition is satisfied
```

Using WHILE the condition is tested at the start of the loop and using UNTIL it is tested at the end of the loop (thus executing the loop at least once).

11.2 The FOR loop which may be implemented using the decrement and branch **(DBRA)** instruction. DBRA has a data register as the source operand and a label as the destination operand. The data register is decremented and if the result is -1 the loop terminates, otherwise a branch is made to the label. DBRA is only one of a range of Decrement and Branch instructions which can also test the condition codes (in Program 11.6 DBEQ is used to exit from a loop (a) if the loop count has decremented to -1, or (b) if a space character was hit on the keyboard).

```
LOOP FOR a number of times
    ..S..
END LOOP
```

Loops may be nested as required (care must be taken in the allocation of data registers for the loop counters, see Exercise 11.2).

12.1 The logical **NOT** instruction inverts or complements the value of every bit in a destination operand (length byte, word or long word specified by a *data alterable addressing mode*). The logical **AND** and **OR** instructions perform the logical operations (of length byte, word or long word) between each bit of a source operand and the corresponding bit of a destination operand. The operands are either:

source in a data register and destination in a data register or specified using a *alterable memory addressing mode*;

or source specified using a *data addressing mode* and **destination in a data register**.

EOR performs exclusive OR (length byte, word or long word) between a source operand in a data register and a destination operand specified by a *data alterable addressing mode.*

```
NOT.W     D0          invert bits in D0 (word)
AND.B     D0,D1       AND byte D0 and D1
OR.W      LABEL,D0    OR memory byte with D0
EOR.L     D0,D2       exclusive OR D0 and D2 long word
```

The immediate forms of the logical instructions (ANDI, ORI and EORI) use an immediate source operand of length byte, word or long word and a destination operand specified using a *data alterable addressing mode.* Special forms of ANDI, ORI and EORI can be used with a destination operand in the CCR.

```
ANDI.B    #$0F,D0        AND byte $0F with D0
ORI.W     #$0FF00,D0     OR word $0FF00 with D0
ORI.L     #$1F,LABEL     OR long word $1F with memory
EORI.W    #$0F,D2        exclusive OR $0F with D2
ANDI.B    #$F7,CCR       clear the N bit
ORI.B     #4,CCR         set the Z bit
EORI.B    #4,CCR         invert the Z bit
```

12.2 described the shift and rotate instructions. The operands may be:

1 Destination operand is a data register. All operand lengths are allowed and the number of bits to shift or rotate are set by:

(a) an immediate source operand in the range 1 to 8.

(b) a source operand in another data register, having a value in the range 0 to 63.

2 Destination operand in memory (specified using a *memory alterable addressing mode*). The operand length is word, and the shift or rotate is by 1 bit position only.

ASL (arithmetic shift left) shifts the most significant bit out into C and X, and shifts in 0s from the right (Z, N and V are set according to the result). This can be used as a multiply by 2 operation.

ASR (arithmetic shift right) extends the sign bit from the left and shifts the least significant bit out into C and X (Z and N are set according to the result and V is cleared). This can be used as a divide by 2 operation.

LSL (logical shift left) is identical to ASL except that V is cleared.

LSR (logical shift right) is identical to ASR except that 0s are shifted in from the left (i.e. not sign extended).

ROL and ROR rotate (left and right) the operand by the specified number of bits (leaving in C a copy of the last bit moved, Z and N are set according to the result, V is cleared and X is not affected).

ROXL and ROXR rotate (left and right via X bit) are similar to ROL and ROR except that the bits are rotated via the extend bit X.

12.3 BCLR (bit clear), BSET (bit set), BCHG (bit change) and BTST (bit test) operate upon a single bit in a destination operand. The bit position is specified by the source operand (either immediate or in a data register) and the destination operand which is:

(a) a long word in a data register;

or (b) a byte in memory specified using a *data alterable addressing mode.*

The operation (i) tests the value of the specified bit and sets the Z bit to this value, and then (ii) sets the specified bit according to the operation (except in the case of BTST which only does the test).

```
BCLR.L    #0,D0          test and clear bit 0 in D0
BSET.B    #3,LABEL       test and set bit 3 of memory
BSET.L    D0,D1          test and set bit in D1
BCHG.L    #23,D0         test and invert bit 23 of D0
BTST.L    #3,D0          test bit 3 of D0
```

12.4 The test and set (TAS) instruction tests a byte operand, specified using a *data alterable addressing mode,* and sets the N and Z bits. The high order bit of the operand is then set to 1. The instruction is indivisible and is used to synchronise several processors.

13

Address Register Addressing

The processor contains eight address registers, A0 through A7, used for storing memory addresses (e.g. as array pointers). Address register A7 serves a special purpose as the system **Stack Pointer** (the stack is a dynamic data area, see Chapter 14). The contents of address registers can be accessed directly using *address register direct addressing mode*, with the restriction that they can only be used for word and long word length operands. There are special versions of move, add and subtract for use when the destination operand is specified using *address register direct addressing mode*, for example:

```
MOVEA.W     LABEL,A0     move contents of LABEL into A0
MOVEA.L     #LABEL,A0    move address of LABEL into A0
SUBA.L      #$0FF,A1     subtract constant $FF from A1
ADDA.W      #$0F,A2      add constant $F to A2
```

The source operand of these instructions can be specified using *any addressing mode*. A major difference between the above instructions and the ordinary MOVE, ADD, SUB, etc., is that the CCR condition code bits are **not affected**. This allows address registers to be loaded without affecting condition codes set by the results of previous calculations. To test the value of an address register (and set the condition code bits) the **CMPA** instruction compares a source operand, specified by *any addressing mode*, with a destination operand in an address register (length word or long word). Remember that addresses should be treated as **unsigned numbers** when using conditional branch instructions.

```
CMPA.L      #$1000,A0    compare $1000 with A0
```

13.1 Address register indirect addressing

In many cases the effective address of an operand is not known until run-time, e.g. to access an array element. The address registers can contain address values to be manipulated and then used as pointers to memory operands. This is called *address register indirect addressing mode* (indicated by enclosing the address register name in brackets) and may be used for all types of operand reference: data, memory, alterable and control.

Address register indirect addressing mode is often used for processing arrays and lists. Program 13.1 (next page) uses address register indirect addressing to sum the elements of an array of word values (stored in a data area at the end of the program). At line:

12 `MOVEA.L #ARRAY,A0`, moves the start address of the array into A0 (which will then be used to access the array elements using *address register indirect addressing mode*)
13 D0 (which will hold the SUM) is cleared
14 Moves the length of the array minus 1 into D1 where it will be used for loop control
15 `ADD.W (A0),D0`, adds the next element of the array to SUM (in D0)
16 Two is added to the address in A0 so it points to the next array element for the next loop, i.e. the elements are words each of length two bytes
17 The DBRA instruction branches to line 15 if the loop is not finished (D1 <> -1)

When the loop is complete (D1=-1) the value of SUM is displayed. It is important to note that when working in assembly language the initial value of the array index is 0.

```
 1                                * Program 13.1 - sum an array of 10 elements
 2                                *
 3                                * sum = 0
 4                                * FOR  index := 1 TO 10 DO
 5                                *     sum := sum + array[index]
 6                                * END LOOP
 7                                * STOP
 8                                *
 9                                * SUM in D0, loop count in D1 & array address in A0
10                                *
11                                          XREF      WRTEXT,WRDECW,WRLINE
12 001000 207C 00001036 START:              MOVEA.L   #ARRAY,A0     array start address
13 001006 4280                              CLR.L     D0            initialise SUM to 0
14 001008 7209                              MOVEQ     #9,D1         D1 holds loop count
15 00100A D050          LOOP:               ADD.W     (A0),D0       add array element
16 00100C 5488                              ADDQ.L    #2,A0         point at next element
17 00100E 51C9     FFFA                     DBRA      D1,LOOP       loop if D1 <> -1
18 001012 4EB9 0002802C                     JSR       WRLINE
19 001018 4EB9 00028038                     JSR       WRTEXT
20 00101E 73 75 6D 20 6F                    DC.W      'sum of array = $'
21 00102E 4EB9 00028074                     JSR       WRDECW        display sum
22 001034 4E4E                              TRAP      #14           !! STOP PROGRAM
23                                *
24                                * data area
25 001036 0014 0023 0043ARRAY:              DC.W      20,35,67,-6,78,-89,4,56,-9,10
26                                          END
```

Program 13.1 Program to sum the elements of an array (Force or Bytronic board)

13.1.1 Postincrement and predecrement addressing modes

Address Register Indirect with Postincrement (also called autoincrement) and *Address Register Indirect with Predecrement* (also called autodecrement) addressing modes are variations of address register indirect addressing.

Postincrement: *after* the operand address has been obtained from the address register, the contents of the address register are incremented by the length of the operand (i.e. 1 for byte, 2 for word and 4 for long word). *Postincrement mode* is indicated by a + symbol after the operand. For example, in the following instruction the source operand is obtained (from the address in A0), A0 is incremented by 2 and the ADD.W performed:

```
ADD.W       (A0)+,D0    add indirect then increment
```

Predecrement: the contents of the address register are decremented by the length of the operand *before* the operand address is obtained (e.g. to process a list backwards). *Predecrement mode* is indicated by a - symbol before the operand. For example, in the following instruction the contents of A0 are decremented by 4, the source operand is obtained (from the address in A0) and the ADD.L performed:

```
ADD.L       -(A0),D0    decrement then add indirect
```

Postincrement and predecrement addressing modes can be used for data, alterable and memory operand references, i.e. they may not be used for a control reference such as:

```
            JMP         (A0)+       this is not allowed
```

Program 13.2 (below), a modified version of Program 13.1 for the XA8 cross assembler, uses *postincrement mode* to sum the array elements. The summation loop is (lines 18 and 19):

```
LOOP:       ADD.W       (A0)+,D0     add array element
            DBRA        D1,LOOP      loop if D1 <> -1
```

In line 18 the source operand of the ADD instruction is specified using *postincrement mode*. After the word length array element has been added to D0, A0 is automatically incremented by 2 to point to the next array element (i.e. the operand size is word).

Note in Program 13.2 that the code and data are in separate program sections specified by the .PSECT pseudo-operator (_text for the code section and _data for the data section).

```
--------------------------------------------------------------------------------
 1                       ; Program 13.2 - sum an array of 10 elements
 2                       ;
 3                       ; sum = 0
 4                       ; FOR index := 1 TO 10 DO
 5                       ;     sum := sum + array[index]
 6                       ; END LOOP
 7                       ; STOP
 8                       ;
 9                       ; SUM in D0, loop counter in D1 & array address in A0
10                       ;
11                               .PROCESSOR  M68000
12                               .EXTERNAL   WRTEXT,WRDECW,WRLINE
13                               .PSECT      _text        ;start code section
14                       ;
15   000000  207C        START:  MOVEA.L     #ARRAY,A0    ;array start address
             00000034
16   000006  4280                CLR.L       D0           ;initialise SUM to 0
17   000008  7209                MOVEQ       #9,D1        ;D1 holds loop count
18   00000A  D058        LOOP:   ADD.W       (A0)+,D0     ;add array element
19   00000C  51C9FFFC            DBRA        D1,LOOP      ;loop if D1 <> -1
20   000010  4EB9                JSR         WRLINE
             00000000
21   000016  4EB9                JSR         WRTEXT
             00000000
22   00001C  73756D20            .TEXT       "sum of array = $"
23                               .EVEN
24   00002C  4EB9                JSR         WRDECW       ;display sum
             00000000
25   000032  4E4E                TRAP        #14          ;!! STOP PROGRAM
26                       ;
27                               .PSECT      _data        ;start data section
28                       ; data area
29   000000  0014        ARRAY:  .WORD       20,35,67,-6,78,-89,4,56,-9,10
30                               .END
--------------------------------------------------------------------------------
```

Program 13.2 Program to sum an array using *postincrement mode* (XA8 assembler)

Exercise 13.1 (*see Appendix G for sample answer*)

Design, code and test a subroutine which compares two ASCII character strings of the same length and displays a message indicating whether or not they are identical. On entry to the subroutine the following parameters should be set up:

D0 (word value) holds the string length in bytes;
A0 (long word) holds the address of the first character of string 1;
A1 (long word) holds the address of the first character of string 2.

Hint; the CMPM instruction compares two operands in memory (length byte, word or long word) when both are accessed using *postincrement addressing mode*, e.g.:

```
        CMPM.B   (A0)+,(A1)+  compare two byte operands in memory
```

13.1.2 Address register indirect with displacement

A variation of indirect address register addressing adds a fixed displacement or offset value to the contents of an address register to form the effective address. For example:

```
            CLR.B       $40(A0)         clear a byte in memory
```

The above instruction adds $40 to the contents of A0 to form the effective address (i.e. if A0 = $2000 then the byte at memory address $2040 would be cleared). The displacement is stored, following the operation-code in memory, as a **16-bit signed value** which is sign extended to 32 bits. This addressing mode can be used for all types of operand reference (data, memory, alterable and control) and is useful for accessing variables from a base address (see below), elements of arrays or records (see Appendix E), or values that are stored in stack structures (see Chapter 14). The displacement is a byte count, therefore it is necessary to allow the correct offsets for word (*2) and long word (*4) operands.

```
-------------------------------------------------------------------------------
1                           * Program 13.3:  SUM=A+B using word data in memory
2                           *  data accessed from a base address using
3                           *  address register indirect with displacement mode
4                           *
5                           * define variables as offsets from base address
6           00000000        A:        EQU       0          A offset from base
7           00000002        B:        EQU       2          B offset from base
8           00000004        SUM:      EQU       4          SUM offset from base
9           00001000                  ORG       $1000
10 001000 207C 00002000 START:    MOVEA.L   #BASE,A0   load base address
11 001006 3028      0000            MOVE.W    A(A0),D0   get value of A into D0
12 00100A D068      0002            ADD.W     B(A0),D0   add B to it
13 00100E 3140      0004            MOVE.W    D0,SUM(A0) put result into SUM
14 001012 4E4E                      TRAP      #14        !! STOP PROGRAM
15                          * data area initialised with test values
16          00002000                  ORG       $2000
17 002000 01C8              BASE:     DC.W      456       A initial value 456
18 002002 FFB2                        DC.W      -78       B initial value -78
19 002004 0000                        DC.W      0         SUM is initialised to 0
20                                    END

A           00000000 E      B         00000002 E     SUM        00000004 E
START       00001000 L      BASE      00002000 L
-------------------------------------------------------------------------------
```

Program 13.3 Memory operands accessed using base addressing techniques

Program 13.3 above (a modified version of Program 8.1) shows the use of *address register indirect with displacement addressing mode* to access variables using base address techniques.

line

6-8	the variable names are defined as offsets from a base (each variable is a word)
10	the base address of the data area is moved in A0
11-13	the calculation SUM=A+B is performed with the variables being accessed with *address register indirect with displacement addressing mode*, e.g. in line 12 A0 = $2000 and B = 2 therefore when the instruction `ADD.W B(A0),D0` is executed the source operand effective address is B(A0) = $2002

Program 13.3 assumed that previous calculations had stored the values of A and B into the data area. In practice the data area would probably be on the stack in which case the offset would be negative values (see Chapter 14 Program 14.2).

13.1.3 Address register indirect with index and displacement

Address register indirect with index and displacement can be used for all types of operand reference (data, memory, alterable and control). The effective address is the sum of three values:

(a) the contents of an address register (long word value);
(b) the contents of an index register which can be either a data or address register (long word or sign extended word value);
plus (c) a displacement which is part of the op-code (sign extended byte value).

This addressing mode may be used to provide access to complex data structures (see Appendix E for examples), e.g. the effective address of an element of a record in an array of records would be the sum of:

(a) an address register which holds the start address of the array of records,
plus (b) an index register which selects the particular record in the array,
plus (c) a displacement which selects the element or field within the record.

The index register is a data or address register with the length being specified using a .W (sign extended word) or .L (long word) suffix. For example:

```
CLR.B       $40(A0,D1.W)      clear a byte
CLR.B       $4(A0,D1.L)       clear a byte
```

If, in the first example, A0 = $20000 and D1 = $1000, the effective address is:

= long word contents of A0 + sign extended word contents of D1 + $40
= 20000 + $1000 + $40 = $21040 (the address of the byte cleared)

In the second example, the effective address is:

= the long word contents of A0 + long word contents of D1 + $4

13.2 The CASE control structure

```
CASE condition of
     1:  S1
     2:  S2
     3:  S3
ENDCASE
```

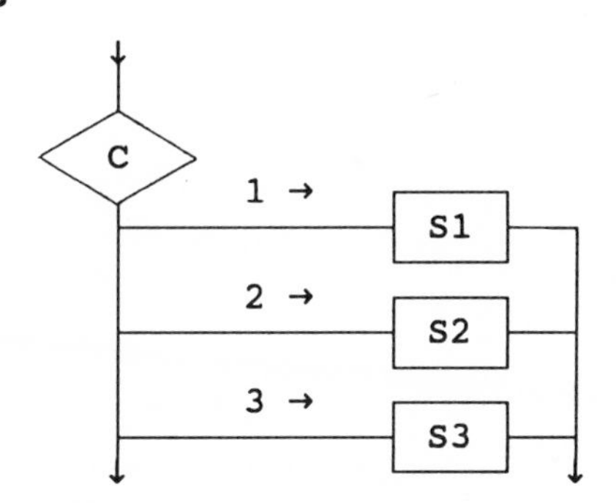

In the CASE control structure a control variable is used to select one of a number of choices (via a series of branches). If, in the diagram, the condition C has the value 1, statement S1 is executed, if 2, statement S2 is executed, etc. It is essential to ensure that all possible values of the condition have a CASE (this may be done by placing an IF condition test before the CASE and if the condition value is invalid it may be ignored or an error message displayed).

Program 13.4 (next page) shows how a CASE structure can be implemented using *address register indirect with index and displacement addressing mode.* The correct CASE statement is accessed via a jump table of CASE addresses which is defined in lines 37, 38, and 39, using the DC.L pseudo-operator. When a label name is used in the operand field of such a statement the memory address equivalent to the label is placed in memory, i.e. at line 37 the address of label S1 (value $001060) is stored starting at memory address $001054. When executed the correct CASE is accessed as follows:

line

26-29	tests that the condition value is in the range 1 to 3; if the test fails the program branches back to line 22 to request input (an error message could be displayed)
30	the condition value (in the range 1 to 3) is moved into D1 (D0 is preserved for calculations when the CASE has been selected)
31	the value is then decremented (the offsets in the JUMP table start from 0)
32	the value is multiplied by 4 using ASL.L (each JUMP address is four bytes long)
33	the address of the start of the CASE jump table (label JUMP) is moved into A0
34	`MOVEA.L 0(A0,D1.W),A0` moves the contents of the effective address into A0, i.e. 0 + JUMP + contents of D1 will be the address of S1, S2 or S3
35	`JMP (A0)` jumps to the correct CASE (label S1, S2 or S3)

In Program 13.4 the start of the CASE jump table is in long word form in A0 (where it is used in line 34) and therefore the algorithm may be used anywhere in memory (note that it is not relocatable). In practice, the condition value would be calculated earlier in the program to determine the CASE to be selected. The condition must be checked to ensure that it is within the CASE range otherwise the address obtained from the CASE table would be incorrect and the program would crash.

A fully relocatable CASE statement using *program counter relative with index and displacement addressing mode* will be described in Chapter 15.6.

```
--------------------------------------------------------------------------------
1                            * Program 13.4 - CASE program structure using
2                            *  address register indirect with index and displacement
3                            *
4                            * LOOP FOR ever
5                            *     WRITE(newline,'enter number to test CASE statement ')
6                            *     READ(NUMBER)
7                            *     IF NUMBER is the range 1 to 3
8                            *         CASE NUMBER
9                            *              1: NUMBER=NUMBER*10
10                           *              2: NUMBER=NUMBER*100
11                           *              3: NUMBER=NUMBER*1000
12                           *         END CASE
13                           *     WRITE(' NUMBER = ',NUMBER)
14                           *     END IF
15                           * END LOOP
16                           *
17                           * D0 holds NUMBER
18                           * D1 holds NUMBER for calculation of index into JUMP table
19                           * A0 holds the start address of the JUMP table of CASEs
20                           *
21                                     XREF      WRLINE,WRTEXT,WRDECW,RDDECW
22 001000 4EB9 0002802C START:         JSR       WRLINE
23 001006 4EB9 00028038                JSR       WRTEXT
24 00100C 65 6E 74 65 72               DC.W      'enter number to test CASE statement $'
25 001032 4EB9 00028088                JSR       RDDECW          read number
26 001038 4A40                         TST.W     D0              IF D0 <= 0 invalid number
27 00103A 6FC4                         BLE.S     START
28 00103C 0C40      0003               CMPI.W    #3,D0           IF D0 > 3 invalid number
29 001040 6EBE                         BGT.S     START
30 001042 2200                         MOVE.L    D0,D1            copy NUMBER to D1
31 001044 5341                         SUBQ      #1,D1            table offsets are from 0 !!
32 001046 E581                         ASL.L     #2,D1            *4 for long word offsets
33 001048 207C 00001054                MOVEA.L   #JUMP,A0         address of JUMP table
34 00104E 2070      1000               MOVEA.L   0(A0,D1.W),A0 get address of CASE
35 001052 4ED0                         JMP       (A0)             jump to correct CASEe
36                           * JUMP table of CASE statement addresses
37 001054 00001060      JUMP:          DC.L      S1              first case statement address
38 001058 00001066                     DC.L      S2              second case statement address
39 00105C 0000106C                     DC.L      S3              third case statement address
40                           * CASE statements
41 001060 C1FC      000A S1:           MULS.W    #10,D0          start of first CASE statement
42 001064 600A                         BRA.S     ENDCASE
43 001066 C1FC      0064 S2:           MULS.W    #100,D0         second CASE statement
44 00106A 6004                         BRA.S     ENDCASE
45 00106C C1FC      03E8 S3:           MULS.W    #1000,D0        third CASE statement
46 001070 4EB9 00028038 ENDCASE:       JSR       WRTEXT          end of CASE statement
47 001076 20 4E 55 4D 42               DC.W      ' NUMBER = $'
48 001082 4EB9 00028074                JSR       WRDECW
49 001088 6000      FF76               BRA.L     START
50                                     END
--------------------------------------------------------------------------------
```

Program 13.4 CASE structure using address register indirect with index & displacement

Problem for Chapter 13

Design and code a subroutine which calculates the mean of an array of integer word values. On entry to the subroutine the following parameters should be set up:

D0 contains the number of word values in the array;

A0 contains the address of the first element of the array.

On exit, return the mean in D0. Test the subroutine with a suitable program and test data.

14

Stack operations and calling subroutines

A stack is a dynamic data structure in which data is added to (or pushed on) and removed from (or pulled off) the top, such that it follows a 'last-in first-out' structure (as opposed to a queue which is a 'first-in first-out' structure). Although a stack structure can be accessed using *postincrement and predecrement addressing modes* with any address register, A7 is used to access the system stack area which is set up by the monitor or operating system.

14.1 Using address register A7 as the stack pointer

In normal practice, when using the MC68000 processor, address register A7 is set up to point to the **top** of the processor system stack area in memory. *Predecrement mode* is then used to move (or push) information onto the stack, and *postincrement mode* used to move (or pull) it off. This means that A7 starts at a high address (the top of the stack) and decrements as information is pushed onto the stack. Generally, before starting user programs, the operating system or monitor allocates an area of memory for the user stack, and sets up the stack pointer A7 to the start address of this area (the top address). It has been assumed, in the sample programs throughout this book, that the stack has been set up in this way before program execution starts. It is possible for user programs to allocate their own stack area and set up the stack pointer A7 to point to this.

There is an important difference between A7 and the other address registers, in that the stack pointer A7 **must always** be aligned to an even address. When byte length operands specify register A7 in *postincrement or predecrement mode*, the value of A7 will be altered by 2 so that it always points to a word boundary. Note that some assemblers allow the name SP, as well as A7, to be used for the stack pointer.

```
1                          * Program  14.1 - pushing data onto the stack
2                          *
3                          * Set A7 (Stack Pointer) to $1000
4                          * Load D0 with some long word test data
5                          * Push a byte, word and long word (D0) onto Stack
6                          *
7                          * after execution use monitor to check stack contents
8                          *
9 001000 2E7C 00001000 START:       MOVEA.L    #$1000,A7      Set Stack Pointer
10 001006 203C 89ABCDEF             MOVE.L     #$89ABCDEF,D0  put data in D0
11 00100C 1F3C     0041             MOVE.B     #'A',-(A7)     Push byte
12 001010 3F3C     1234             MOVE.W     #$1234,-(A7)   Push word
13 001014 2F00                      MOVE.L     D0,-(A7)       Push long word
14 001016 4E4E                      TRAP       #14            !! STOP PROGRAM
15                                  END
```

Program 14.1 Program pushing data onto the stack (Force or Bytronic board)

Program 14.1 shows how to push data onto the stack. In line 9 the stack pointer A7 is set to point to the top of the stack+2 (memory address $1000 in this case). D0 is loaded with a long word value (line 10), and then data is pushed onto the stack (a character (byte) in line 11, a word in line 12 and the long word from D0 in line 13). As the information is pushed onto the stack, A7 will take the values $1000, $0FFE, $0FFC and $0FF8 (i.e. the stack area starts at $0FFE). After the executing line 13 of Program 14.1 (i.e. use a breakpoint to stop the program at address $1016), A7 would contain the value $0FF8 (pointing to the last word pushed), and the contents of the stack area would appear thus:

	address	contents		comments on contents
A7 after line 13→	000FF8	89	AB	long word from D0
	→ 000FFA	CD	EF	"
A7 after line 12→	000FFC	12	34	word $1234
A7 after line 11→	000FFE	41	??	byte 'A'
A7 after line 9 →	001000	2E	C7	first program instruction

Note that in line 11, `MOVE.B #'A',-(A7)`, which pushes a byte, the stack pointer A7 is incremented by 2 to maintain the word boundary alignment (A0 to A6 would be incremented by 1). Remember, the stack pointer always points to the last word pushed.

After register contents have been pushed onto the stack, the registers can be used for other calculations, and then the original information retrieved by pulling data from the stack in the **opposite order** to the pushes, thus:

```
MOVE.L    (A7)+,D4    pull long word into D4
MOVE.W    (A7)+,D3    pull word into D3
MOVE.B    (A7)+,D0    pull byte into D0
```

Data pushed onto the stack does not have to be pulled back into the same register. It is important, however, to maintain the integrity of the stack by pushing and pulling the correct **amount** of information of the correct **length**. Otherwise data on the stack will not be as expected, and the program will fail. It is very easy to pull more data off than has been pushed, and vice versa. In particular, when writing subroutines, care should be taken since the JSR and RTS instructions make use of the stack, and getting the stack in a mess will cause the program to fail on execution of the RTS instruction (JSR and RTS were described briefly in Chapter 9.2, and will be described in more detail in section 2 below).

Although address register A7 is the general stack pointer, used by several MC68000 instructions, any address register can be used as a stack pointer to a stack area within the user program. Care should be taken when pushing and pulling mixtures of byte, word and long word (only A7 automatically maintains even address alignment).

Exercise 14.1 *(see Appendix G for sample answer)*

Values on the stack may be accessed directly using the *address register indirect with displacement addressing mode*. Modify Program 14.1 to move the following values:

1 the byte 'A' on the stack into D1;
2 the word value $1234 on the stack into D2;
3 the long word value on the stack (originally from D0) into D3.

14.2 The JSR, BSR and RTS instructions

The jump to subroutine (**JSR**) and branch to subroutine (**BSR**) instructions are used to call subroutines, e.g. (SUB is the label associated with the first instruction of the subroutine):

```
        JSR       SUB       call subroutine at label SUB
        JSR       (A0)      call subroutine (address in A0)
        BSR.S     SUB       call subroutine (short form)
        BSR.L     SUB       call subroutine (long form)
```

The operations involved in executing a JSR or BSR instruction are:

(a) using *predecrement mode*, the contents of the PC (i.e. the address of the next instruction) are pushed onto the stack as a long word address,

then (b) a jump or branch is made to the start of the subroutine by moving the operand effective address (the address equivalent to the label SUB in the above examples) into the PC.

Thus JSR could be considered as equivalent to the following two operations combined:

```
        MOVEA.L   #RETURN,-(A7)    save return address
        JMP       SUB              jump to start of routine
RETURN: ....                       next instruction
```

The JSR instruction uses *control addressing modes* (see Appendix B) and the BSR instruction, in its .S and .L forms, uses *Program Counter relative addressing mode.* Because BSR is relocatable it should be used as much as possible, i.e. to call subroutines with the same source file and, if possible, subroutines in other files (depends on the assembler and linker). Many of the sample programs in this book use the subroutine library of Appendix D which was in ROM on the target single board computers. It was therefore necessary to use JSR with *absolute addressing mode* to call them.

The return from subroutine (RTS) instruction uses *postincrement mode* to pull the long word value from the top of the stack into the PC. This value **should be** the address pushed by the corresponding JSR or BSR instruction (i.e. the address of the instruction following the JSR or BSR). It is recommended that a subroutine contains only one RTS instruction, so that it is clear where the subroutine exit point is. If a return is required from within the body of the subroutine a branch to the RTS instruction may be used.

One way of passing parameters to subroutines is on the stack (see Chapter 16.5). On exit from the subroutine the parameters have to be removed or deallocated, either before the RTS (see Program 16.2) or on return to the calling program (see the C program Program 29.1 and the generated assembly language code Fig. 29.2). The MC68010/20/30/40 processors have a return from subroutine instruction which also performs the parameter deallocation function:

```
        RTD       #displacement       return and deallocate
```

The RTD instruction:

(a) uses *postincrement mode* to pull the long word value from the top of the stack into the PC (i.e. the subroutine return address),

then (b) adds the sign extended 16-bit displacement value to the stack pointer A7.

14.3 The move multiple (MOVEM) instruction

When a program calls a subroutine it may have important values in the data/address registers. High-level languages tend to have a convention for register use when calling subroutines in which some registers may not be used at all, some may be used but the original values restored and some may be used at will (see Chapter 29 for an example).

A subroutine must therefore ensure that any registers used internally are saved onto the stack on entry, and restored on exit (remembering not to save and restore registers used for returning parameter values to the calling program). This could be done by a number of separate MOVE operations, but a more efficient method is to use the move multiple instruction MOVEM, by which a number of registers can be moved in a single operation.

The source operand of the **move multiple registers** to **memory** instruction is a list of registers, and the destination operand is specified using either *predecrement or a control alterable addressing mode*:

```
MOVEM.L    D0-D7/A0-A6,-(A7)              save to stack
MOVEM.L    D0-D2/D4/A1/A4-A6,-(A7)        save to stack
MOVEM.W    D1-D5/A1/A3,LABEL              save to LABEL
```

The operand length for MOVEM may be word .W or long word .L. The register list is specified to the assembler in a special form (cross assemblers may be more restrictive):

1. the character slash, /, is used to separate names of registers;
2. the character minus, -, is used to indicate groups of registers;
3. the register numbers are in ascending order with the data registers specified first followed by the address registers.

The first example pushes long words D0 to D7 and A0 to A6 onto the stack. The second example pushes long words D0, D1, D2, D4, A1, A4, A5 and A6 onto the stack. The third example moves words D1 to D5, A1 and A3 to memory starting at location LABEL. The order in which the registers are moved when the instruction is executed is from D0 to D7 then A0 to A7. The exception is *predecrement mode* which pushes A7 to A0 then D7 to D0 onto the stack (to be pulled off in the reverse order using *postincrement mode*).

The **move multiple registers** from **memory** instruction has a source operand specified using either *postincrement or a control addressing mode*, and the destination operand is a list of registers. For example, the following instructions restore the register values moved in the corresponding **individual** instructions in the previous examples of MOVEM.

```
MOVEM.L    (A7)+,D0-D7/A0-A6         restore from stack
MOVEM.L    (A7)+,D0-D2/D4/A1/A4-A6   restore from stack
MOVEM.W    LABEL,D1-D5/A1/A3         restore from LABEL
```

The destination register list for MOVEM is written in the same order as the original source list (i.e. **not** in the reverse order). The order in which the registers are moved when the instruction is executed is from D0 to D7, then A0 to A7 (when pushing and pulling using a sequence of separate MOVE operations, the reverse order must be used).

It is good practice to use the MOVEM instruction to save and restore registers which are used for working values within subroutines. If this is done the calling program will not have to save important values stored in registers before calling the subroutine. All the library subroutines of Appendix D save and restore registers in this way.

14.4 Saving and restoring condition codes

If, when a subroutine is called, the values of the condition codes are important, they may be saved by pushing the contents of the Status Register onto the stack on entry to the subroutine, and restored it on exit. RTR is a special form of RTS which:

(a) pulls a word off the top of the stack and uses the five least significant bits to set the Condition Code bits,

then (b) pulls the long word value off the top of the stack into the PC.

For example:

```
SUB:     MOVE.W    SR,-(A7)       save condition codes on stack
         MOVEM     D0-D5,-(A7)    then save registers
         .....
         MOVEM     (A7)+,D0-D5    restore registers
         RTR                      restore condition codes return
```

When using a **MC68010** the first instruction of SUB should be replaced by:

```
SUB:       MOVE.W     CCR,-(A7)       save condition codes
```

14.5 The LINK and UNLK instructions

Chapter 7.3 stated that one of the ways to store variables was on the stack (used extensively by high-level languages) and the MC68000 has the LINK and UNLK instructions to facilitate this. The LINK instruction is used to allocate space on the stack which is then available for storing variables, e.g.:

```
           LINK       An,#m         allocate m bytes on the stack
```

The operation of the LINK instruction is:

1 The contents of address register An are moved onto the stack as a long word value.
2 The updated stack pointer (A7) is moved into the address register An.
3 The sign extended value of the immediate word operand m is then added to the stack pointer (because the stack works down memory, m should be a negative value).

This instruction can be used:

(a) in the main program to allocate data areas which exist until the program terminates,

and (b) in subroutines to allocate local data storage which will be deallocated when the subroutine terminates.

Program 14.2 (next page), a modified version of Program 13.3, shows a main program which uses LINK to allocate a data area on the stack. To access the data on the stack *address register indirect with displacement addressing mode* is used with A0 holding the base address of the data area. Because A0 points to the top of the data area the displacements are negative values.

```
--------------------------------------------------------------------------------
1                        * Program 14.2:  SUM=A+B using word data on stack
2                        *  data on stack accessed from a base address using
3                        *  address register indirect with displacement mode
4                        *
5                        * define variables as offsets from base address
6         FFFFFFFE       A:         EQU       -2          A offset from base
7         FFFFFFFC       B:         EQU       -4          B offset from base
8         FFFFFFFA       SUM:       EQU       -6          SUM offset from base
9         00001000                  ORG       $1000
10 001000 2E7C 00001000 START:      MOVEA.L   #$1000,A7   set stack
11 001006 4E50      FFFA            LINK      A0,#-6      reserve storage
12 00100A 317C      01C8            MOVE.W    #456,A(A0)  set A value
                    FFFE
13 001010 317C      FFB2            MOVE.W    #-78,B(A0)  set B value
                    FFFC
14 001016 3028      FFFE            MOVE.W    A(A0),D0    get value of A into D0
15 00101A D068      FFFC            ADD.W     B(A0),D0    add B to it
16 00101E 3140      FFFA            MOVE.W    D0,SUM(A0)  put result into SUM
17 001022 4E58                      UNLK      A0
18 001024 4E4E                      TRAP      #14         !! STOP PROGRAM
19                                  END
--------------------------------------------------------------------------------
```

Program 14.2 Using LINK to allocate a data area (Force/Bytronic board)

At line 17 of Program 14.2 (before executing UNLK, see below) the stack would appear (A0 is assumed to contain 0 initially):

	address	contents		comments on contents
A7 after line 11→	000FF6	01	7A	SUM = A + B
	000FF8	FF	B2	B value -78 decimal
	000FFA	01	C8	A value 456 decimal
A0 points here →	000FFC	00	00	A0 most significant word
	000FFE	00	00	A0 least significant word
Original A7 →	001000	2E	7C	first program instruction

The program sequence is:

6-8 the variable names are defined as offsets from a base (each variable is a word and A0 points to the top of the data area thus the offsets are negative values)

10 the stack pointer A7 is initialised to $1000

11 `LINK A0,#-6` allocates a data area of 3 words (6 bytes) on the stack; after executing this instruction A0 = $FFC the start of the data area and A7 = $FF6 the top of the stack

12-13 initial values are assigned to A and B

14-16 SUM = A+B is performed: the variables being accessed using *address register indirect with displacement addressing mode*, e.g. in line 15 A0 = $FFC and B = -4 so when `ADD.W B(A0),D0` is executed the source operand effective address B(A0) = $FF8

When the data area is finished with it must be deallocated (e.g. before the RTS instruction to deallocate the local variables of a subroutine) using the UNLK instruction:

```
        UNLK        An
```

The stack pointer is (a) loaded with the contents of the address register, and (b) the address register is loaded with the long word from the top of the stack. UNLK reverses the operation of the corresponding LINK, restoring the address register and stack pointer to the original values (assuming the address register and stack have not been damaged). The data area will be deallocated and its contents lost (also any information which had been moved onto the stack after LINK will also be lost). It is important that UNLK is used within the same level of subroutine (or main program) as the corresponding LINK instruction. If a number of data areas have been allocated they must be deallocated in the reverse order.

Problem for Chapter 14

1 A main program sets the stack pointer to $1000 then calls subroutine SUB1. SUB1 uses LINK to allocate a data area of 8 bytes (using A0) then calls subroutine SUB2. SUB2 allocates a data area of 10 bytes on the stack (using A1). Draw a diagram of the stack and list the contents of the stack pointer, A0 and A1 at the end of this process. Write a simple test program (similar to Program 14.2) to check the answer.

2 One advantage of languages that have stack operations is that recursive functions can be written, i.e. a function which calls itself. For example, although factorial can be implemented using a FOR loop (see Problems for Chapter 11) it can also be written using recursion. Program 14.3 shows a listing of a MC68000 assembly language factorial function using recursion. Explain how it works and show the stack contents on each function call for N=4.

```
--------------------------------------------------------------------------------
14                      * Factorial using recursion: Pascal equivalent
15                      *
16                      * FUNCTION FACTOR(N:INTEGER):INTEGER;
17                      * BEGIN
18                      * IF N=1 THEN FACTOR=1
19                      *        ELSE FACTOR:=N*FACTOR(N-1);
20                      * END;
21                      *
22                      * on entry: D0 holds word value of N (assumed N > 0)
23                      * on exit:  D0 holds N!
24                      *
25 001040 3F00          FACTOR:  MOVE.W   D0,-(A7)    move value of N to stack
26 001042 5340                   SUBQ.W   #1,D0       decrement N
27 001044 6706                   BEQ.S    FACT1       if 0 end of recursion
28 001046 61F8                   BSR.S    FACTOR      ELSE call FACTOR again
29 001048 C0DF                   MULU     (A7)+,D0    FACTOR = N*(N - 1)
30 00104A 6002                   BRA.S    FACT2       return
31 00104C 301F          FACT1:   MOVE.W   (A7)+,D0    end of recursion N = 1
32 00104E 4E75          FACT2:   RTS                  return
33                               END
--------------------------------------------------------------------------------
```

Program 14.3 Factorial function using recursion

15

Relocatable code and PC relative addressing modes

Many of the sample programs presented in this book contained references to absolute addresses in memory, i.e. either the instructions or the data were specified to be at particular memory locations. When working on small programs targeted at single board computer systems this is usually no problem. However, large programs are made up of a number of modules, i.e. a main program and subroutines, and it is important that the code generated by the assembler or compiler is position independent or relocatable in memory. If a subroutine is relocatable, it can be executed anywhere in memory without affecting the results. It is worth noting that high-level language compilers generate relocatable machine code.

15.1 Relocatable code

There are a number of factors to consider when writing relocatable code:

(a) is the module code specified to start at an absolute address, i.e. using an ORG pseudo-operator or equivalent ?

and/or (b) are there references which will be resolved as absolute addresses, i.e. operand references, memory references in tables (e.g. the addresses in the CASE table in lines 38 to 40 of Program 13.4).

15.1.1 Generation of code

The object code of a module or set of modules may be assigned to absolute physical addresses in memory at various stages:

1 During the assembly stage by specifying an ORG pseudo-operator or equivalent. This should be avoided except when necessary, e.g. writing system monitors or small embedded control systems (even then it is best to specify the absolute addresses at the link stage).

2 During the link stage by specifying code and/or data sections to start at particular absolute addresses in memory, e.g. in Chapter 8.3 the XA8 linker linkx was directed to generate absolute code with the code section (_text) starting at address $1000 and the data section (_data) starting at address $4000, i.e.:

```
linkx -a -db 0x4000 -tb 0x1000 p8_1a.o
```

If linkx is not given absolute specifications it links the modules and leaves the result in relocatable format.

3 If the linker generates linked relocatable code the operating system loads this into absolute addresses in memory and resolves any absolute references (possibly under the control of a memory management unit). Generally in this type of environment the user has no control over where the code is placed in physical memory.

The way modules are generated, linked and loaded depends upon the assembler, linker and operating system environment and users have to conform to the rules (see manuals), e.g.:

(a) when generating code for single board computers or small systems with a simple monitor, the user can position code in memory as required (either by specifying ORG directives or, preferably, by directing the linker to generate absolute code),

or (b) under an operating system which looks after the organisation of memory the user may have no control over final position of the code.

Even when the resultant code is loaded into absolute physical memory locations it may still be relocatable at run time if it contains no absolute references (see next section).

15.1.2 Avoiding absolute memory references

Absolute memory references are either operand references using absolute addressing mode or other references to absolute memory locations (e.g. in tables), see Fig. 15.1 below.

When using a disk based assembler absolute references will, if no ORG pseudo-operators are used, be generated within each file from a base of 0, e.g. see Programs 8.1a and 8.1b. The references to absolute physical addresses will be finally set up by the linker or operating system (in a virtual memory system these will be logical addresses mapped to physical addresses using a memory management unit).

By avoiding absolute memory references it is possible to generate code which is in effect totally relocatable and may be moved about memory as required. For example, the problem at the end of Chapter 17 is to write a memory test program which copies itself throughout memory testing block by block.

```
          MOVEA.L   #CASE,A0     load CASE table address
          LEA.L     CASE,A0      load CASE table address
          MOVE.W    LABEL,D0     move from absolute address
          MOVE.W    D0,LABEL     move to absolute address
          JSR       SUB          call subroutine
CASE:     DC.L      CASE1        CASE routine absolute address
```

Fig. 15.1 Examples of absolute address specifications

15.2 Re-entrant code

In a multi-processing environment many programs may be waiting to run. At any instant the operating system decides which process can execute (i.e. use the processor) either by using a scheduling algorithm and/or on a priority basis (Tanenbaum 1987). In such an environment common systems routines may be used by many programs and the efficiency of the overall system can be improved if only one copy of each routine is in memory (which may be 'called' as required by the executing programs). As processes switch (e.g. a higher priority process becomes ready to run) a suspended process may have been in the middle of using such a routine which is now required by the currently executing process. By writing re-entrant code it is possible to have subroutines which may be interrupted, used by another processes, and then resumed. When writing MC68000 code the main rule to be followed when implementing re-entrant subroutines is to avoid the use of absolute data in memory. Variables used by the routine should be either on the stack or in registers (both of which are preserved when processes switch).

15.3 The LEA and PEA instructions

The effective address of an operand (defined in Chapter 6.1) is the memory address where the operand is eventually found. To determine the effective address, the processor uses one of a number of *addressing modes* which are defined by the operand field of an instruction. A useful instruction which can be used by programs which process complex data structures is LEA (Load Effective Address). Rather than loading an operand, LEA calculates the **effective address** of the operand (as a long word value) and then moves it into an address register. The resultant address can then be used with *address register indirect addressing modes*. The source operand of LEA is specified using a *control addressing mode* and the destination operand must be an address register: e.g.:

```
LEA         $40(A0,D1.W),A1    load address into A1
CLR.B       (A1)               clear the memory byte
```

If A0 = $20000 and D1 = $1000, the value $21040 is loaded into A1 and then the contents of that address cleared. These two statements are in effect equivalent to:

```
CLR.B       $40(A0,D1.W)       clear a byte
```

PEA calculates the effective address of an operand (as a long word value) and then pushes it onto the stack. The source operand is specified using a *control addressing mode* (the destination operand is assumed to be on the stack), e.g.:

```
PEA         $40(A0,D1.W)       push effective address
```

This can be used, for example, to push the effective addresses of parameters onto the stack before calling a subroutine (see Chapter 16.5).

15.4 Position independence with the MC68000

When using a disk based system to generate position independent MC68000 object code the assembly language programmer must ensure that:

1 No assembler directives are used which force instruction and data generation at an absolute address (i.e. there are no ORG statements). The assembler will generate the code from a base address of zero and the linker or operating system will place it as required in physical memory.

2 No absolute addresses are specified as immediate operands, effective addresses for operands and in data declarations, see Fig. 15.1. The only exceptions are when absolute addresses are used to access the exception vector table (see Chapter 23.2), I/O register addresses (see Chapter 18.4), etc.

15.5 Program Counter relative addressing

One technique to produce relocatable code is to use operand references which are relative to the current instruction, i.e. by specifying a displacement to be added to the PC (Program Counter). This addressing mode is called *Program Counter relative with displacement* (or PC relative for short). For example, the **only** addressing mode used by branch instructions (see Chapter 10.2) is PC relative and thus they are automatically relocatable.

15.5.1 Program Counter with displacement addressing mode

Program Counter with displacement addressing mode is similar to *address register indirect with displacement*, but uses the PC instead of one of the address registers. It is specified thus:

```
        LEA.L     HELLO(PC),A0    load message address into A0
        BSR       WRTXTA          display message
        PEA.L     BYE(PC)         push message address
        BSR       WRTXTS          display message
         .....
         .....
* constant data area in code section
HELLO: DC.W       'hello user$'
BYE:   DC.W       'goodbye user$'
```

The (PC) following the label indicates to the assembler that the references to the operands are to be generated using *Program Counter with displacement addressing mode.* The displacement calculated by the assembler is stored as a signed 16-bit value in the word following the operation-code in memory. If the displacement cannot be resolved into a 16-bit value (the maximum that can be handled by this addressing mode), the assembler will display an error message. At execution time the displacement is sign extended and added to the PC to form the operand effective address (i.e. similar to the long form of the branch instruction).

In the above example the subroutines and the message addresses (assumed to be constant data in the code section - see below) are referred to using PC relative addressing. Note:

1 PC relative references must be within the same program section (see Chapter 8.3).
2 Some assemblers are more restrictive in that PC relative references must be within the same source file, e.g. the XA8 cross assembler. In such cases JSR with absolute addressing will have to be used to call subroutines in other files.
3 Some assemblers, e.g. XA8, will attempt to generate PC relative references even when given operands in absolute format.

PC relative addressing modes may not be used to access alterable operands, i.e.:

```
        MOVE.W        D0,A(PC)     this is NOT allowed !
```

This restriction is because PC relative addressing modes may only reference operands in the program code section and data should not be written into this.

15.5.2 Program Counter with index and displacement addressing mode

This is similar to *address register indirect with index and displacement addressing mode*, and is specified thus:

```
          MOVE.W       TABLE(PC,Rn),D0
          ......
          ......
TABLE:    DC.W         10,20,30
```

The displacement is generated by the assembler is placed as a signed 8-bit value within the operation-code of the instruction. If the displacement (the number of bytes between the instruction and TABLE above) will not fit into 8 bits the assembler will display an error message. The index register Rn (which gives an index into TABLE above) may be an address or data register holding a value of length word .W (sign extended) or long word .L. For example:

```
        MOVE.W      TABLE(PC,D1.W),D0      index is word
        MOVE.W      TABLE(PC,D1.L),D0      index is long word
```

15.6 Relocatable CASE control structure

Chapter 13.2 described the CASE control structure and how to implement it using *address register indirect with index and displacement addressing mode* (Program 13.4).

Program 15.1 (next page) shows how a CASE structure can be implemented using *program counter relative with index and displacement addressing mode.* If the structure is to be totally relocatable not only must the operand references be relative, but also the CASE addresses in the jump table. In Program 15.1 the correct CASE statement is accessed via a table of CASE addresses which are held as **displacements relative to the start of the table itself**, i.e. defined in lines 37, 38 and 39 using the DC.L pseudo-operator. For example, in line 39 the value of S3-JUMP is the displacement (the number of bytes) between the label JUMP and the label S3, i.e. $18 bytes. When the program is executed the correct CASE is accessed as follows:

line

26-29	tests that the condition value is in the range 1 to 3; if the test fails the program branches back to line 22 to request input (an error message could be displayed)
30	the condition value (in the range 1 to 3) is moved into D1
31	the value is then decremented (i.e. the displacements start from 0)
32	the value is multiplied by 4 (the relative addresses in the CASE jump table are four bytes long)
33	`LEA.L JUMP(PC),A0` loads the base address of the start of the CASE jump table (address $1052) into A0 (using LEA with *PC relative addressing mode*)
34	`ADDA.L 0(A0,D1.W),A0` adds the relative displacement (from the CASE table) to the base address (label JUMP) to obtain the address of S1, S2 or S3
35	`JMP (A0)` then jumps to the correct CASE (label S1, S2 or S3)

For example, if the user had entered 2 to select the CASE at label S2:

(a) in line 34 the values would be A0 = $1052 (address of JUMP) D1 = $4, hence the **source operand effective address** 0(A0,D1.W) = $1056

(b) the contents of address $1056 (value $12) would be added to the address of the JUMP table in A0 (value $1052) making the result A0 = $1064 (the address of label S2).

(c) A0 is then used as the JMP address (to access S2) in line 35

This version of the CASE structure is totally relocatable and may be moved anywhere memory and executed (e.g. using the memory copy function of the system monitor). Note that the subroutines calls using JSR are to routines in ROM and are to absolute addresses (subroutines within the program would be called using BSR making the calls relocatable).

```
-----------------------------------------------------------------------------------
 1                         * Program 15.1 - relocatable CASE structure using
 2                         *    PC relative with index and displacement
 3                         *
 4                         * LOOP FOR ever
 5                         *     WRITE(newline,'enter number to test CASE statement ')
 6                         *     READ(NUMBER)
 7                         *     IF NUMBER is the range 1 to 3
 8                         *         CASE NUMBER
 9                         *             1: NUMBER=NUMBER*10
10                         *             2: NUMBER=NUMBER*100
11                         *             3: NUMBER=NUMBER*1000
12                         *         END CASE
13                         *     WRITE(' NUMBER = ',NUMBER)
14                         *     END IF
15                         * END LOOP
16                         *
17                         * D0 holds NUMBER
18                         * D1 holds NUMBER for calculation of index into JUMP table
19                         * A0 holds the start address of the JUMP table of CASEs
20                         *
21                                   XREF      WRLINE,WRTEXT,WRDECW,RDDECW
22 001000 4EB9 0002802C START:       JSR       WRLINE
23 001006 4EB9 00028038              JSR       WRTEXT
24 00100C 65 6E 74 65 72             DC.W      'enter number to test CASE statement $'
25 001032 4EB9 00028088              JSR       RDDECW          read number
26 001038 4A40                       TST.W     D0              IF D0 <= 0 invalid number
27 00103A 6FC4                       BLE.S     START
28 00103C 0C40      0003             CMPI.W    #3,D0           IF D0 > 3 invalid number
29 001040 6EBE                       BGT.S     START
30 001042 2200                       MOVE.L    D0,D1           copy NUMBER to D1
31 001044 5341                       SUBQ      #1,D1           table offsets are from 0 !!
32 001046 E581                       ASL.L     #2,D1           *4 for long word offsets
33 001048 41FA      0008             LEA.L     JUMP(PC),A0     load JUMP table address
34 00104C D1F0      1000             ADDA.L    0(A0,D1.W),A0   add offset to base address
35 001050 4ED0                       JMP       (A0)            jump to correct CASE
36                         * JUMP table of CASE addresses (long word relative values)
37 001052 0000000C      JUMP:        DC.L      S1-JUMP        first case statement offset
38 001056 00000012                   DC.L      S2-JUMP        second case statement offset
39 00105A 00000018                   DC.L      S3-JUMP        third case statement offset
40                         * CASE statements
41 00105E C1FC      000A S1:         MULS.W    #10,D0         first CASE statement
42 001062 600A                       BRA.S     ENDCASE
43 001064 C1FC      0064 S2:         MULS.W    #100,D0        second CASE statement
44 001068 6004                       BRA.S     ENDCASE
45 00106A C1FC      03E8 S3:         MULS.W    #1000,D0       third CASE statement
46 00106E 4EB9 00028038 ENDCASE:     JSR       WRTEXT         end of CASE statement
47 001074 20 4E 55 4D 42             DC.W      ' NUMBER = $'
48 001080 4EB9 00028074              JSR       WRDECW
49 001086 6000      FF78             BRA.L     START
50                                   END
-----------------------------------------------------------------------------------
```

Program 15.1 Relocatable CASE structure using PC relative with index and displacement (Force or Bytronic board)

Review of addressing modes

Examples of addressing modes using MOVE are presented at the end of this review.

Data register direct addressing mode: (Chpater 6.2 page 40) the operand is in a data register (used for data source operands and data alterable destination operands):

```
CLR.W      D0         clear data register D0
TST.W      D0         test D0 and set CCR bits
MOVE.L     D0,D1      copy D0 to D1
```

Immediate addressing mode: (Chapter 6.3 page 41) is used for **source** operands which are constants:

```
MOVE.W     #10,D0     move 10 decimal into D0
CMP.B      #'X',D0    is character in D0 = 'X'
```

The operand *immediately* follows the instruction in memory.

Absolute (or direct) addressing mode: (Chapter 8.2 page 60) is used for accessing operands at absolute memory locations (used for all types of operand reference: data, memory, alterable and control):

```
CLR.W      $1000           clear memory address $1000
CMPI.B     #10,$4000       test address $4000 = 10
MOVE.B     $1000,$2000     copy memory to memory
```

Address register direct addressing mode: (Chapter 13 page 100) the operand is in an address register (word and long word only):

```
MOVEA.L    A0,A1          copy A0 to A1
ADDA.L     #$1000,A0      Add constant $1000 to A0
CMPA.L     #$4000,A0      test A0 = $4000 ??
```

Note the use of MOVEA, ADDA and CMPA not MOVE, ADD and CMP.

Address register indirect addressing mode: (Chapter 13.1 page 100) an address register contains the address of the operand, i.e. a pointer to an operand in memory (used for all types of operand reference: data, memory, alterable and control):

```
CLR.B      (A0)          clear contents of address in A0
CMP.B      #10,(A0)      compare memory with 10
MOVE.L     (A0),(A1)     copy memory to memory
```

Address Register Indirect with Postincrement addressing mode: (Chapter 13.1 page 101) *after* the operand address has been obtained from the address register, the contents of the address register are incremented by the length of the operand (i.e. 1 for byte, 2 for word and 4 for long word, used for all types of operand reference):

```
CLR.B      (A0)+           clear indirect then increment
TST.W      (A0)+           test contents then increment
MOVE.L     (A0)+,(A1)+     copy memory to memory
```

Address Register Indirect with Predecrement addressing mode: (Chapter 13.1 page 101) the contents of the address register are decremented by the length of the operand *before* the operand address is obtained (used for all types of operand reference):

```
        CLR.B      -(A0)          decrement then clear indirect
        TST.W      -(A0)          decrement then test contents
        MOVE.L     -(A0),-(A1)    copy memory to memory
```

Address register indirect with displacement: (Chapter 13.1 page 103) a fixed displacement or offset value is added to the contents of an address register to form the effective address (used for all types of operand reference):

```
        CLR.B      $40(A0)           clear a byte in memory
        CMPI.W     #1000,$20(A1)     memory contents = 1000 ??
        MOVE.L     $2(A0),$10(A1)    copy memory to memory
```

If A0 = $2000 the effective address of CLR.B would be $2040.

Address register indirect with index and displacement: (Chapter 13.1 page 104) the effective address is the sum of three values (used for all types of operand reference):

(a) the contents of an address register (long word value);
(b) the contents of an index register which can be either a data or address register (long word or sign extended word value);
plus (c) a displacement which is part of the op-code (sign extended byte value).

```
        CLR.B      $40(A0,D1.L)                  clear memory byte
        CMPI.B     #'A',$20(A2,D2.W)             memory = 'A' ??
        MOVE.L     $2(A0,D0.W),$10(A1,D1.L)      memory to memory
```

If A0=$20000 and D1=$1000 the effective address of CLR.B would be $21040.

Program Counter with displacement addressing mode: (Chapter 15.5 page 117) a fixed displacement or offset value is added to the contents the Program Counter (PC) to form the effective address (may be used for data, memory and control references, i.e. it may **not** be used for **alterable destination** operands):

```
        ADD.W      X(PC),D0      add memory to D0
        LEA.L      X(PC),A0      load data address into A0
        .....
X:      DC.W       10
```

Program Counter with index and displacement addressing mode: (Chapter 15.5 page 117) the effective address is the sum of three values (may be used for data, memory and control references, i.e. it may **not** be used for **alterable destination** operands):

(a) the contents of the Program Counter (PC),
(b) the contents of an index register which can be either a data or address register (long word or sign extended word value);
plus (c) a displacement which is part of the op-code (sign extended byte value).

```
        MOVE.W     TABLE(PC,D1.L),D0    move memory to D0
        ......
TABLE:  DC.W       10,20,30
```

Examples of addressing modes using the MOVE instruction

The MOVE instruction copies the contents of the source operand specified using *any addressing mode* to a destination operand specify by a *data alterable addressing mode* (the contents of the source are left intact). The following table shows the various combinations of source operands with the destination operand D0 (see Appendix B for full details)

MOVE instruction	Source addressing mode and source effective address	page
MOVE.W D1,D0	data register direct mode data register D1	40
MOVE.W $1000,D0	Absolute (memory) mode contents of memory location $1000	60
MOVE.W #$1000,D0	Immediate mode immediate constant value $1000	41
MOVE.W A0,D0	Address register direct mode address register A0	100
MOVE.W (A0),D0	Address register indirect A0 contains address (a pointer)	100
MOVE.W (A0)+,D0	Indirect with postincrement increment A0 after operand obtained	101
MOVE.W -(A0),D0	Indirect with predecrement decrement A0 before operand obtained	101
MOVE.B $10(A0),D0	Indirect with displacement address in A0 + $10	103
MOVE.W $4(A0,D1.L),D0	Indirect with index and displacement address in A0 + contents of D1 + $4	104
MOVE.W $100(PC),D0	Program Counter with displacement $100 + PC (program counter)	117
MOVE.W $10(PC,A1.L),D0	PC with index & displacement $10 + PC + contents of A1	117

The following table shows the various combinations of source operand D0 with destination operands (see Appendix B for full details)

MOVE instruction	destination effective address	page
MOVE.W D1,D0	data register D0	40
MOVE.W D0,$1000	move to memory location $1000	60
MOVE.W D0,(A0)	A0 contains address of destination	100
MOVE.W D0,(A0)+	" with postincrement	101
MOVE.W D0,-(A0)	" with predecrement	101
MOVE.B D0,$10(A0)	address in A0 + $10	103
MOVE.W D0,$4(A0,D1.L)	address in A0 + contents of D1 + $4	104
MOVEA.W D0,A0	address register A0*	100

* note the use of MOVEA when the destination is address register direct mode

Table B.2 shows that any of the source operand shown in the first table may be used with any of the destination operands shown in the second table when the instruction is MOVE (other instructions tend to be much more restrictive; see Appendix B).

Review of MC68000 instructions

Below is a summary of the MC68000 instruction set grouped as to use (hence the same instruction may appear in more than one place). The examples use a range of addressing modes (see Review on Addressing Modes for a description of each mode with examples).

General data manipulation instructions

instruction		information moved	page
CLR.B	D0	clear bits 0 to 7 of D0 to 0	40
EXG	D0,D1	exchange registers (all 32 bits)	51
SWAP	D0	swap lower and upper 16 bits of data register	52
MOVE.W	D1,D0	move (copy) data register*	41
MOVE.W	$1000,D0	move contents of memory address $1000*	64
MOVE.W	#$1000,D0	move constant value $1000*	41
MOVEQ	#10,D0	move constant (-128 to +127) to data register	42
MOVEA.L	#LABEL,A0	move to address register (CCR not effected)	100
MOVEM.L	D0-D3/A0,X	move multiple to memory (label X)	110
MOVEM.L	Z,D0/A0-A4	move multiple from memory (label Z)	110
MOVEP.W	D0,0(A0)	move peripheral data (to peripheral)	152
MOVEP.L	1(A0),D0	move peripheral data (from peripheral)	152
LEA.L	X(PC),A0	load effective address	116
PEA.L	X(PC)	push (on to stack) effective address	116

* a full list of the combinations of source and destination operands for the MOVE instruction is presented at the end of the Review of Addressing Modes (on page 122).

Arithmetic operations

instruction		operation performed	page
ADD.W	D0,D1	add binary	50
SUB.L	D3,LABEL	subtract binary	50
ADDI.W	#50,D0	add constant (immediate operand)	50
SUBI.W	#50,LABEL	subtract constant (immediate operand)	50
ADDQ.W	#2,D0	add constant in range 1 to 8	51
SUBQ.L	#6,D1	subtract constant in range 1 to 8	51
ADDX.L	D2,D0	add with extend (via X in CCR)	56
ADDA.W	#$0F,A2	add to address register (CCR not effected)	100
SUBA.L	D0,A1	subtract from address register "	100
NEG.W	D0	negate binary	51
EXT.W	D0	extend sign of byte value to word	52
EXT.L	D0	extend sign of word value to long word	52
MULS.W	#30,D0	signed binary multiply	52
MULU.W	#$4F,D5	unsigned binary multiply	52
DIVS.W	#10,D0	signed binary divide	52
DIVU.W	VALUE,D1	unsigned binary divide	52

Subroutine, exception processing and stack manipulation instructions

instruction		operation performed	page
JSR	SUB	call subroutine at label SUB	109
JSR	(A0)	call subroutine (address in A0)	109
BSR.S	SUB	call subroutine (short form)	109
BSR.L	SUB	call subroutine (long form)	109
RTS		return from subroutine	109
RTR		return and restore condition codes	111
RTE		return from exception	211
RTD	#displacement	return and deallocate (68010)	109
PEA.L	X(PC)	push (to stack) effective address	116
MOVE.L	D0,-(A7)	move (push) on to stack	107
MOVE.L	(A7)+,D4	move (pull) off the stack	108
MOVEM.L	D0-D7/A0-A6,-(A7)	move multiple to stack	110
MOVEM.L	(A7)+,D0-D7/A0-A6	move multiple from stack	110
LINK	An,#m	allocate m bytes on the stack	111
UNLK	An	deallocate (previous LINK)	113

Program control instructions

instruction		operation performed	page
BRA.S	TEST	unconditional branch -126 to +129 locations	77
BRA.L	TEST	unconditional branch -32766/32769 locations	77
JMP	TEST	jump (unconditional)	78
DBRA	D1,LOOP	decrement and branch if D1 < > -1	87
Bcc.S	TEST	conditional branch, cc values are in	80
Bcc.L	TEST	(Table 10.1 and 10.2)	80
TST.W	D0	test operand and set condition codes	81
CMP.W	D0,D1	compare operands and set condition codes	82
CMPI.B	#'X',D0	compare constant with destination	82
CMPA.L	#$1000,A0	compare with address register	100
CMPM.B	(A0)+,(A1)+	compare operands using postincrement mode	82
Scc	X	Set according to condition codes	84
BTST.L	D0,D1	test bit in D1 specified by D0	95
TAS.B	D0	test & set operand	96
CHK	#9,D0	check array index bounds	201

RESET and TRAP instructions

instruction		operation performed	page
TRAP	#n	TRAP instruction n in range 0 to 15	201
TRAPV		Trap on overflow	201
RESET		Issue internal RESET	151
STOP	#n	Stop and wait for exception	212

Logical, shift/rotate and single bit manipulation instructions

instruction		operation performed	page
NOT.B	D0	logical NOT (invert bits)	91
AND.B	D0,D1	logical AND	91
OR.W	#$0FF00,D0	logical OR	91
EOR.B	D0,D1	logical exclusive OR	91
ANDI.B	#$0F,D0	logical AND constant with destination	92
ORI.L	#$1F,LABEL	logical OR constant with destination	92
EORI.W	#$0F,D2	logical exclusive OR constant with destination	92
ASL.B	#3,D0	arithmetic shift left (constant 1-8)	93
ASR.W	D1,D0	arithmetic shift right (D1 in range 0-63)	93
LSL.L	#3,D0	logical shift left	94
LSR.W	Z	logical shift right memory (word) by 1 bit	94
ROL.B	#3,D0	rotate left	94
ROR.L	D1,D0	rotate right	94
ROXL.B	#3,D0	rotate left via X	95
ROXR.W	Z	rotate right via X	95
BCLR.L	#0,D0	test & clear bit	95
BSET.B	D1,LABEL	test & set bit in memory	95
BCHG.L	#23,D0	test & change bit	95
BTST.B	#1,LABEL	test bit	95
BTST.L	D0,D1	test bit in D1 specified by D0	95

Operations on 68000 processor registers

MOVE instruction		information moved	page
MOVE.W	#0,CCR	move to condition code register	79
MOVE.W	CCR,D0	move from condition code register (68010)	79
ANDI.B	#$F7,CCR	logical AND constant with CCR	92
ORI.B	#4,CCR	logical OR constant with CCR	92
EORI.B	#4,CCR	logical exclusive OR constantt with CCR	92
MOVE.W	#$2000,SR	move to MC68000 Status Register	199
MOVE.W	SR,D0	move from MC68000 Status Register	199
ANDI.W	#$0DFFF,SR	logical AND constant with SR	199
ORI.W	#$2000,SR	logical OR constant with SR	199
EORI.W	#$2000,SR	logical exclusive OR constant with SR	199
MOVE.L	A0,USP	move to user stack pointer	200
MOVE.L	USP,A1	move from user stack pointer	200
MOVEC	D1,VBR	move to control register (68010)	204
MOVEC	VBR,A0	move from control register (68010)	204

16

Subroutines, variable usage and parameter passing

16.1 The XDEF and XREF pseudo-operators

The sample programs in this book are very short and can be held within a single disk file. Practical programs of any size soon become too large to be managed within a single file and sets of subroutines which perform logical tasks are separated out and placed in their own disk file, i.e. modules and sets of modules as described in Chapter 9.1. Each file is assembled or compiled separately and then linked to form the executable program.

If the main program and subroutines are to reference subroutines in other files there must be some means of informing the assembler that:

(a) a label defined within the file may be referenced by other files, e.g. XDEF;
and (b) a label reference within a file is defined in another file, e.g. XREF.

```
XDEF        label list
XREF        label list
```

The XDEF directive specifies that labels in the list (which are defined in the current file) are to be passed to the linker as symbols which may be referenced by code in other files. The XREF directive specifies that labels in the list are not defined within the file, but are external, to be satisfied at link time. The subroutine library in Appendix D shows the use of XDEF and sample programs use XREF reference the subroutines.

In addition to subroutine names any label may be passed between files in this way, i.e. labelled variables in memory may be made accessible (global) to all files in the program (see next section).

The equivalent pseudo-operators for the XA8 cross assembler are .PUBLIC and .EXTERNAL (see Program 9.2).

16.2 The scope of variables in a program

The scope of a variable determines where it can be accessed in a program. In the discussion on modular programming in Chapter 9.1 it was stated that a module (i.e. subroutines) should only be allowed access to variables which it actually uses (if any subroutine can access any data errors can be very difficult to track down).

16.2.1 Variables in registers

The data and address registers should only hold information which is being used by the section of code currently executing. Information should be loaded from memory into registers as needed and saved (assuming it has been modified) when no longer required.

16.2.2 Labelled variables in memory.

A labelled memory location (see Chapter 8) can be referenced by any instruction in the current program source file, i.e. is *global* to that file. XDEF and XREF can then be used to make such labels *global* to all files that make up the complete program, i.e. the names of subroutines need to be global to all modules which *call* them.

A requirement of modular programming is that an individual module should only be allowed access to variables which it actually uses. Making variables *global* therefore violates this principal and in practice global variables should only be used where absolutely necessary, e.g. when communicating with interrupt service routines (see Chapter 23.7 and the sample programs in Chapter 24).

16.2.3 Variables on the stack

The use of the LINK and UNLK instructions (see Chapter 14.5) allows the stack to be used for the storage of variables:

(a) the main program can allocate variables which exist until the program terminates,
and (b) subroutines can allocate local variables which are discarded on return to the calling program.

High-level languages tend to make extensive use of the stack in this way and assembly language programs which are called from high level language programs should maintain the practice (see Chapter 29 for an example of a C program).

16.3 Passing parameters to subroutines

Although data may be passed in and out of subroutines via external global variables (see section 2.2 above) this practice is not recommended (it becomes very difficult to maintain track of which subroutines are accessing which variables).

The standard way to pass information in and out of subroutines is via parameters which can be strictly controlled in terms of only passing data which is actually required. As previously described in Chapter 9.2.1, subroutine parameters are:

(a) the data or address values passed into the subroutine for processing;
and/or (b) values returned to the calling program on exit from the subroutine.

The RDDECW library subroutine, previously used in sample programs, returns a single parameter in data register D0. Subroutines may be very complex with several parameters used for passing information in and out. Parameters may be passed in a number of ways:

1 via the data and/or address registers (see section 4 below);
2 on the stack (see section 5 below);
3 placed immediately after the JSR or BSR instruction in memory (see section 6 below).

In general the sample programs in this book have used registers for parameter passing, and for most assembly language programs this will normally suffice. When interfacing assembly language subroutines with high-level languages parameters are usually passed on the stack (see Chapter 29). The following sections describe the parameter passing techniques using, as an example, a subroutine to sum the elements of an array.

```
--------------------------------------------------------------------------------
1                        ; Program 16.1, subroutine parameters in registers
2                        ;
3                        ; Set up parameters of first array in A0 & D0
4                        ; WRITE(newline,'sum of first array = ',SUM)
5                        ; Set up parameters of second array in A0 & D0
6                        ; CALL SUMARR to sum the second array
7                        ; WRITE(newline,'sum of second array = ',SUM)
8                        ;
9                                  .PROCESSOR M68000    ;68000 processor
10                                 .EXTERNAL  WRLINE,WRTEXT,WRDECW
11                                 .PSECT     _text     ;open code section
12   000000  207C        START:    MOVEA.L    #ARR1,A0  ;address of first array
             00000076
13   000006  700A                  MOVEQ      #10,D0    ;length of first array
14   000008  6156                  BSR        SUMARR    ;SUM the first array
15   00000A  4EB9                  JSR        WRTEXT
             00000000
16   000010  0A                    .BYTE      10,13     ;newline
17   000012  73756D20              .TEXT      "sum of first array = $"
18                                 .EVEN
19   000028  4EB9                  JSR        WRDECW    ;display the SUM
             00000000
20   00002E  207C                  MOVEA.L    #ARR2,A0  ;address of second array
             0000008A
21   000034  7005                  MOVEQ      #5,D0     ;length of second array
22   000036  6128                  BSR        SUMARR    ;SUM the second array
23   000038  4EB9                  JSR        WRTEXT
             00000000
24   00003E  0A                    .BYTE      10,13
25   000040  73756D20              .TEXT      "sum of second array = $"
26                                 .EVEN
27   000058  4EB9                  JSR        WRDECW    ;display the SUM
             00000000
28   00005E  4E4E                  TRAP       #14       ;!! STOP PROGRAM
29                       ; data area
30                                 .PSECT     _data     ;open data section
31   000000  0014        ARR1:     .WORD      20,35,67,-6,78,-89,4,56,-9,10
32   000014  0038        ARR2:     .WORD      56,-78,68,45,-7
33                       ;
34                       ; subroutine to sum elements of an array of word integer values
35                       ;
36                       ; SUBROUTINE SUMARR
37                       ; SUM = 0
38                       ; LOOP FOR number of elements in array
39                       ;     SUM = SUM + next array element
40                       ; ENDLOOP
41                       ; RETURN
42                       ;
43                       ; on entry:-    A0 holds address of start of array
44                       ;               D0 holds number of elements in array
45                       ; on exit:-     D0 holds SUM of elements of array
46                       ;
47                                 .PSECT     _text       ;reopen code section
48   000060  48A74080 SUMARR:      MOVEM      D1/A0,-(A7) ;save registers
49   000064  2200                  MOVE.L     D0,D1       ;number of elements
50   000066  5341                  SUBQ.W     #1,D1       ;subtract 1 for DBRA
51   000068  4280                  CLR.L      D0          ;clear SUM
52   00006A  D058        SUMAR1:   ADD.W      (A0)+,D0    ;add next element
53   00006C  51C9FFFC              DBRA       D1,SUMAR1   ;finished ?
54   000070  4C9F0102              MOVEM      (A7)+,D1/A0 ;restore registers
55   000074  4E75                  RTS                    ;return
56                                 .END
--------------------------------------------------------------------------------
```

Program 16.1 Passing subroutine parameters in registers (XA8 assembler)

16.4 Passing parameters to subroutines via registers

When using registers, the parameters passing in or out can be:

(a) values in data registers (e.g. WRDECW expects to find the number to be displayed in D0, and RDDECW returns the number entered in D0);

and/or (b) parameter addresses in address registers (e.g. WRTXTA expects the address of the text message to be in A0).

Parameters passed in this way can be accessed immediately by the subroutine. This is the most simple means of passing parameters and is used in most of the assembly language programs presented in this book. Its disadvantages are:

1 The number of parameters is limited by the number of registers.
2 The calling program may have important values in registers which have to be saved before the parameters are loaded.

Program 16.1 (previous page) calls subroutine SUMARR to sum the elements of an array, where the parameters are passed in registers. When subroutine SUMARR is called A0 contains the start address of the array and D0 contains the number of array elements and, on exit, the value of the sum is returned in D0. Note that registers used within the subroutine are saved and restored using MOVEM and, in line 50, the array length is decremented by 1 for use as a DBRA loop counter. The assembler listing was generated by the XA8 cross assembler; note the use of the .PSECT pseudo-operator to open (line 11) and reopen (line 47) the code section and open the data section (line 30).

If a large number of parameters are to be passed, an address register can point to a parameter block where the actual parameters are stored. Examples of a parameter blocks are shown in Program E.1b and E.2B (Appendix E).

16.5 Passing parameters to subroutines on the stack

When passing parameters via the stack, the parameter values (or the addresses of the parameters) are pushed onto the stack and the subroutine is called using JSR or BSR. The subroutine then has to retrieve the parameter information before it can be used.

If the stack is used within the subroutine (e.g. by MOVEM) care must be taken to ensure that the stack integrity is maintained. Otherwise when RTS is executed the value on the top of the stack will not be the return address and the program will crash.

Program 16.2 (next page but one) is a modified version of Program 16.1 which pushes the subroutine parameters onto the stack before the JSR instruction (lines 12 and 19 push the address of the array onto the stack, and lines 13 and 20 push the length of the array). The subroutine SUMARR then pulls the parameters off the stack and carries out the addition. When the subroutine terminates the parameters must be removed from the stack either within the subroutine or by the calling program after return (in the case of a high-level language refer to the compiler manual for details).

The following diagram shows the stack pointer and stack contents at various points during the first call to SUMARR (e.g. put a breakpoint at address $1098, the value on the stack are long words unless otherwise stated):

	address	contents		comments
line 50, A7 after MOVEM	→ 0FEE	??	??	saved D1
	→ 0FF0	??	??	"
	→ 0FF2	??	??	saved A0
	→ 0FF4	??	??	"
line 49, A7 after call	→ 0FF6	00	00	return address
	→ 0FF8	10	14	"
line 14, A7 before call	→ 0FFA	00	0A	array length (word)
	→ 0FFC	00	00	array address
	→ 0FFE	10	76	"
line 12, original A7	→ 1000	2E	7C	first instruction

The sequence of Program 16.2 is (values are for the first call of SUMARR):

11 sets up the stack pointer A7 to $1000

12 `PEA.L ARR1` pushes the start address of array ARR1 onto the stack, long word value $00001076, resultant A7 = $0FFC; note that `MOVE.L #ARR1,-(A7)` or `PEA.L ARR1(PC)` could have been used (the latter being relocatable)

13 pushes the array length onto the stack, word value 10, resultant A7 = $0FFA

14 subroutine called, pushes return address ($00001014) on stack, resultant A7 = $0FF6

49 subroutine entry, uses MOVEM to save D1 and A0 onto stack, resultant A7 = $0FEE

50 `MOVE.W 12(A7),D1` gets the array length off the stack at $0FFA twelve bytes above current stack pointer, i.e. A7 = $0FEE so the effective address of 12(A7) is $0FEE + 12 = $0FFA

51 subtracts one from the array length for use as the DBRA loop counter

52 `MOVE.L 14(A7),A0` gets the array start address off the stack (at $0FFC fourteen bytes above current stack pointer)

53 clear the SUM (in D0)

54 performs the summation of the array using *postincrement addressing mode*

55 program loop control

56 uses MOVEM to restore D1 and A0, A7 now equals $0FF6

57 `MOVE.L (A7),6(A7)` moves the subroutine return address (from $FF6) up the stack (to $FFC), overwriting the first parameter (the array address)

58 `ADDQ.L #6,A7` resets A7 to point at the new position of the return address at $0FFC

59 returns to the calling program with the parameters removed from the stack

When calculating the position of the parameters on the stack (in lines 50 and 52) allowance has to be made not only for space used by the return address but the values pushed by MOVEM, i.e. four bytes are used for every long word pushed by MOVEM.

Lines 56 and 57 of Program 16.2 remove the parameters from the stack. An alternative is that the subroutine leaves the stack intact (as on entry) and after return the calling program removes the parameters, i.e. in this case by adding 6 to A7. High-level languages which use the stack to pass parameters will have rules for passing them, i.e. the order in which they are passed and in what form (values, pointers, etc.), and how and when they are removed (see Chapter 29 for an example using the C language).

```
------------------------------------------------------------------------------
 1                            * Program 16.2, subroutine parameters on the stack
 2                            *
 3                            * Push parameters of first array onto stack
 4                            * CALL SUMARR to sum the first array
 5                            * WRITE(newline,'sum of first array = ',SUM)
 6                            * Push parameters of second array onto stack
 7                            * CALL SUMARR to sum the second array
 8                            * WRITE(newline,'sum of second array = ',SUM)
 9                            *
10                                     XREF      WRLINE,WRTEXT,WRDECW,WRREGS
11 001000 2E7C 00001000 START:         MOVEA.L   #$1000,A7     set up A7
12 001006 4879 00001076                PEA.L     ARR1          first array address
13 00100C 3F3C     000A                MOVE.W    #10,-(A7)     array length
14 001010 6100     0082                BSR       SUMARR        SUM the array
15 001014 4EB9 0002802C                JSR       WRLINE
16 00101A 4EB9 00028038                JSR       WRTEXT
17 001020 73 75 6D 20 6F               DC.W      'sum of first array = $'
18 001036 4EB9 00028074                JSR       WRDECW        display SUM
19 00103C 4879 0000108A                PEA.L     ARR2          second array address
20 001042 3F3C     0005                MOVE.W    #5,-(A7)      second array length
21 001046 6100     004C                BSR       SUMARR        SUM the array
22 00104A 4EB9 0002802C                JSR       WRLINE
23 001050 4EB9 00028038                JSR       WRTEXT
24 001056 73 75 6D 20 6F               DC.W      'sum of second array = $'
25 00106E 4EB9 00028074                JSR       WRDECW        display SUM
26 001074 4E4E                         TRAP      #14           !! STOP PROGRAM
27                            * data area
28 001076 0014 0023 0043 ARR1:         DC.W      20,35,67,-6,78,-89,4,56,-9,10
29 00108A 0038 FFB2 0044 ARR2:         DC.W      56,-78,68,45,-7
30                            *
31                            * subroutine to sum elements of a word array
32                            * SUBROUTINE SUMARR
33                            * SUM = 0
34                            * LOOP FOR number of elements in array
35                            *     SUM = SUM + array element
36                            * ENDLOOP
37                            * RETURN
38                            *
39                            * on entry: top of stack   =  return address
40                            *           top of stack+4 = number of array  elements
41                            *           top of stack+6 = start address of array
42                            *
43                            * on exit:  D0 holds SUM of the elements of the array
44                            *
45                            * Other internal registers:-
46                            * D1 is used to store number of elements in array
47                            * A0 holds the address of the array
48                            *
49 001094 48E7      4080  SUMARR:  MOVEM.L  D1/A0,-(A7)       save registers
50 001098 322F      000C           MOVE.W   12(A7),D1         get array length
51 00109C 5341                     SUBQ.W   #1,D1             -1 for DBRA
52 00109E 206F      000E           MOVE.L   14(A7),A0         get array address
53 0010A2 4280                     CLR.L    D0                SUM = 0
54 0010A4 D058            SUMAR1:  ADD.W    (A0)+,D0          add next element
55 0010A6 51C9      FFFC           DBRA     D1,SUMAR1         finished ?
56 0010AA 4CDF      0102           MOVEM.L  (A7)+,D1/A0       restore registers
57 0010AE 2F57      0006           MOVE.L   (A7),6(A7)        return address
58 0010B2 5C8F                     ADDQ.L   #6,A7             reset stack
59 0010B4 4E75                     RTS                        and return
60                                 END
------------------------------------------------------------------------------
```

Program 16.2 Passing subroutine parameters on the stack

16.6 Subroutine parameters following the JSR or BSR in memory

Another way to pass parameters into subroutines is to place the parameters or parameter addresses immediately after the subroutine call instruction (JSR or BSR) in memory, e.g. subroutines WRTEXT and DELAY in the library of Appendix D use this technique. After the JSR or BSR the top of the stack contains the address of the first parameter which can then be obtained by using *address register indirect with displacement addressing mode.*

Program 16.3 (next page) shows the sum array elements SUMARR subroutine using this technique to pass the parameters. In Program 16.3, the address pushed onto the stack by the JSR instructions (lines 10 and 17) is the address of the start of the parameters (a word which contains the number of elements in the array, followed by words containing the array data), not the return address. The parameters are then accessed within SUMARR as follows:

43 MOVEM is used to push D1 and A0 onto the stack

44 `MOVE.L 8(A7),A0` moves the start address of the parameters into A0 (i.e. A7 is pointing to the last word pushed onto the stack, MOVEM has just pushed two long words, therefore the start address of the parameters is 8 bytes above the top of the stack)

45 `MOVE.W (A0)+,D1` moves the number of elements in the array into D1 (after the execution of this instruction A0 will contain the address of the first element of the array)

46 subtracts one from the array length for use as the DBRA loop counter

47 the value of SUM is set to 0 (held in D0)

48 on each iteration of the loop the next array element is added to the SUM in D0 (and 2 is automatically added to A0)

49 DBRA controls the loop

50 on exit from the loop A0 will contain the address of the memory location following the last array element; `MOVE.L A0,8(A7)` moves this address onto the stack into the memory location from where line 44 obtained the parameter address, ready for RTS

51 MOVEM restores D1 and A0

56 RTS returns to the calling program; execution continues from the instruction following the end of the array data (lines 13 and 20)

Parameter passing in this way requires great care, e.g. if the parameter specifying the number of array elements is incorrect the program will crash as the RTS return address will be incorrect at line 53. An additional problem is that the parameter values are fixed when the program is written. This can be overcome by following the JSR with the address of a parameter block where the data values are held (or this could be passed in an address register). The contents of the parameter block can then be changed at run time.

In practice, this technique would only be used for small systems coded totally in assembly language, e.g. the sample programs in this book. This technique has disadvantages in that (a) it mixes code and data and (b) it is never used in high level languages. It is, however, a useful example of stack manipulation techniques and it is worthwhile being aware that it exists, e.g. the M68 and Kaycomp monitor calls via the TRAP #11 instruction use this technique to pass the command (see Chapter 9.4).

```
------------------------------------------------------------------------------
 1                         * Program 16.3 - subroutine parameters follow JSR
 2                         *
 3                         * CALL SUMARR to sum the first array
 4                         * WRITE(newline,'sum of first array = ',SUM)
 5                         * CALL SUMARR to sum the second array
 6                         * WRITE(newline,'sum of second array = ',SUM)
 7                         *
 8                         *
 9                                   XREF      WRLINE,WRTEXT,WRDECW
10 001000 6100      007C START:      BSR       SUMARR      SUM first the array
11 001004 000A                       DC.W      10          number of elements, etc
12 001006 0014 0023 0043             DC.W      20,35,67,-6,78,-89,4,56,-9,10
13 00101A 4EB9 0002802C              JSR       WRLINE
14 001020 4EB9 00028038              JSR       WRTEXT
15 001026 73 75 6D 20 6F             DC.W      'sum of first array = $'
16 00103C 4EB9 00028074              JSR       WRDECW      display SUM
17 001042 6100      003A             BSR       SUMARR      SUM second array
18 001046 0005                       DC.W      5           number of elements, etc
19 001048 0038 FFB2 0044             DC.W      56,-78,68,45,-7
20 001052 4EB9 0002802C              JSR       WRLINE
21 001058 4EB9 00028038              JSR       WRTEXT
22 00105E 73 75 6D 20 6F             DC.W      'sum of second array = $'
23 001076 4EB9 00028074              JSR       WRDECW      display SUM
24 00107C 4E4E                       TRAP      #14         !! STOP PROGRAM
25                         *
26                         * subroutine to sum elements of a word array
27                         * SUBROUTINE SUMARR
28                         * SUM = 0
29                         * LOOP FOR number of elements in array
30                         *     SUM = SUM + array element
31                         * ENDLOOP
32                         * RETURN
33                         *
34                         *  on entry top of stack points to:-
35                         *          word holding number of elements in array
36                         *          followed by elements of array (words)
37                         * on exit: D0 holds sum of elements of array
38                         *
39                         * Other internal register used:
40                         * D1 is used to store number of elements in array
41                         * A0 is used to hold the array pointer
42                         *
43 00107E 48E7      4080 SUMARR:     MOVEM.L   D1/A0,-(A7)      save working registers
44 001082 206F      0008             MOVE.L    8(A7),A0         get return address
45 001086 3218                       MOVE.W    (A0)+,D1         get number of elements
46 001088 5341                       SUBQ.W    #1,D1            subtract 1 for DBRA
47 00108A 4280                       CLR.L     D0               initialise sum to 0
48 00108C D058           SUMAR1:     ADD.W     (A0)+,D0         add array element
49 00108E 51C9      FFFC             DBRA      D1,SUMAR1
50 001092 2F48      0008             MOVE.L    A0,8(A7)         adjust return address
51 001096 4CDF      0102             MOVEM.L   (A7)+,D1/A0      restore registers
52 00109A 4E75                       RTS                        return to calling program
53                                   END
------------------------------------------------------------------------------
```

Program 16.3 Subroutine parameters following JSR in memory

16.7 TRAP instructions

TRAP instructions take the form:

```
        TRAP      #n          system TRAP, n = 0 to 15
```

In principle a TRAP instruction is similar to a subroutine call. In the *exception vector table* at the bottom of memory (fully explained in Chapter 23) there are sixteen long word entries corresponding to the sixteen possible TRAP instructions (i.e. n can take the value 0 to 15). These table entries hold the start addresses of exception service routines, which, are similar in structure to a subroutine (the final instruction is RTE instead of RTS).

Program 16.4 shows how to program `TRAP #0`. In line 5, `MOVE.L #TRP_0,$80` moves the start address of the service routine (label `TRP_0`) into the exception vector table entry corresponding to `TRAP #0` (memory address $80 - see Table 23.2). When `TRAP #0` is executed in line 6 the processor obtains the address of the service routine from memory address $80 and calls the routine. Program 16.4 assumed that the exception vector table was in RAM memory (not ROM) and that the program was executing with the processor in Supervisor mode (see Chapter 23 for a full discussion of these topics).

Information can be passed into TRAP service routines using any of the parameter passing techniques described for subroutines (note that TRAP pushes more information onto the stack than JSR or BSR). In practice, TRAPs tend to be used for fixed system wide facilities, e.g.:

1. Some TRAP entries may be reserved for use by the operating system or monitor for system functions (see manuals), e.g. M68 uses `TRAP #12` for breakpoints.
2. Some TRAP entries may be used to enable programs to access monitor functions at run time, e.g. M68 provides various functions via `TRAP #11` - see Chapter 9.4.

The remainder may be available for user programmed functions (depends upon the operating system or monitor - see manuals).

```
--------------------------------------------------------------------------------
1                             * Program 16.4 -  on a TRAP #0 displays a message
2                             * assumes that the exception vector table is in RAM
3                             *
4                                       XREF      WRLINE,WRTEXT
5 001000 21FC 0000100C START:           MOVE.L    #TRP_0,$80     set up TRAP #0 table entry
                   0080
6 001008 4E40                           TRAP      #0             execute the TRAP
7 00100A 4E4E                           TRAP      #14            !! STOP PROGRAM
8                             *
9                             * exception service routine for TRAP #0
10 00100C 4EB9 00028038 TRP_0:          JSR      WRTEXT          display a message
11 001012 61 20 54 52                   DC.W     'a TRAP #0 occurred $'
12 001026 4EB9 0002802C                 JSR      WRLINE
13 00102C 4E73                          RTE
14                                      END
--------------------------------------------------------------------------------
```

Program 16.4 Subroutine parameters following JSR in memory

17

Medium to large programs: design, coding and testing

The program implementation problems set at the end of previous chapters have been fairly simple, e.g. a page or so of source code, but we are now getting to the stage where more realistic programs may be attempted. This chapter discusses the problems of implementing large software systems and introduces some useful techniques.

17.1 Software as a component of total system cost

Until the late 1970s hardware was the major contributor to the cost of a computer system, requiring programmers to produce compact fast and efficient code with little thought given to future enhancements or maintenance. Problems occurred when modifications were required to correct faults or add facilities. The code could not be understood, even by the original programmer, and total rewrites were often required. Fig. 17.1 (Chikofsky and Rubenstein 1988) shows how the contributions of hardware and software towards the total cost of a complex computer based product have changed over the years. In particular software maintenance costs are predominating; Chikofsky and Rubenstein give the example of the development cost of software for the USA Air Force's F-16 jet fighter being $85 million with expectations to spend about $250 million on software maintenance. Although the exact percentage contributions will vary from product to product, e.g. the cost of a mass produced washing machine controller would still be mainly hardware, the cost of software development and maintenance is clearly of major importance.

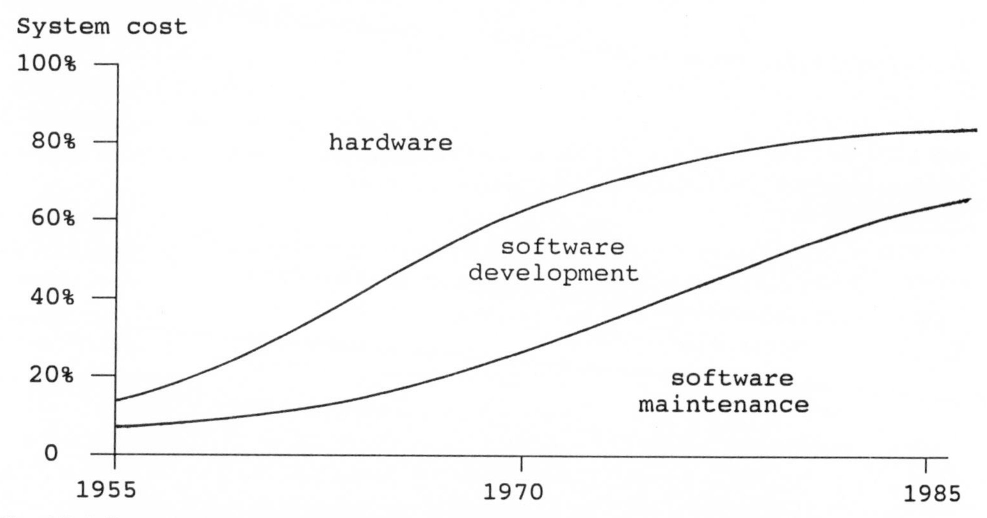

Fig. 17.1 Contribution of hardware and software to total system cost (Chikofsky and Rubenstein 1988) (c) IEEE Computer 1988

As software costs increased and applications became more critical (e.g. in a process control system where a fault could destroy a chemical plant), it was realised that scientific methods had to be applied to the design and implementation of programs. These techniques are called **software engineering** (Steward 1987) and are used to speed up the design, coding and testing of programs, and give a better quality product that is easy to maintain and use.

17.2 General requirements for software systems

As end-user requirements become more exacting system software grows in complexity and size. Systems consisting of 200,000 to 500,000 lines of code, once beyond imagination, are now commonly found in a single product. In general software systems have a number of overall requirements:

Correctness and reliability, which outweigh all the other requirements. Faults in programs can range in effect from being costly, for example, incorrectly ordered stock for a supermarket, to disastrous, for example, a fault that could lead to an explosion in a computer controlled chemical process.

Flexibility and reusability. To be commercially viable components of software systems must be capable of being used across a range of products and computer systems (Bramer 1988). This flexibility is more often provided by the use of high level languages than assembly language, but all programs are more easily adapted to changing requirements (and they will change), if written in a modular manner and well documented.

Efficiency. As a software system grows in size and complexity it also tends to run slower. Slow response time may be annoying to an engineer using a CAD system but it could prove disastrous in a jet aircraft travelling at Mach 2. Efficiency is the usual reason for coding in assembly language, but it must never be pursued at the expense of correctness and only in extreme cases at the expense of flexibility. Today it is generally simpler and cheaper to purchase faster hardware than spending time gaining marginal improvements in software efficiency.

Maintainability. This deals with the problem of updating and improving software regardless of complexity. Modifying systems consisting of hundreds of thousands of lines of code written by software engineers who have since moved on is a daunting task. Will new staff be able to understand the code as originally written ? If not, product maintenance will be very costly, time consuming and prone to error.

In an era of increasing system size and complexity an important issue is productivity. A software system consisting of 500,000 lines of code may take 150 man-years to implement. Without use of modern techniques and tools such systems may never become operational, e.g. CASE (Computer Aided Software Engineering) tools aid with software specification, design, debugging, integration, performance analysis, verification, and maintenance (Chikofsky and Rubenstein 1988, Wallace & Fujii 1989, Oman 1990). The functionality offered by software development tools depends to a large extent on the complexity and size of the project and hence the funds available to purchase tools and support software and hardware. In practice this can range from a simple editor, compiler and run-time system on a PC to a CASE system running on a high powered professional workstation (Oman 1990).

17.3 Software engineering techniques

A full discussion of Software Engineering techniques is beyond the scope of this book but in general the stages in the development of a software product are:

Functional specification. Using System Analysis techniques a functional specification of the system is developed.

System design. The system is designed in terms of structure charts and specifications of modules.

Detailed design and coding. Working from the design specification modules are designed, coded and tested.

Hardware integration. Modules are executed in any specialist target hardware environment to check for hardware/software interaction problems.

Unit tests. Modules which form units or components of a system are integrated and tested.

System test. The complete system is integrated and tested, often at end-user sites.

Maintenance. Correction of existing errors and updating the functionality of the product.

The process can be iterative, but the earlier in the sequence that errors are found or updated requirements specified, the better. Fig. 17.2 shows the relationship between the cost of correcting an error and the stages in product development where it is found. For example, if modifications are required during system test (e.g. due to the analysis being incorrect), the design, coding and testing phases will have to be repeated.

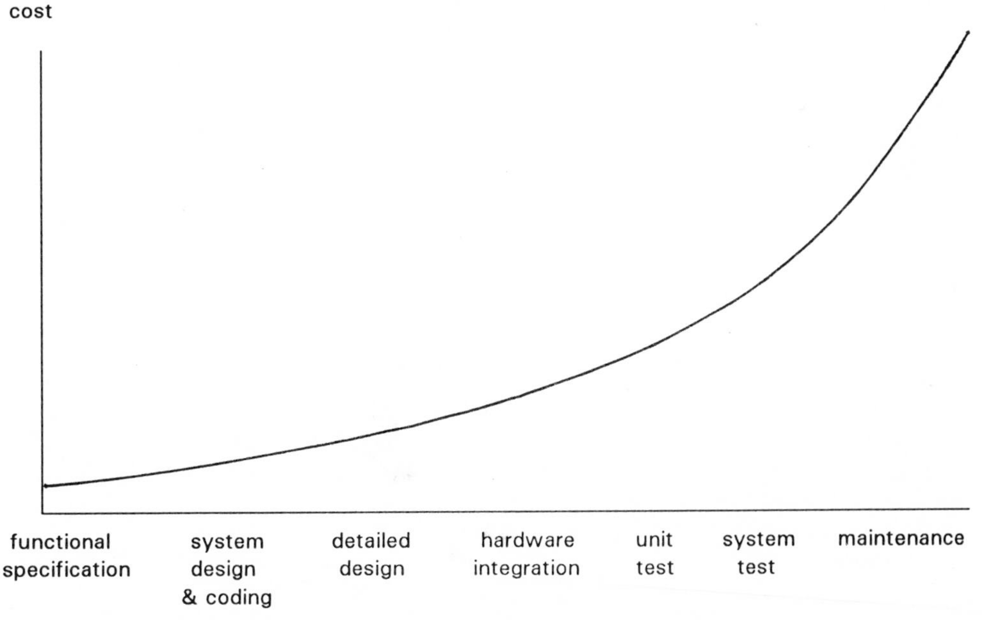

Fig. 17.2 Relationship between the cost of correcting an error and the stages in product development where it is found.

17.4 Documentation

Documentation is vitally important if a program is to be understood, not only by others, but by the original programmer at a later stage. In particular, sequences of assembly code statements that appear 'obvious' in meaning when written, unless thoroughly commented, can be incomprehensible when read a week later. In the sample program listings in this book, the final design of a module, produced as a **Structured English** description, is always typed into the source program as comments at the head of each section of code. In addition, comments alongside the code should say **why** a particular operation is being carried out (i.e. how the design is being implemented). It is not to be expected that there will be a one-to-one relationship between the Structured English and sequences of assembly code statements. Assembly code is written for efficiency and the Structured English is a design aid, not a strait-jacket.

17.5 System analysis

The object of the analysis phase is to find out exactly what the end-user wants the final program to do (the programmer himself may be the end-user). A problem is that often users do not know what they want because they do not know the capabilities of modern computer systems. It is the job of full time professional analysts to analyse complex problems.

From the analysis phase, a requirements specification is drawn up, which specifies exactly what the program will do. When complete, the finished program can be checked against this document.

All the exercises and problems presented in this book are in effect specifications of programs to be written. It may appear from these specifications that the end product is obvious, and no analysis was required. Even simple problems, however, do require some analysis if only to find out what type of data is to be processed and how much there will be, e.g. will a file contain 100 items or 10000000000000 (which will not fit onto a floppy disk).

17.6 Program design - stepwise refinement

It is recognised that human beings can only carry between five and ten separate operations in their heads at a time. If a program is more than 10 or 20 lines of code (depending upon experience and complexity of problem), any attempt to design and code the entire program in one go will lead to a badly conceived and incorrect program. One design technique is stepwise refinement, whereby a task is visualised initially at a very simple level and subsequently refined in more and more detail.

17.6.1 Design exercise - a system monitor for a single board computer system

A system monitor is required for a single board computer based on the MC68000 (the facilities of a typical systems monitor were outlined in Chapter 3.1), e.g. the following is an outline design of parts of the M68 system monitor (see the answers to the exercises of Chapter 3 and 4 in Appendix G for sample runs of M68).

M68 system monitor, level 1 refinement: power up or RESET processing

```
initialise system data areas and stack
initialise exception vectors
initialise I/O devices
check CPU type (68000 or 68010) and size of RAM, etc.
display system start up message
perform hardware checks and report any errors found
display system prompt
LOOP for ever
    IF a key has been pressed
        process user command
    ELSE
        IF 10 minutes have elapsed since last command
            LOOP
                perform non-destructive memory tests
            UNTIL a key is hit
ENDLOOP
```

These statements represent the tasks which are executed when the power is switched on or the reset button is hit (each task would probably be a subroutine, which calls further subroutines, etc). Statements in line are executed sequentially, i.e. the system data areas and stack are initialised, followed by the exception vectors, I/O devices, etc. Indented statements are executed according to LOOP control or conditions tested, i.e. IF a key has been hit the user command is processed, otherwise, IF nothing has happened for ten minutes non-destructive memory tests are performed UNTIL a key is hit.

This shows the overall tasks but gives no idea of how to implement initialise exception vectors or process a user command, etc. Further refinement is therefore required.

level 2 refinement: initialise exception vectors (see Chapter 23)

```
IF exception vectors are in ROM
   set up JMP table to exception routines in RAM
ELSE (exception vectors are in RAM)
   load exception routine addresses into vector table in RAM
```

level 2 refinement: check CPU type (68000 or 68010) and size of RAM, etc.

```
 attempt to execute MC68010 instruction, e.g. MOVE.W CCR,D0
 IF illegal instruction exception occurs
     CPU is a MC68000
 ELSE
     CPU is a MC68010
determine size of RAM fitted
determine I/O devices fitted, MC6821, MC6840, MC68230, etc.
```

level 3 refinement: determine size of RAM fitted

```
IF Bytronic or Force board
   LOOP
       writing words of data into RAM
   UNTIL bus error exception occurs
   size of RAM equals the bus exception address - 2
ELSE
   IF Kaycomp board
       write test_value to last word of first 16K RAM
       IF last word of 32K of RAM equals test_value
          write test_value_2 to last word of first 16K RAM
          IF last word of 32K of RAM test_value_2
              size of RAM is 16K
       ELSE
          write test_value to last word of first 32K RAM
          IF last word of 64K of RAM equals test_value
              write test_value_2 to last word of first 32K
              IF last word of 64K of RAM equals test_value_2
                  size of RAM is 32k
              ELSE
                  size of RAM is 64K
```

The technique used to determine the size of the RAM will depend upon the target system hardware (with some systems it may not be possible to find the RAM size). Two techniques are presented in the above design:

1 On the Bytronic and Force board a bus error exception is generated if non RAM is written to; thus the RAM size is found by writing data into RAM until an exception occurs (i.e. when ROM is written to).
2 The Kaycomp can be fitted with 16, 32 or 64K of RAM. It does not have bus error exception circuits. However, if less than 64K is fitted the RAM is mirrored, i.e. 32K of RAM appears twice in RAM addresses 0 to 32K and 32 to 64K, thus if the same data appears in both halves 32K is fitted (16K is mirrored four times).

level 2 refinement: process user command (user has hit a key)

```
read(key_hit)
CASE key_hit
     'R': register command processing;
     'M': memory command processing;
     'G': program execute command processing;
     'T': program trace command processing;
     'B': program breakpoint command processing;
ELSE
     display error message and a list of valid commands
```

Note that commands to M68 are sequences of single key hits (see Appendix G for sample answers to Exercise 3.1, 4.1, etc).

Level 3 refinement: register command processing.

```
write('Register ')
read(key_hit)
CASE key_hit
    'I': write('initialise'); initialise D0-D7 & A0-A6 to 0
    'D': write('display'); display register values on screen
    'S': register set command processing
ELSE
    display error message and a list of valid commands
```

Level 4 refinement: register set command processing

```
write('set ',newline)
write(' Enter D or A (data/address register) and number ?')
read(key_hit)
IF key_hit <> 'D' or 'A'
    display error message
ELSE
    read(number)
    IF number <> '0' to '7'
        display error message
    ELSE
        write('new value ? ')
        read(number)
        set specified register to number
```

Numeric values may be entered as decimal or hexadecimal numbers (provided by common subroutines) and a command sequence may be aborted at any time by hitting the <ESC> key (the code for this is not shown so the above design requires further enhancement).

17.6.2 Design exercise - discussion

The refinement can stop when the design is of sufficient detail that it can be translated directly into program instruction statements, or calls to existing modules. For example, the level 2 and 3 refinement of **check CPU type (68000 or 68010), size of RAM, etc.** could be coded directly from the above design. In practice, how the monitor is implemented depends upon the software facilities, if any, which exist on the board:

1. If there is no software for the target board (i.e. a new design) the first task is to implement part of the **level 1 refinement: power up or RESET processing**. A reset exception vector routine (see Chapter 23.4) would have to be implemented which, after some initialisation, sets up the I/O device which controls serial communications with the host terminal or computer. Routines could then be implemented to do basic I/O (read and write characters, text strings, numeric values, etc.) and the rest of the system could be built up from this. The major problem is that successive modifications of the software would have to be burned into ROM to be tested (a very time consuming task). As soon as basic monitor facilities have been implemented and it is possible to load and execute programs one can proceed as follows.

2 If basic monitor facilities already exist on the board the new monitor can be tested by loading it into the user RAM memory area and executing it as a *normal* user program, i.e. by jumping to the start of the reset exception processing routine. The new monitor would take over control of the board and its facilities can be tested. If the exception vector table is in RAM exceptions can be tested as well. When satisfied the new monitor can be burned into ROM and fitted into the board.

The design process is generally iterative and modifications to existing stages may be needed. This can occur when a new situation arises that requires a backtrack in the design process. If possible, modifications should be carried out at the design stage; once a program has been coded, modification becomes much more expensive (see Fig. 17.2).

In practice, critical situations may occur where there is no means of recovering from an error condition, and the program would terminate with a **fatal error** or **unrecoverable error** message to the user. Such critical situations should be discovered during the specification and design phases, and some means to recover from them found.

Structured English is a tool. In some situations, for example to maintain a Company's documentation standards, a set form of Structured English must be adhered to. In general, however, Structured English should be written in a manner that the programmer is comfortable with, and that suits the application area and programming language being used. It can often be convenient to number Structured English statements with statement numbers and refinement level.

17.7 Testing

Although methods of thorough testing are beyond the scope of this book, the subject is introduced in the following section. Unless a program has been formally proved, confidence in it can only be determined by thorough testing. Two approaches are:

Black Box Testing. All reasonable combinations of input to a module are tried out, the results are predicted **in advance** and compared with the actual results. Such test data should, if possible, be devised at the program specification and design stage before any coding is carried out.

White Box Testing. Test data is devised to follow every possible path through the module. Again, results must be predicted in advance; it is easy to convince oneself that incorrect results are correct once a program has produced them.

In practice, a combination of these methods is often used. For example, white box testing (together with the display of internal variables) is often used when homing in on a fault. When testing any particular input value, it is important to test:

(a) normal values;
(b) ends of ranges;
(c) other special values (e.g. 0 often causes trouble);
and (d) possible invalid values.

Suppose, for example, valid numerical data input to a module should be -7 to +7 inclusive. A suitable set of test data could be -7, -1, 0, +1, +7, -8 and +8 where the latter two values should be rejected as invalid input.

Problem for Chapter 17

One of the tasks performed when a system is powered up is a RAM memory check. Write a program to test **part** of the user RAM area of your computer system, e.g.:

```
test_pattern = first_test_pattern
REPEAT
   WRITE('testing memory block with pattern ',test_pattern)
   FOR test_address = start_address TO end_address
      memory[test_address] = test_pattern
      IF memory[test_address] <> test_pattern
          WRITE('write error at address ',test_address,
                ' contents were ',memory[test_address],
                ' expected ',test_pattern)
   DELAY for 25mSec
   FOR test_address = start_address TO end_address
      IF memory[test_address] <> test_pattern
          WRITE('refresh error at address ',test_address,
                ' contents were ',memory[test_address],
                ' expected ',test_pattern)
   test_pattern = next_test_pattern
UNTIL all test_patterns completed
```

1 `memory[test_address]` indicates the contents of memory at address `test_address`.

2 The start_address and end_address should be in an area of RAM unused by the monitor or the program, e.g. for the Force, Bytronic and Kaycomp boards:
Force/Bytronic: program origin = $1000, start_address=$4000 end_address=$8000
Kaycomp: program origin = $400400, start_address=$404000 end_address=$408000

2 Two tests are performed for each test pattern written:
(a) a read after write check to ensure that the basic write and read circuits are OK;
(b) after 25mSec an additional read check ensures that dynamic memory refresh is OK.

4 Use the following test patterns which should show up errors due to shorts to 0 or 1, shorts between pins, crossed pins, etc.

```
0000       $FFFF      $5555      $AAAA
rotating 1 pattern -
     0001 0002 0004 0008 0010 0020 to 2000 4000 8000
marching 1's pattern -
    0001 0003 0007 000F 001F 003F 007F 00FF 01FF 03FF
       to 1FFF 3FFF 7FFF FFFF FFFE FFFC FFF8 FFF0
       to FC00 F800 F000 E000 C000 8000
```

To test the program generate an error, e.g. at a particular address don't write `test_pattern` but some other value. **Beware** that if the program goes wrong it could overwrite all of memory. For example, if using a ROM based editor/assembler the editor text buffer could be overwritten and the program source code lost. In such cases save the program source code before executing the program.

Testing for 'mirrored' blocks. One problem with the above test is that it does not test for mirrored blocks, i.e. memory which appears more than once in the memory map (due to an address decoding fault). Modify the program as follows:

```
WRITE('Testing with first test pattern ',first_pattern)
FOR test_address = start_address TO end_address
   memory[test_address] = first_pattern
   IF memory[test_address] <> first_pattern
       WRITE('write error at address ',test_address,
             ' contents were ',memory[test_address],
             ' expected ',first_pattern)
old_pattern = first_pattern
REPEAT
   WRITE('Checking ',old_pattern,' writing ',new_pattern)
   new_pattern = next_pattern
   DELAY for 25mSec
   FOR test_address = start_address TO end_address
      IF memory[test_address] <> old_pattern
          WRITE('refresh error at address ',test_address,
                ' contents were ',memory[test_address],
                ' expected ',old_pattern)
      memory[test_address] = new_pattern
      IF memory[test_address] <> new_pattern
          WRITE('write error at address ',test_address,
                ' contents were ',memory[test_address],
                ' expected ',new_pattern)
   old_pattern = new_pattern
UNTIL all test_patterns completed
```

Relocating memory test program. In certain cases memory will appear to be working correctly using the above techniques but fail when a program is executed in it. Implement a self relocating memory test, e.g. on the Bytronic or Force boards:

1. The program starts at location $1000
2. It tests a block starting at $1800 length $800 bytes using the above technique.
3. The program relocates (copies) itself to the block starting at $1800
4. The program restarts, executing the copy at $1800 (make sure that any data initialisation code in the original program is not re-executed)
5. It tests a block starting at $2000 length $800 bytes
6. The program relocates to location $2000 and tests the block starting at $2800, etc.

When the program has tested the area of memory (e.g. up to $8000) it should test the block $1000 to $1800 and then relocate itself to $1000, etc. Remember to generate relocatable code (see Chapter 15), i.e. use PC relative addressing modes. A common fault is that the program appears to work until it has relocated to the top of the test area and crashes when overwriting the original program (i.e. when testing block $1000 to $1800). If so it is likely that either the program has not correctly relocated (the original was being executed all the time) or absolute addressing modes have been used (accessing data in the original).

18

Introduction to Input/Output programming

18.1 The I/O device interface

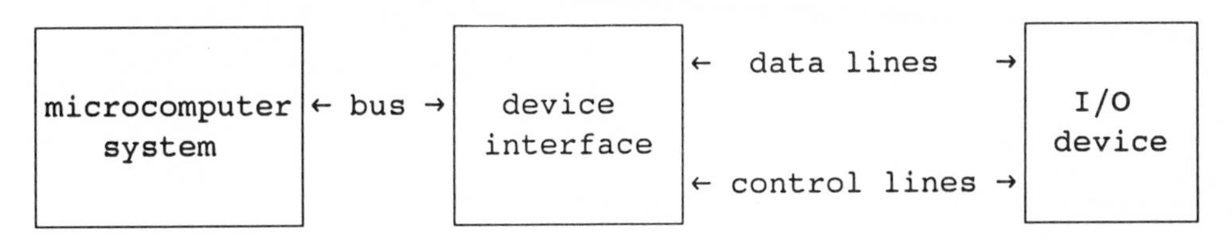

Fig. 18.1 Microcomputer, I/O device and its interface

The diagram of Fig. 18.1 represents an I/O device connected to a microcomputer. To enable a program running in the microcomputer to control an input/output device (e.g. disk, terminal, printer, etc.), an interface is required which is connected to the system bus. The bus carries the address, data and control information between the microprocessor, the memory, and the I/O device interfaces connected to the system. The device interface will contain the address decoding and control circuits which determine when the device is being addressed, and **data**, **status** and **control** registers which enable a program to control the device and transfer data:

Status Register(s). Contains information about the state of the I/O device, e.g. if it is ready to transfer data or is busy (do not confuse this with the MC68000 Status Register SR).

Control Register(s). By loading data into this register the program can control the action of the device (e.g. set the line speed of a serial interface).

Data Register(s). Contains the data being passed between the program and the I/O device.

The number of registers associated with an I/O interface is a function of the complexity of the device. A complex device (e.g. MC68230 PIT) may have several status and control registers in its interface. On the other hand, a simple device (e.g. the MC6821 PIA) may have a combined control and status register, with certain bits used for control and others for reporting status.

In general, the device interface is connected to the external device by a number of lines:

Data Lines. These lines are used for the transfer of data between the interface and the external device.

Control Lines. These lines carry signals which enable a program to determine the status of the device and control it.

18.2 Polled I/O programming

Before a program can transfer data to an I/O device it must ensure that the device is not busy (e.g. a printer may be in the process of printing a character). The simplest I/O programming technique is to use a polling loop, in which the program polls or examines the interface status register to determine if the device is ready for a data transfer. If the device is busy the program loops back to check the status register again. If the status register indicates that the device is ready for more data, the program carries out the next data transfer. In general, with a simple device such as a printer, a single bit in the status register indicates the state of the device (i.e. if it is 0 the device is busy or if it is 1 the device is ready for the data transfer).

```
Initialise interface and printer
LOOP
    LOOP
        test 'printer ready bit' in interface status register
    UNTIL printer is ready for next data byte
    Transmit next data byte to printer
UNTIL all characters have been printed
terminate program
```

Above is the outline pseudo-code of a program printing a sequence of characters on a printer. After initialising the printer the program continues thus:

1. The interface status register is polled to see if the printer is ready for the next character (e.g. a BTST bit test instruction is used to test the status bit).
2. If it is not ready, the program loops back to step 1 and tests the status register again.
3. If it is ready, the program sends the next character to the printer.
4. If all the characters have been printed the program stops, if not, it loops back to step 1 to print the next character.

If the device is slow (a printer may print 200 characters per second which is very slow in computer processor speed terms), the program spends most of its time in the polling loop where it is doing nothing useful. Operating systems generally have a program which allow a user to list a program on the printer, e.g. the PRINT command in MS-DOS. If polled I/O is used, the user, after requesting the printout, may well have to wait many minutes while a large listing is being printed. This time could well be spent editing another program or doing some other useful work. It is possible, with a simple output device such as a printer, for the operating system to allow the user to run other programs while every now and again polling the printer to see it is ready for the next character. The problem with such a technique is that some devices are time critical, in that data transfer can only take place within a particular time period, e.g. high speed disk I/O. If the operating system is using polled I/O to such a device, while at the same time running a program that is doing a large amount of processing, it can easily miss such a time period. I/O programming using interrupts to control data transfer allows such I/O events to take place in parallel with other processing.

18.3 Interrupt I/O programming

This technique allows programs to be run in parallel with an I/O data transfer. When the I/O device interface is ready for the next data transfer it sends a signal to the processor which interrupts the program being executed, and transfers control to an **interrupt service routine** (which is similar in format to a normal program subroutine). The interrupt service routine code performs the data transfer to the I/O device and terminates with an RTE (Return from Exception) instruction (full details in Chapter 23). The RTE instruction causes the processor to resume execution of the program that had been interrupted. Consider the following:

1 The user requests the operating system to print a file.
2 The operating system initialises the printer and then sends the first character.
3 The operating system then prompts the user for the next command.

Then in parallel:

4 The user continues to edit, assemble and run other programs.
5 When the printer is ready for the next character the interface sends an interrupt signal to the processor:
   (a) the current program (of 4 above) is halted;
   (b) the printer interrupt service routine is called;
      (i) the routine sends the next character to the printer;
      (ii) the routine is terminated with RTE,
   and (c) the interrupted program (of 4 above) is resumed.

The interrupt driven I/O technique allows the user to continue working while the program listing is being printed. In a multi-processing environment (i.e. an operating system which allows a number of processes or tasks to be run concurrently (Tanenbaum 1987)) all input/output devices would need to be interrupt driven, i.e. all other processes could not be stopped while I/O was being performed. The drawbacks with interrupt driven I/O systems is that the interface is more complex and it is more difficult to write and test the I/O driver programs than when using polled I/O techniques.

Another complication when using interrupts is that many processors allow a number of levels of interrupt priority. The priority system is built into the computer hardware with each device assigned a priority depending upon its importance and speed (i.e. fast devices having a high priority and slow devices a low priority). Interrupts from devices are then dealt with in order of priority, with high priority devices able to interrupt the servicing of lower priority devices.

Consider the case where an operating system is carrying out a transfer of information to a high speed disk and a printer concurrently. The disk data transfers are very time critical and, once the device is ready, the transfer must be carried out within a very short time period (tens of microseconds) otherwise the disk will have rotated past the required data. In such a case interrupts from the disk are assigned a higher priority than interrupts from the printer and may interrupt the printer interrupt routine as well as normal programs. For example:

1 A program is running in the processor.
2 The printer is ready for the next character and the printer interface interrupts the current program.
3 Control is transferred to the printer interrupt service routine which prepares to send the next character to the printer.
4 The disk becomes ready for a data transfer and the disk interface interrupts the current program (i.e. the interrupt service routine of the printer).
5 Control is transferred to the disk interrupt service routine which performs the data transfer and is terminated with RTE.
6 The printer interrupt service routine resumes, finishes the data transfer to the printer and terminates with RTE.
7 The original program is resumed.

18.4 Accessing the I/O device interface registers

Each I/O device interface contains a number of registers which pass data and status/control information between a program running in the processor and the I/O device circuits. There are two techniques for accessing I/O device registers:

(a) by means of special I/O instructions;
or (b) by having the I/O registers appear as part of the primary memory of the processor.

18.4.1 Special I/O instructions

Some processors have a set of special instructions for accessing I/O device registers. Each register is assigned an I/O port number (built into the hardware of the interface) and this is used as an operand in the I/O instructions. For example, the Intel 8086 family of microprocessors uses the I/O instructions IN and OUT, thus:

```
IN    AL,n   read contents of I/O port n to register AL
OUT   n,AL   output contents of register AL to port n
```

Clearly, before any I/O programming can be carried out, the port addresses and the format of the information within them must be known (from hardware manuals). This technique has the advantage that I/O instructions stand out in the program code but its disadvantage is that special I/O instructions are required (with associated control signals).

18.4.2 Memory mapped I/O registers

In this case the I/O registers appear as part of the primary memory of the computer. To program the device the programmer must know the memory addresses assigned to the registers and the format of the data within them. Although the registers are mapped as part of the memory address space of the computer, they are **not** normal memory for program and data storage. The main advantage of this techniques is that the standard memory manipulation instructions can be used to access the I/O registers. The main disadvantage is that I/O instructions in the program code appear no different from normal memory accesses. The use of modular programming techniques to isolate I/O routines together with good comments can overcome this disadvantage.

The MC68000 uses memory mapped I/O, For example, the sample programs in Chapter 25 use the MC6850 ACIA asynchronous serial interface (the next chapter will discuss parallel and serial I/O techniques) where the I/O registers of port A (length byte) are mapped at the following addresses on the Force board (Force 1984):

```
$50040: ACIA Control Register (to control the ACIA)
$50040: ACIA Status Register (reports the status of the ACIA)
$50042: Data Register (transmit and receive)
```

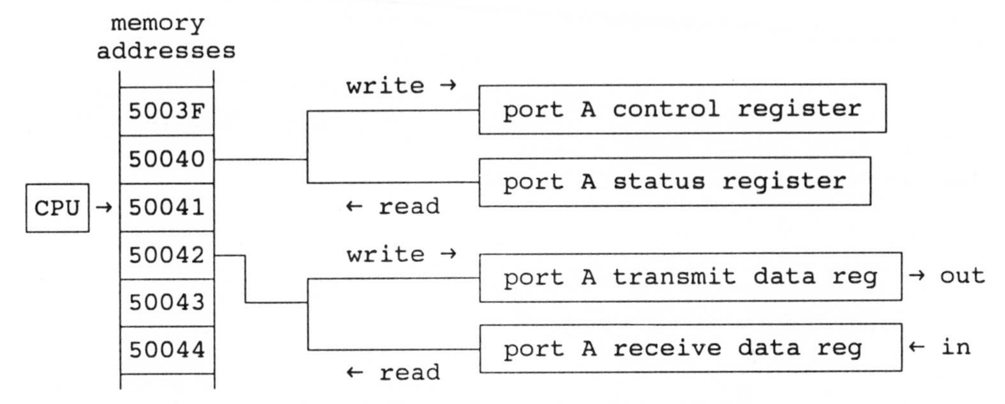

Fig. 18.2 Diagram showing the I/O registers of the MC6850 ACIA

Fig. 18.2 is a diagram showing the address mapping of the I/O registers of the MC6850 on the Force board:

1 **Two** separate registers, Control and Status, are accessed via address $50040:
   (a) on a **write** a byte is written to the control register;
   (b) on a **read** a byte is read from the status register.
2 **Two** separate data registers are accessed via address $50042:
   (a) on a **write** a byte is written to the **transmitter data register** (this holds the character being transmitted down the serial line);
   (b) on a **read** a byte is read from the **receiver data register** (holds the last character received from the serial line).

Thus it is **not possible** to read the control register and the transmitter data register or write to the status register and the receiver data register.

The I/O registers may be accessed by instructions as though they are memory, e.g. (using *absolute* addressing mode):

```
MOVE.B    D0,$50042     transmit character in D0
MOVE.B    $50042,D1     read received character into D1
MOVE.B    #$35,$50040   move 35 hex to ACIA Control Reg
MOVE.B    $50040,D0     move ACIA Status Reg into D0
```

The first instruction **writes** the byte from D0 to the transmitter data register, the second **reads** the byte from the receiver data register, the third **writes** $35 to the Control Register and the fourth **reads** the Status Register into D0.

It is **not** possible to read the contents of the Control Register (or transmitter data register) or write to the Status Register (or the receiver data register). In cases where different I/O registers are accessed on a memory read and a memory write, *memory to memory* operations cannot be used (i.e. AND, OR, BCHG, etc.), e.g.:

```
        ORI.B       #$35,$50040     ???????????
```

If location $50040 was normal RAM memory its contents would be read, ORed with $35, and the result written back. In this case the Status Register is read, ORed with $35, and the result written into the Control Register; a totally meaningless operation. It is possible, however, to use *memory to memory* operations when the same I/O register is accessed both on read and write. The sample programs in Chapter 20 use the MC6821 PIA parallel interface, where the port A registers are mapped to the following byte sized addresses:

```
$5CEF1: Port A Data Register
$5CEF3: Port A Control and Status Register
```

In this case:

1 there is a single data register used for transmit and receive at $5CEF1
2 there is a combined Control/Status register at $5CEF3.

As the **same** physical register is accessed on memory read and write, instructions such as OR, AND, BSET, BCHG, etc., can be used on these registers. Note, however, that some of the bits in the Control/Status Register are read only (data written to them is ignored).

In practice, when writing programs the numeric addresses of I/O registers should not be used in instruction operands. The EQU pseudo-operator (or similar) should be used to define the I/O register names and their equivalent addresses, e.g.:

```
* Define PIA memory mapped addresses for FORCE  board
PIA:        EQU         $5CEF1       base address of PIA
ADATA:      EQU         PIA          Data Register A
ACTRL:      EQU         PIA+2        Control Register A
BDATA:      EQU         PIA+4        Data Register B
BCTRL:      EQU         PIA+6        Control Register B
```

`PIA: EQU $5CEF1` defines the base address of the PIA and the I/O registers are specified relative to this base. To modify the program for a similar system it should then be possible to just edit the base address definition (interrupt handling and service routines are ignored for now). Once defined the I/O register names can then be used as instruction operands, e.g.:

```
        MOVE.B      #4,ACTRL        set up A Control Register
        MOVE.B      #$0F,ADATA      write to A data register
        MOVE.B      ADATA,D0        read A data register
        MOVE.B      D0,BDATA        write to B data register
        MOVE.B      BDATA,TEST      read B data reg to memory
        MOVE.B      ADATA,BDATA     copy A data to B data
```

In the above examples *absolute addressing mode* was used to access the I/O registers; and this is common practice when writing small programs. However, sometimes it is necessary to use *addresses register indirect with displacement addressing mode*, e.g.:

```
* Define PIA memory mapped addresses for FORCE  board
PIA:       EQU       $5CEF1       base address of PIA
ADATA:     EQU       0            Data Register A
ACTRL:     EQU       2            Control Register A
BDATA:     EQU       4            Data Register B
BCTRL:     EQU       6            Control Register B
```

The base address of the PIA is defined followed by the I/O registers specified as **offsets** from the base. In the program the base address is loaded into an address register and then *address register indirect with displacement* addressing mode used to access the I/O registers, e.g.:

```
        MOVEA.L   #PIA,A0                load base address of PIA
        MOVE.B    #4,ACTRL(A0)           set up A Control Register
        MOVE.B    #$0F,ADATA(A0)         write to A data register
        MOVE.B    ADATA(A0),D0           read A data register
        MOVE.B    D0,BDATA(A0)           write to B data register
        MOVE.B    BDATA(A0),TEST         read B data to memory
        MOVE.B    ADATA(A0),BDATA(A0)    copy A data to B data
```

This method should be used when writing common subroutines for a system which contains a number of devices of the same type, e.g. a small control system may have several MC6821 PIAs. To program a particular device its base address is loaded into a specified address register then the appropriate common subroutine called. When using *address register indirect with displacement addressing mode* take care not to corrupt the address register. In particular when writing interrupt routines remember to load an address register with the base address on entry. The interrupt service routine should not assume, because the main program loaded the base value into a particular address register, that it is available (the main program may have called a subroutine which is using the register for something else). Note that even if I/O registers are specified as offsets it is still possible to use absolute addressing to access them, e.g.:

```
            MOVE.B    PIA+BDATA,D0      read B data register
```

It is worth noting at this point that the MC68000 processor can operate either in Supervisor Mode or User Mode (in User Mode instructions that interfere with the overall operation of the computer cannot be used; see Chapter 23.1 for details). In the case of multi-user and multi-processing environments the operating system runs in Supervisor Mode and the user programs in User Mode. In addition, such computer systems will generally not allow user programs running in User Mode to access I/O registers and other critical parts of memory. It is assumed in the following chapters that a single user program development microcomputer is being used which runs all programs in Supervisor Mode. Students on formal courses will be given guidance by tutors, otherwise refer to the microcomputer manuals (see Appendix C for comments on different microcomputer configurations).

18.5 The RESET instruction

The processor has a RESET line (connected to the reset inputs of I/O device interfaces) which can be used for input (from the RESET button) or output via the RESET instruction:

```
            RESET
```

When RESET is executed a pulse is output on the RESET line which will reset the interface circuits (i.e. clear registers, interrupt conditions, etc.). This instruction is normally used on monitor or operating system start-up, or after a serious fault condition.

18.6 The MOVEP (move peripheral data) instruction

In general, any instructions which are used to read and write memory can be used to access the I/O registers. To simplify address decoding, sequences I/O registers for a particular device often use alternate (even or odd) memory addresses, e.g. the ACIA port addresses described above were $50040 and $50042. In such a case .B (byte) length operands may be used to transfer individual bytes to these registers, or the MOVEP instruction may be used to transfer words or long words. MOVEP transfers two or four bytes of data between a specified **data register** and **alternate byte locations in memory**. The start address in memory is specified using *address register indirect with displacement addressing mode* and the high-order byte is moved **first**, from/to the start address, followed by the low-order byte(s) from/to the following **alternate** addresses. For example (assume A0 = $50000):

```
MOVEP.W     D0,0(A0)
MOVEP.L     1(A0),D0
```

The first instruction moves the bits 8 to 15 of D0 to address $50000 and the bits 0 to 7 to address $50002. The second instruction will transfer the bytes:

(a) $50001 to bits 24 to 31 of D0;
(b) $50003 to bits 16 to 23 of D0;
(c) $50005 to bits 8 to 15 of D0;
and (d) $50007 to bits 0 to 7 of D0.

18.7 DMA (Direct Memory Access)

In the examples considered in the last section, the transfer of data between the I/O device interface and the memory of the computer was carried out by program instructions. If a large amount of data is to be transferred at high speeds (e.g. from a hard disk), a technique called DMA (Direct Memory Access) can be used. When a data transfer is to use DMA:

1. The program loads the interface with the address in memory where the data is to be transferred to/from, how much data to transfer, and then tells it to start the transfer.
2. The program can then carry on to do other operations.

Then:

3. When the interface has data to transfer it requests the processor for the use of the bus.
4. At the end of the current bus cycle the processor will:
   (a) release the bus and inform the interface it is available;
   then (b) wait for the interface to signal that it has finished with the bus.
5. The interface uses the bus to transfer the data (this may be one byte or several).
6. When the transfer is complete, the interface signals to the processor that it has finished with the bus.
7. The processor regains control of the bus and resumes execution of instructions.

The interface transfers data directly to/from memory at the full speed of the bus/memory combination, and requires no program intervention. The I/O device status register will indicate when the **whole** data transfer is complete.

This technique, enabling the I/O interface to gain control of the bus and carry out the data transfer independently of the processor, is called Direct Memory Access. I/O interfaces using DMA are very complex, and are only used to interface devices that transfer large volumes of data at very high speed (possibly where interrupt driven I/O may be too slow).

18.8 Problems with programming a new interface

Programming I/O interfaces is not simple; in particular, modern general purpose interface chips (e.g. the MC68230) can contain twenty or thirty I/O registers, be used in a variety of different modes, and in general be very complex and difficult to operate.

In general, when faced with implementing software to drive an I/O device interface the programmer works from manufacturers data sheets. These define:

(a) its functionality (what it will do), to enable system designers to select an appropriate device for a particular application or system;

(b) its hardware characteristics, to enable an electrical engineer to build it into a microcomputer system;

and (c) the I/O registers and their functions, to enable a programmer to program the device.

Manufacturers data sheets tend to be written by engineers (hardware and/or software) who are experts on the device. The data sheets are highly factual, written in very technical terms and tend to be obscure and difficult to understand (for a complex device the technical documentation may be fifty pages long). In general, no programming examples are given, making it very difficult for an inexperienced programmer to get started. In such circumstances the recommended approach is to program the device to do simple things, building up to more complex as experience is gained (this approach is taken in this book). When a thorough understanding of the device is attained it may then be linked to the application environment for testing (which may sustain damage if incorrectly operated).

For example, in the case of a parallel interface, simple circuits can be connected which give an immediate indication of program operation. The next chapter will introduce parallel I/O programming techniques and Appendix F describes some sample test circuits. Readers should then continue to the chapter which introduces to the parallel I/O interface for their microcomputer (Chapter 20 for the MC6821 PIA or Chapter 21 for the MC68230 PIT).

Problem for Chapter 18

Examine the manuals for any microcomputer systems you have access to and in each case determine (if possible):

1 The I/O devices connected to the computer, the associated registers and memory mapped addresses.

2 The clock speed of the processor, and if the memory uses WAIT states.

19

Introduction to parallel and serial input/output

19.1 Parallel input/output

The Motorola MC6821 PIA and MC68230 PIT chips contain parallel interfaces which can be used to interface microprocessor systems to external devices, such as printers, control systems, etc. (IBM/PC and compatible microcomputers tend to use the Intel 8255 Programmable Peripheral Interface chip which is similar in function to the MC6821). This section discusses parallel input/output techniques and their application and Chapters 20 and 21 introduce the MC6821 and MC68230 PIT respectively (Appendix F describes various simple circuits which can be attached to such interfaces). It is assumed that readers have a knowledge of basic digital interface circuits (Loveday 1984).

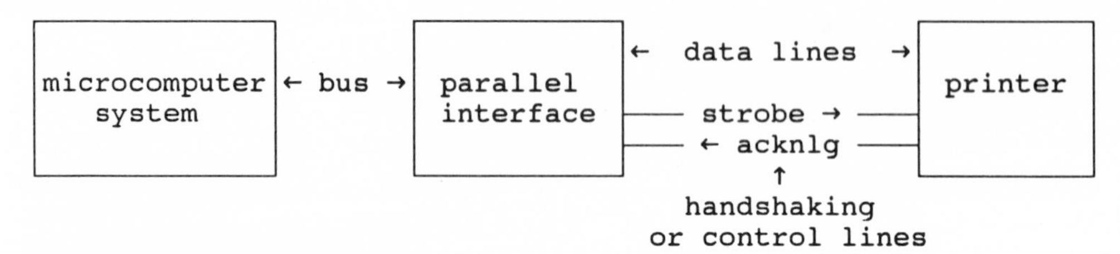

Fig. 19.1 Microcomputer connected to a printer via a parallel I/O interface

Fig. 19.1 shows a microcomputer system connected to a printer via a parallel interface. The interface is connected to the microcomputer system bus which carries the address, data and control information between the microprocessor, the memory and I/O device interfaces. The parallel interface will contain the address decoding and control circuits which determine when the device is being addressed, and data, status and control registers which enable a program to control the interface and transfer data.

For example, parallel printers are generally connected to a computer using a Centronics compatible parallel interface which has eight data lines and two handshaking or control lines called **STROBE** (for data available) and **ACKNLG** (for data acknowledged). A timing diagram showing the transmission of two characters is shown in Fig. 19.2.

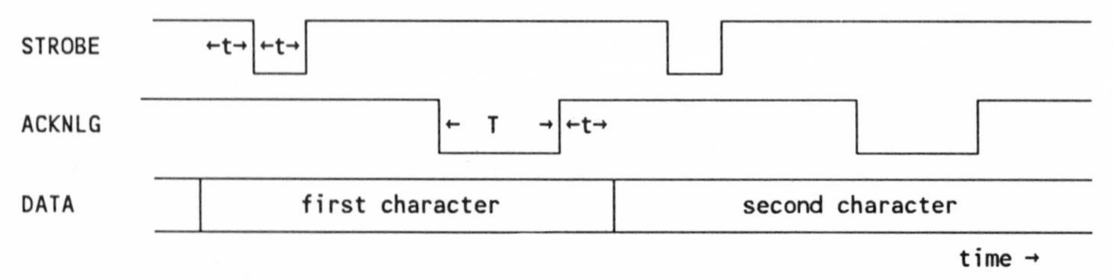

Fig. 19.2 Centronics parallel printer timing diagram, $t>0.5$ & $T>1.5$ microseconds

DATA LINES. These eight lines carry the data bits from the parallel interface to the printer with the data signals being active high, i.e. logical 1 is +5 volts and logical 0 is 0 volts. The data must be placed on the data lines at least 0.5 microseconds before the start of the STROBE pulse and be maintained at least 0.5 microseconds after the end of the ACKNLG pulse (to allow the electrical levels on the cables to settle and hence ensure that data can be read correctly).

STROBE (data available signal output from the parallel interface) is normally high (+5 volts). When data has been placed on the data lines a low pulse (at least 0.5 microseconds in length) is transmitted to the printer to inform it that data is available.

ACKNLG (data acknowledge signal from the printer) is normally high (+5 volt). After the printer receives the STROBE signal it reads the data and processes it (buffers it or prints a line). An acknowledge pulse (low) is then sent to the parallel interface to indicate that it can transmit the next data byte. The width of the ACKNLG pulse is typically 1.5 microseconds. Once the parallel interface senses the ACKNLG pulse the next byte can be loaded onto the data lines and the STROBE pulse transmitted.

This system of **STROBE** (for data available) and **ACKNLG** (for data acknowledged) is called **hardware handshaking** (the pulse width times are typical and many modern printers are less critical). If a program is to output several lines of text, each character must be transmitted using this sequence. For example, Fig. 19.2 shows the transmission of two characters and the corresponding handshake signals. In addition there are often other signals that report printer error conditions or are used to control printer actions.

Depending upon the application the direction of data transfer may be one way (e.g. input from a keyboard or output to a printer), or bidirectional, with the direction of data transfer depending upon instructions from the program or signals from the I/O device (e.g. a pair of computers carrying out a two way information exchange over a single parallel communications channel). With bidirectional transfer the control lines will switch roles when the transmitting and receiving ends change over. In addition, both ends must agree, at any instant, which is transmitting and which is receiving, otherwise both could be trying to transmit or both receive at the same time. In simple interfacing applications, a particular parallel interface is usually used for either input or output with additional parallel interfaces added as required (the names given to the handshaking signals also depend upon the application). The main advantages of parallel data transfer is that it is simple and fast, being limited by:

1 The bandwidth of the transmission lines, i.e. which determines the maximum rate at which data can be physically transferred.
2 The rate at which the transmitter can supply data for transmission, and the rate at which the receiver can process it (signals on the handshaking lines control this data flow to ensure that the transmitter does not send data faster than the receiver can deal with it). In practice the data transfer rate between two computers, or a computer and a control system, may be several hundred thousand bytes per second.

In general, the electrical levels used by parallel interfaces are the standard digital circuit levels where a 0 volt level represents logical 0 and a +5 volt level represents logical 1 (Loveday 1984); these levels can be changed when connecting to external equipment by using line drivers, amplifiers, relays, etc.

19.1.1 The parallel input/output (PIO) interface

Parallel I/O chips usually contain more than one 8-bit parallel interface port and associated control lines, e.g. the MC6821 has two ports called A and B and the MC68230 has three ports called A, B and C. The MC6821 and MC68230 allow bidirectional data transfer on all ports with individual bits of each port being either input or output as selected by the program.

19.1.1.1 Data registers

Associated with each parallel I/O port of the MC6821/MC68230 is an 8-bit **Data Register** which contains output and/or input data to be transferred to and from the eight data lines connecting the port to external devices. If data is written (by the program) to any bits in a data register which have been specified as input lines, it will be ignored, i.e. only output lines will be effected.

19.1.1.2 Data direction registers

Associated with the **Data Register** of each parallel I/O port of the MC6821/MC68230 is an 8-bit **Data Direction Register**. The bit pattern in the data direction register determines which bits in the associated Data Register are used for input and which for output. Thus the bits in the Data Register (and the associated data lines) can be all input, or all output, or a mixture, as determined by the corresponding Data Direction Register bits.

19.1.1.3 Control registers

Associated with each parallel I/O port of the MC6821 and MC68230 is a **Control Register** which controls the operation of the port, e.g. handshake lines, whether interrupts are enabled, etc. (the MC68230 also has general control registers which control the overall device).

19.1.1.4 Status information

Each parallel I/O port has status information which enables the program to determine the state of the control lines, etc. The MC6821 presents this within a combined control/status register for each port while the MC68230 has a separate port status register which indicates the status of all the parallel port control lines.

19.1.1.5 Parallel interface test circuits

The simplest way to approach the programming of a new I/O device, in particular one as complex as the MC68230 PIT, is to implement a series of test programs (see discussion in Chapter 18.8). Appendix F presents a number of simple circuits which can be used to test programs driving parallel interfaces. After reading this chapter the reader should continue with the chapter on the parallel interface relevant to their microcomputer environment (Chapter 20 for the MC6821 and Chapter 21 for the MC68230).

19.2 Asynchronous serial communications

The major disadvantage of parallel communications systems is that, due to the large number of communications lines required, the physical distance between devices is limited. A serial communications system uses two communications lines, one for receive and one for transmit, making it useful over long distances and even via telephone lines.

In a serial communications system a byte of data is transferred between the transmitter and receiver bit by bit over a single communications line. Bit 0 is transmitted, then bit 1, etc., through to bit 7. When the first byte is finished, the next byte, if any, can be sent. The data is transferred at an agreed rate of a number of bits per second, which is called the baud rate. To separate the data bytes a **START bit** is transmitted before the first data bit and one or two **STOP bits** are transmitted after the last data bit.

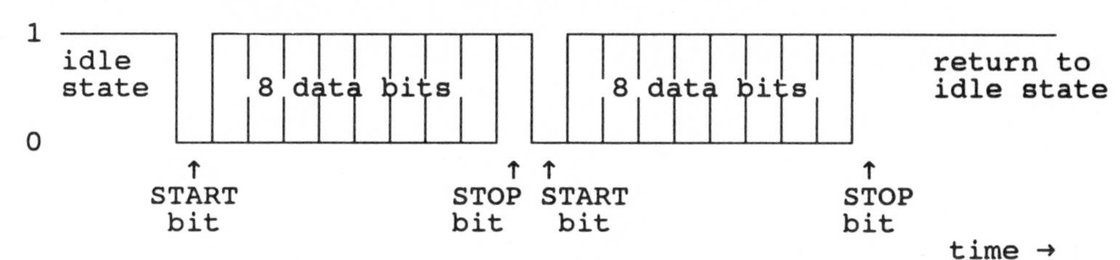

Fig. 19.3 Format of data on an asynchronous serial line

Fig. 19.3 shows the format of the data on a serial line when two 8-bit (byte) characters are transmitted (plus one START bit and one STOP bit in each case). When no data is being transmitted the line is in the idle state at logical 1, which corresponds to a nominal voltage level of -6 volt. When the serial interface is ready to transmit a data byte, it first transmits a START bit of logical 0 (nominal level +6 volt). The START bit serves to separate the idle state from the first data bit which could be a 0 or a 1. The data bits are then transmitted, bit 0 first, one after the other. After the data bits, one or two STOP bits, at logical 1, are transmitted. If, after the STOP bit(s), more data is available the next START bit is transmitted, otherwise the line returns to the idle state. Thus if another data byte is to be immediately transmitted (as in Fig. 19.3), the STOP bit(s) serve to separate the last data bit of a data byte, from the START bit of the next data byte.

To check for errors when transmitting characters over a noisy communications channel, a parity check bit can be generated which can replace bit 7 of the character or be appended on the end of it to form a 9-bit code (Appendix A presents details of parity generation). Thus for each data byte transmitted, 10, 11 or 12 bits are actually transmitted (START + data bits + parity bit + STOP bit(s)).

The baud rate can be from 50 up to 19.2K baud (bits/second). If, for example, the baud rate is 1200, each bit takes 0.8333 milliseconds to transmit with 120 bytes per second transferred if one STOP bit is used.

The main application of asynchronous serial communication is the transmission of character data at random intervals (for example, when a user is hitting the terminal keyboard to send characters, one per byte, to a computer, and the computer is sending characters to the terminal display). Another form of serial communications is synchronous, which is used where large blocks of data are to be transferred (e.g. file transfer).

19.2.1 The serial input/output (SIO) interface

Within the computer system, data transmission is via a high speed parallel bus system for the transfer of address, data and control signals. The SIO interface converts the parallel data from the data bus into a serial bit stream at the correct baud rate, and adds parity, START and STOP bits as required. Another SIO interface at the other end converts the data back to a parallel form to be placed on the data bus. Fig. 19.4 shows a diagram of the major components.

```
computer ← bus → SIO ← data lines → SIO ← bus → computer

               ↑               ↑               ↑
           parallel          serial         parallel
             data             data            data
```

Fig. 19.4 A serial communications system

The SIO interface is generally called a UART (Universal Asynchronous Receiver Transmitter). The MC6850 ACIA (see Chapter 25) contains one serial interface and the MC68681 DUART (see Chapter 26) contains two serial interfaces. Associated with each serial interface are two ports, one for transmission and one for receive allowing *full duplex* simultaneous data transfer in both directions (*simplex* is one way transmission and *half duplex* is two way over a single communications line).

It is vital that the transmitter and receiver agree on the baud rate, number of data bits (normally eight but seven if bit 7 is used as the parity check bit), parity and number of STOP bits (otherwise the information received will be incorrect).

19.2.2 Modems

Serial communications can be used over very large distances, via telephone lines. A device called a modem (modulator/demodulator) converts the digital -6 and +6 volt signal levels to analogue signals, which can be transmitted over a telephone line. Another modem at the other end converts these signals back to the -6 and +6 digital signal levels. Fig. 19.5 shows a diagram of such a system.

```
computer ←bus→ SIO ↔ modem ← telephone line → modem ↔ SIO ←bus→ computer
```

Fig. 19.5 Serial communications via a telephone line

19.2.3 The EIA RS232C standard

The EIA RS232C standard (EIA RS-232-C 1981, Cambel 1984) was originally developed to foster data communications via public telephone networks. It defines the interface between a DTE (Data Terminal Equipment, i.e. a user terminal), and a DCE (Data Communications Equipment, i.e. a modem), using serial binary data interchange. The nominal signal levels are -6 volt for logical 1 (sometimes called MARK), but any level between -3 and -12 volt is accepted, and +6 volt for logical 0 (sometimes called SPACE), but any level between +3 and +12 volt is accepted. Signal levels between -3 and +3 volt are not valid and data would probably be corrupt.

The facilities of particular serial I/O chips vary but the most common signals are (the following assumes a DTE communicating with a DCE using the standard RS232 25 way D type connector):

TXD (pin 2) and RXD (pin 3) are the serial transmit and receive data lines.

$\overline{RTS}$ (Request To Send - DTE output to DCE on pin 4) is the **transmit handshaking output** which the DTE asserts to indicate that it wishes to transmit data.

$\overline{CTS}$ (Clear To Send - DTE input from DCE on pin 5) is the **transmit handshaking input** to the DTE. The DCE (e.g. modem) asserts $\overline{CTS}$, in response to $\overline{RTS}$, to indicate that it is ready to accept data.

$\overline{DSR}$ (Data Set Ready - DTE input from DCE on pin 6) indicates that the remote device (e.g. modem) is switched on and ready.

Signal ground pin 7 is the common signal return path.

$\overline{DCD}$ (Data Carrier Detect - DTE input from DCE on pin 8) is the **receive handshaking input.** If the external device is able to send characters this signal is asserted (e.g. a modem would negate $\overline{DCD}$ if the data carrier was lost due to a telephone line fault).

$\overline{DTR}$ (Data Terminal Ready - DTE output to DCE on pin 20) indicates that the host device (a terminal in the original RS-232C standard) is switched on and ready.

The $\overline{RTS}$, $\overline{CTS}$, $\overline{DSR}$, $\overline{DCD}$ and $\overline{DTR}$ are all active low. In such a case, if the signal is negated the line is at logical 1 (a -6 volt level), and if it is asserted or active the value is logical 0 (a +6 volt level). For example, if the DCE is not ready to accept data the $\overline{CTS}$ line will be logical 1, otherwise, if it is ready $\overline{CTS}$ will be at logical 0. In particular, when working over half duplex lines, $\overline{CTS}$ and $\overline{RTS}$ are used to control which end of the communications link can transmit at any instant.

The problem arises that microcomputers are not consistent in the wiring of serial interfaces. Some serial interfaces are wired as DTEs, some as DCEs, and some as neither. In many cases, only some of the signals are connected or used by the software and this leads to confusion in linking terminals to computers, printers to computers, and computers to computers (reference to hardware manuals and sometimes circuit diagrams is therefore required; see also Cambel 1984).

19.2.4 Handshaking with serial communications systems

Parallel communications system require handshaking signals to inform the receiver when valid data is available and the transmitter when the next data can be transmitted. In the case of asynchronous serial communications the START bit informs the receiver when data is about to arrive and it can read the bits in at the appropriate baud rate. Thus in applications when the receiver can read the data off the line and process it at the same speed or faster than the transmitter can send there is no problem. However, serial communications are often used to connect slow devices to a computer. For example, if a printer which can print 100 characters per second is connected to a computer running at 9.6 Kbaud (approximately 900 characters per second) data would be lost as soon as the printer buffer was full, i.e. the first page or so is printed correctly then the text becomes garbled.

19.2.4.1 Hardware handshaking

Hardware handshaking makes use of the $\overline{\texttt{RTS}}$, $\overline{\texttt{CTS}}$, $\overline{\texttt{DCD}}$ and/or $\overline{\texttt{DTR}}$ lines of the SIO interface. For example, serial printers have a *printer ready for data* or *printer busy* signal which is used for handshaking. Fig. 19.6 shows the connection of a Force board (Force 1984) to an Epson FX-80 printer. The FORCE board uses $\overline{\texttt{CTS}}$ (Clear to Send input) input as its handshaking input and the Epson uses the RS232C signal $\overline{\texttt{DTR}}$ (Data Terminal Ready on pin 20) for the *printer ready* signal (it is at logical 0 when the printer can accept another character). In the case of Fig. 19.6 the RXD, $\overline{\texttt{RTS}}$ and $\overline{\texttt{DCD}}$ signals are ignored.

FORCE computer		Epson printer	
Signal name	pin	pin	Signal name
ground	1	1	ground
transmit data	→ 2	3 →	receive data
clear to send	← 5	20←	printer ready
data common	7	7	data common

Fig. 19.6 Epson FX-80 printer attached to a FORCE microcomputer

19.2.4.2 Software handshaking

The most common software handshaking technique uses the XON/XOFF protocol (this protocol is used to stop and start character display on a terminal connected to a computer). The protocol requires a full duplex communications system in which the receiver transmits the XOFF character (transmit off - ASCII DC3 or CTRL/S) to stop the transmitter (e.g. when the printer buffer is full) and the XON (transmit on - ASCII character DC1 or CTRL/Q) to restart transmission.

Often printers and other devices which can be attached to computer via serial lines allow a range of handshaking options both hardware and software (clearly the computer and the device must agree on which to use or difficulties arise).

Exercise 19.1 (*see Appendix G for sample answer*)

List the advantages and disadvantages of parallel and serial communications systems.

20

Introduction to the MC6821 PIA

The MC6821 PIA (Peripheral Interface Adaptor - Motorola 1985a) chip contains a pair of 8-bit bidirectional parallel interface ports, called A and B, which are **usually** used for input and output respectively. Each 8-bit port has associated with it:

Data Lines. Eight bidirectional data lines. The port A data lines are called PA0 to PA7 and port B data lines PB0 to PB7.

Control Lines. Two control lines per port for handshaking; one for input only (called CA1 and CB1), and the other which is programmable to be input or output (CA2 and CB2).

Data Register (8 bits). Contains output and/or input data to be transferred to/from the eight data lines.

Data Direction Register (8 bits). The bit pattern in this register determines which bits in the Data Register are used for input and which for output. If a bit in the Data Direction Register is:

0: the corresponding bit in the Data Register is used for input;

or 1: the corresponding bit in the Data Register is used for output.

Thus the bits in the Data Register (and the associated data lines) can be all input, or all output, or a mixture, as determined by the corresponding Data Direction Register bits.

Control Register (8 bits). This contains control bits that enable a program to control the device and status bits that are set by signals on the control or handshaking signals.

The MC68000 uses memory mapped I/O addresses to access the MC6821 PIA registers, with each PIA connected to the system having four byte sized addresses associated with it. Table 20.1 lists the PIA registers and the associated memory addresses used by the FORCE microcomputer (Force 1984) which was used to test the programs in this chapter.

address	MC6821 register
$5CEF1	Port A - Data Register and Data Direction Register
$5CEF3	Port A - Control Register
$5CEF5	Port B - Data Register and Data Direction Register
$5CEF7	Port B - Control Register

Table 20.1 Memory mapped registers of the MC6821 PIA for the Force board

Table 20.1 shows that the Data Register and the Data Direction Register of each port share the same memory address. Bit 2 in the associated Control Register determines which is accessed when reading from or writing to the address (more details below). Also note that alternate (odd) addresses are used to simplify address decoding (Clements 1987, Wilcox 1987). Fig. 20.1 is a diagram showing the MC6821 port A registers and their memory mapping. Note how $5CEF1 is used to access the Data Register or the Data Direction

Register depending upon the switch setting (controlled by bit 2 of the Control Register).

The programs in this chapter define the FORCE microcomputer register addresses by means of the EQU pseudo-operator, which are then accessed using MOVE and other instructions, with .B (byte) length operands. Readers implementing the programs may have to alter these addresses to suit the microcomputer being used (refer to hardware manuals).

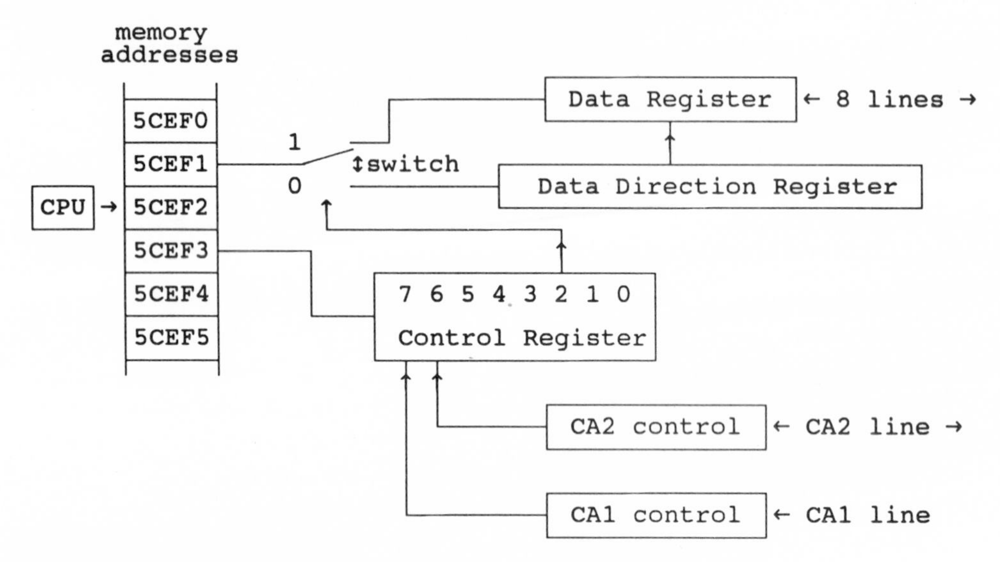

Fig. 20.1 Diagram of the MC6821 PIA Port A registers (Port B is similar)

read only Status bits		read/write Control bits					
7	6	5	4	3	2	1	0
IRQA1	IRQA2		CA2 Control		DDR/DR	CA1 Control	
Cleared by reading the data register	Cleared by reading the data register	**0** CA2 is an input line	**0** bit 6 set by a high to low transition on CA2	**0** CA2 interrupt disabled	**0** access data direction register	**0** bit 7 set by a high to low transition on CA1	**0** CA1 interrupt disabled
			1 bit 6 set by a low to high transition on CA2	**1** CA2 interrupt enabled			
Set by a transition on CA1 input line	Set by a transition on CA2 as input line	**1** CA2 is an output line	**0** automatic handshake	**0**=acknlg **1**=strobe	**1** access data register	**1** bit 7 set by a low to high transition on CA1	**1** CA1 interrupt enabled
			1 CA2 output = bit 3	**0 or 1** output on CA2			

Table 20.2 Control Register for MC6821 PIA port A (port B is similar)

Table 20.2 shows the format of the bits in the port A Control Register of the MC6821 PIA (port B Control Register is similar). At this stage do not worry about the complexity of the table, the various bits will be introduced fully (with practical examples) as they are used. Note that bits 6 and 7 are read only status bits and any data written to them will be ignored.

Bit 2 (DR/DDR access) of the Control Register determines which register is accessed at the memory address associated with the Data Register and Data Direction Register, i.e.:

bit 2 is 0: the Data Direction Register is accessed
or bit 2 is 1: the Data Register is accessed

For example, the following code would set bits 0 to 3 of the port B Data Register for output and bits 4 to 7 for input:

```
MOVE.B    #0,$5CEF7      clear port B control register
MOVE.B    #$0F,$5CEF5    set B Data Direction Register
```

To program the MC6821 PIA the general procedure is:

1. Clear the Control Registers of ports A and B to initialise the interface (by clearing bit 2 this also has the effect that the Data Direction Register may now be addressed).
2. Load the Data Direction Registers to set which bits in the Data Register are input and which are output.
3. Set Control Register bit 2 to access the Data Register and set any other bits required, e.g. to enable interrupts, if CA2 is input or output and how it operates, etc.
4. Load or read the Data Registers as required.

20.1 Simple input and output with the MC6821

Program 20.1 (next page) uses the simple switches and lights circuit described in Appendix F.3 (eight switches attached to port A for input and eight LEDs attached to port B for output display). Note that *absolute addressing mode* is used to access the MC6821 I/O registers. The program sequence is:

line	
14-20	`PIA: EQU $5CEF1` defines the base address of the PIA memory mapped registers for the Force board followed by definitions of the port A and B registers relative to this base (for another system change the base address)
23-24	the Control Registers are cleared to initialise the PIA (thus bit 2 is cleared to 0, so that the Data Direction Registers can be addressed)
25	`MOVE.B #0,ADDR` sets all bits of port A for input, i.e. a 0 in the Data Direction Register sets the corresponding bit in the Data Register for input, and 1 for output
26	`MOVE.B #$0FF,BDDR` sets all bits of port B for output
27-28	`MOVE.B #4,ACTRL` sets bit 2 in the A (and B) Control Register to enable the Data Register to be addressed (an alternative is `BSET.B #2,ACTRL`)
31	`MOVE.B ADATA,D0` reads data from port A Data Register (the switches)
32	`MOVE.B D0,BDATA` outputs data to port B Data Register (to light the LEDs)
34-36	the data value is then displayed on the screen using WRHEXB
38-39	if the switch value was not 0 the program LOOPs back to line 31 otherwise
40	the program terminates

```
--------------------------------------------------------------------------------
1                          * Program 20.1 - MC6821 PIA simple input/output
2                          *   Program which reads switch settings from port A
3                          *    and outputs value to LEDs connected to port B.
4                          *
5                          * initialise PIA: A for input, B for output
6                          * LOOP
7                          *    READ(switch values from port A into D0)
8                          *    WRITE(D0 to port B to LEDs)
9                          *    WRITE('Port A Data Register is ',display D0)
10                         * UNTIL switch values = 0
11                         * STOP program
12                         *
13                         * Define PIA memory mapped addresses for FORCE  board
14         0005CEF1        PIA:      EQU      $5CEF1      base address of PIA
15         0005CEF1        ADATA:    EQU      PIA         Data Register A
16         0005CEF1        ADDR:     EQU      PIA         Data Direction Register A
17         0005CEF3        ACTRL:    EQU      PIA+2       Control Register A
18         0005CEF5        BDATA:    EQU      PIA+4       Data Register B
19         0005CEF5        BDDR:     EQU      PIA+4       Data Direction Register B
20         0005CEF7        BCTRL:    EQU      PIA+6       Control Register B
21                         *
22                                   XREF     WRLINE,WRTEXT,WRHEXB
23 001000 13FC      0000 START:      MOVE.B   #0,ACTRL      clear the Control
              0005CEF3
24 001008 13FC      0000             MOVE.B   #0,BCTRL        Registers
              0005CEF7
25 001010 13FC      0000             MOVE.B   #0,ADDR        port A all bits input
              0005CEF1
26 001018 13FC      00FF             MOVE.B   #$0FF,BDDR     port B all bits output
              0005CEF5
27 001020 13FC      0004             MOVE.B   #4,ACTRL       set bit 2 to access the
              0005CEF3
28 001028 13FC      0004             MOVE.B   #4,BCTRL        A & B Data Registers
              0005CEF7
29                         *
30                         * LOOP reading switches from A and lighting LEDs on B
31 001030 1039 0005CEF1 LOOP:        MOVE.B   ADATA,D0       read the switches
32 001036 13C0 0005CEF5              MOVE.B   D0,BDATA       output to LEDs
33 00103C 4EB9 0002802C              JSR      WRLINE         display value in hex
34 001042 4EB9 00028038              JSR      WRTEXT
35 001048 50 6F 72 74 20             DC.W     'Port A Data Register is $'
36 001062 4EB9 00028050              JSR      WRHEXB
37                         * UNTIL switch values = 0
38 001068 4A00                       TST.B    D0             if <> 0 continue loop
39 00106A 66C4                       BNE.S    LOOP
40 00106C 4E4E                       TRAP     #14            terminate program
41                                   END
--------------------------------------------------------------------------------
```

Program 20.1 Program to read switches and set LEDs via an MC6821 PIA

In Program 20.1 the updated switch value is displayed on the LEDs and as a hexadecimal number on the display screen. In this case no handshaking is required to control the data transfer, i.e. it is a simple program loop which reads the switches and sets the LEDs. In other applications handshaking may be required to control the data flow between the transmitter and receiver (handshaking will be explained in detail later).

Exercise 20.1 *(see Appendix G for sample answer)*

Enter Program 20.1, modify as required until operational. Alter the program to display a rotating bit on the LEDs (i.e. LEDs light in order bit 0 1 2 3 4 5 6 7 0 1 2 3 4 etc.). The bit should move approximately once every second (use subroutine DELAY in the library).

20.2 Using the control lines CA1 and CB1 for input

The MC6821 PIA ports A and B have an input control line, CA1 and CB1 respectively, which can be used for handshaking. The PIA detects **transitions or changes** in the voltage levels on the lines (i.e. the voltage on the line is either 0 or + 5 volt, and it is the changes between these levels that are detected, not the absolute levels themselves). Control Register bits 0, 1 and 7 are used with CA1 (CB1 is similar):

Bit

7: is a **read only status bit** which set by transitions on the CA1 input line, i.e. this bit is normally 0 and when a transition occurs it is set to 1. The program can read the Control Register to check the value of this bit or use interrupts (covered in Chapter 23 and 24.1).

Bit 7 is reset (cleared) by reading the associated Data Register, i.e. a write to the Control Register does not affect bit 7 (nor bit 6), which are set by signals on CA1 (and CA2), and cleared **only by** reading the corresponding Data Register.

1: determines how transitions on CA1 affect bit 7:
- 0: Bit 7 is set to 1 by a high to low transition on CA1, i.e. +5 to 0 volt level change
- 1: Bit 7 is set to 1 by a low to high transition on CA1, i.e. 0 to +5 volt level change

0: is used for program interrupt control:
- 0: Interrupts are disabled
- 1: Interrupts are enabled. When a signal on CA1 sets bit 7 an interrupt signal is sent to the processor.

Thus bits 0 and 1 determine the effect of transitions on the handshake line CA1, and the program can examine bit 7 (see Program 20.2 below) or use interrupts (see Chapters 23 and 24.1) to see whether a transition has occurred.

20.3 Using control lines CA2 and CB2 for input

Control line CA2 (and CB2) can be used for input or for output. If **bit 5** of the associated Control Register is 0 then CA2 is an input line which acts in a similar way to CA1:

Bit

6: is a **read only status bit** which is set by transitions on the CA2 line when it is used for input (bit 6 cleared by reading the associated Data Register).

4: determines how transitions on CA2 affect bit 6; 0 set by high/low or 1 set by low/high.

3: is used for program interrupt control; 0 disabled or 1 enabled.

Thus CA2 and CB2 can be used in a similar way to CA1 and CB1 by clearing bit 5 of the Control Register, manipulating bits 3 and 4 and examining bit 6 (or using interrupts).

```
-------------------------------------------------------------------------------
1                           * Program 20.2 - MC6821 PIA Program
2                           *   Program using a PIA to display a rotating LED pattern.
3                           *    Pattern reverses when switch attached to CA1 is hit.
4                           *
5                           * initialise PIA: port B for output
6                           * WRITE('hit CA1 switch to reverse LED rotation')
7                           * set test_pattern = 1
8                           * set initial direction = left
9                           * LOOP FOR ever
10                          *     LOOP
11                          *         output test_pattern to port B (to LEDs)
12                          *         IF direction = left
13                          *             rotate test_pattern one bit left
14                          *             blink LED attached to CA2
15                          *         ELSE
16                          *             rotate test_pattern one bit right
17                          *             blink LED attached to CB2
18                          *         delay for a time period
19                          *     UNTIL CA1 switch is hit
20                          *     INVERT direction
21                          * END LOOP
22                          *
23         0005CEF1         PIA:      EQU     $5CEF1       PIA base address for FORCE
24         00000000         ADATA:    EQU     0            Data Register A
25         00000002         ACTRL:    EQU     2            Control Register A
26         00000004         BDATA:    EQU     4            Data Register B
27         00000004         BDDR:     EQU     4            Data Direction Register B
28         00000006         BCTRL:    EQU     6            Control Register B
29                          *
30                                    XREF    WRLINE,WRTEXT,WRHEXB,DELAY
31 001000 4EB9 00028038     START:    JSR     WRTEXT
32 001006 68 69 74 20 43              DC.W    'hit CA1 switch to reverse LED rotation$'
33 00102E 207C 0005CEF1               MOVEA.L #PIA,A0          base address of PIA
34 001034 117C 0000 0002              MOVE.B  #0,ACTRL(A0)     clear the Control
35 00103A 117C 0000 0006              MOVE.B  #0,BCTRL(A0)       Registers
36 001040 117C 00FF 0004              MOVE.B  #$0FF,BDDR(A0)   port B all bits output
37 001046 117C 0034 0002              MOVE.B  #$34,ACTRL(A0)   access data registers &
38 00104C 117C 0034 0006              MOVE.B  #$34,BCTRL(A0)   CA2 & CB2 for output
39 001052 1628      0000              MOVE.B  ADATA(A0),D3     clear ACTRL bit 7 if set
40 001056 7201                        MOVEQ   #1,D1            initial test_pattern
41 001058 4282                        CLR.L   D2               direction = left
42                          *
43                          * LOOP rotating left or right
44 00105A 1141      0004    LOOP:     MOVE.B  D1,BDATA(A0)   output test_pattern to LEDs
45 00105E 4A02                        TST.B   D2             direction ?
46 001060 660A                        BNE.S   RIGHT
47 001062 E319                        ROL.B   #1,D1          rotate test_pattern left
48 001064 0868 0003 0002              BCHG.B  #3,ACTRL(A0)   invert CA2 to blink LED
49 00106A 6008                        BRA.S   WAIT
50 00106C E219              RIGHT:    ROR.B   #1,D1          rotate test_pattern right
51 00106E 0868 0003 0006              BCHG.B  #3,BCTRL(A0)   invert CB2 to blink LED
52 001074 4EB9 0002809A     WAIT:     JSR     DELAY          wait for a delay time
53 00107A 0064                        DC.W    100            100 milliseconds
54 00107C 0828 0007 0002              BTST.B  #7,ACTRL(A0)   has CA1 switch been hit ?
55 001082 67D6                        BEQ.S   LOOP           if not rotate again
56                          *  UNTIL CA1 switch is hit
57 001084 1628      0000              MOVE.B  ADATA(A0),D3   read A data to clear bit 7
58 001088 4602                        NOT.B   D2             and invert direction
59 00108A 60CE                        BRA.S   LOOP
60                                    END
-------------------------------------------------------------------------------
```

Program 20.2 Direct program control of the MC6821 PIA control lines CA1, CA2 & CB2

20.4 Using control lines CA2 and CB2 for direct programmed output

If **bits 4 and 5** of the associated Control Register are set (to 1) the CA2 (CB2) line can be used for the output of high or low levels (or pulses) under direct program control via **bit 3**, i.e. if bit 3 is binary 0 the voltage output will be 0 volt and if it is 1 it will be + 5 volts.

Program 20.2

Program 20.2 (previous page) (an extension of Exercise 20.1) makes use of a switch attached to the CA1 handshaking line and LEDs attached to CA2 and CB2 (see sample circuit Appendix F.3). The program displays a rotating pattern on the LEDs attached to the data lines and blinks the LED attached to the CA2 or CB2 line. When the switch attached to CA1 is pressed, the rotation of the pattern displayed on the LEDs is reversed and CA2 or CB2 change over. The sequence of events in Program 20.2 is:

23 `PIA: EQU $5CEF1` defines the absolute base address of the PIA memory mapped registers for the Force board

24-28 the port A and B registers are defined as offsets from the base; *address register indirect with displacement addressing mode* will then be used (see Chapter 18.4)

33 loads the base address of the PIA into A0

34-35 the Control Registers are cleared to initialise the PIA (thus bit 2 is cleared to 0, so that the Data Direction Registers can be addressed)

36 all bits of port B are set for output (port A is not used for I/O)

37-38 `MOVE.B #$34,ACTRL(A0)` sets up the port A (and B) Control Register:
- (a) bit 0 is cleared to 0 disabling CA1 (and CB1) interrupts
- (b) bit 1 is cleared to 0, so that bit 7 will be set on a high to low transition of CA1
- (c) bit 2 is set to 1 to address the Data Register.
- (d) bit 5 is set to 1 to make CA2 (and CB2) output lines
- (e) bit 4 is set to 1 to output the value of bit 3 on CA2 (and CB2)
- (f) bit 3 is cleared to 0 to output a 0 initially on CA2 (and CB2)

39 `MOVE.B ADATA(A0),D3` reads the port A Data Register to clear bit 7 of Control Register (if set by previous program); the result in D3 is not used

40 moves the bit pattern into D1 (to be rotated)

41 the initial rotation direction is set to left (D2=0)

44 `MOVE.B D1,BDATA(A0)` displays the bit pattern on the LEDs (via port B)

45-46 IF the direction is left
- 47 the pattern is rotated left
- 48 the CA2 output line is inverted to blink the yellow LED

ELSE
- 50 the pattern is rotated right
- 51 the CB2 output line is inverted to blink the green LED

52-53 DELAY for a time period

54 `BTST.B #7,ACTRL(A0)` tests Bit 7 of the port A Control Register

55 IF bit 7 was 0 (i.e. switch not pressed) loop back to line 44 to rotate the again.

57 IF bit 7 was set (switch hit) read the port A Data Register to clear bit 7 of the Control Register (if this was not done bit 7 would stay set at 1 and the program would not work correctly, i.e. the direction would invert on each loop)

58 invert the rotation direction

59 loop back to line 44 to rotate again

The rotation reverses when the switch attached to CA1 is pressed (a high to low transition); the release has no effect. To amend the program so that the press of the switch is ignored, and the reverse of rotation occurs when the switch is released, the initialisation statement on line 37 should be changed so that bit 1 of the A Control Register is set to 1 so that a low to high transition on CA1 will set bit 7:

```
        MOVE.B      #$36,ACTRL(A0)
```

Thus it is possible to program the handshaking lines directly (CA1 (CB1) & CA2 (CB2)) to control data transfers (automatic handshaking is covered in the next section).

20.5 Using CA2 and CB2 for automatic handshaking

It would be possible to handshake a parallel communications system directly under program control by toggling CA2 (or CB2) to generate output pulses (see data ready STROBE in Fig. 19.2) and polling CA1 (or CB1) for the reply from the external device (see data acknowledge ACKNLG in Fig. 19.2). The MC6821, however, provides a facility for automatic handshaking (using CA2 and CB2) which is enabled by setting bit 5 of the Control Register to 1 (making CA2/CB2 an output line) and resetting bit 4 to 0.

20.5.1 Automatic Output Acknowledge and strobe on CB2

When using port B for output with Control Register **bit 5** set to 1 and **bit 4** cleared to 0, the setting of **bit 3** determines how CB2 behaves.

0: **Automatic Output Acknowledge:** CB2 is normally high; it goes **low** after a write to the Data Register of port B, and **high** when a transition on CB1 sets bit 7 of the Control Register.

1: **Automatic Output Strobe:** CB2 is normally high. A low pulse is output after a write to the Data Register of port B.

Both these can be used to send a *Data Ready Strobe signal* to an external device (the most appropriate is selected to suit the device handshaking requirements). For example, a Centronics compatible parallel printer can be attached to the MC6821 (see Appendix F.4).

1 the MC6821 GND (ground) to printer GND;
2 the eight data lines of port B to the printer input data lines;
3 STROBE (data ready from interface) handshaking line to CB2;
4 ACKNLG (printer data acknowledge) handshaking line to CB1.

In addition signals from the printer which indicate printer ON/OFF line, out of paper, etc., can be connected to port A to enable a program to determine its current status.

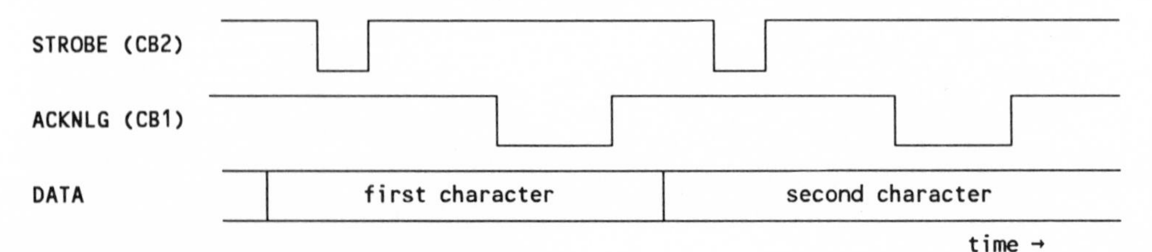

Fig. 20.2 Parallel printer timing diagram

Fig 20.2 shows the timing diagram of an MC6821 operated using *Automatic Output Strobe* handshaking. The sequence of events when a program is ready to output a character (byte) of data to the parallel printer is:

1 The program writes the character to port B Data Register; the program then waits for an ACKNLG signal by polling CB1.
2 On the write to port B Data Register, the *Automatic Output Strobe* sends a low pulse from CB2 to the STROBE input of the printer (normally this line is high).
3 After the printer receives the STROBE pulse, it reads the character byte from the data lines. It may now buffer the character or print a line of data.
4 When the printer is ready to receive the next character, it sends the low ACKNLG pulse on CB1 (this line is normally high), and then waits for the next STROBE pulse.
5 The program has been waiting for an acknowledge signal (ACKNLG) by testing bit 7 of the Control Register. When the ACKNLG pulse from the printer arrives on CB1 it sets bit 7, and the program can return to step 1 to send the next character.

Program 20.3 (end of chapter) uses **Automatic Output Strobe** to print text on a parallel printer (use an oscilloscope to observe the handshaking). The program sequence is:

22-24 initialise PIA and set all bits of port B for output
25 `MOVE.B #$2C,BCTRL` enables *Automatic Output Strobe* by setting B Control Register bit 5 = 1, 4 = 0 and 3 = 1 (bit 2 = 1 to address the Data Register).
26 read B Data Register to clear bit 7 of Control Register (if set)
27 moves the long word address of the start of the text into A0.

Then a program loop transmits the text character by character:

30 the next character is moved into D0
31 if the character was 0 goto line 27 to reset A0 to the start of the text
32 the character is moved into port B Data Register for transmission to the printer; also the *Automatic Output Strobe* causes a low pulse to be transmitted on the CB2 line to inform the printer that data is available
33 the character is displayed on the screen
34 bit 7 of B Control Register is tested to see if an ACKNLG pulse has been received
35 IF bit 7 is not set (no ACKNLG yet) the program loops back to line 34 to try again; if it is set the program continues to line 36
36 ACKNLG pulse received; the port B Data Register is read to clear bit 7 (if bit 7 is not cleared the program will not wait for the next ACKNLG pulse)
39 the program loops back to line 30 to get the next character, etc.

When executing Program 20.3 using a printer with a buffer the screen will display a block of characters (transmitted until the buffer is full) followed by small blocks of characters (as the printer empties the buffer by printing characters and it is refilled). Switching the printer OFF LINE will also stop the transmission of characters.

The handshaking can also be seen visually by using the switches and lights board of Appendix F.3 with the data output and STROBE signal shown on the LEDs and a press switch serves as the ACKNLG signal. The *Automatic Output Strobe* pulse will be too fast to see, so line 25 (of Program 20.3) should be modified to use *Automatic Output Acknowledge* and the comments removed from lines 38 and 39. The sequence is then:

1 when the data is sent to port B it appears on the eight LEDs;
2 the green LED attached to CB2 goes out as the *Automatic Output Acknowledge* high to low transition occurs;
3 the switch attached to CB1 is then hit to emulate the printer data acknowledge printer;
4 when the PIA senses the transition bit 7 in the Control Register will be set and the green LED attached to CB2 will light (as *automatic output acknowledge* is cleared);
4 after the DELAY time the next character appears on the LEDs and screen and the green LED goes out as the *Automatic Output Acknowledge* signal is sent.

20.5.2 Automatic Input Acknowledge and Strobe on CA2

When using port A for input with Control Register bit 5 set to 1 and bit 4 cleared to 0, the setting of bit 3 determines how CA2 behaves:

0: **Automatic Input Acknowledge:** CA2 goes **low** after a read from the Data Register of port A, and **high** when a signal on CA1 sets bit 7 of the Control Register.
1: **Automatic Input Strobe:** CA2 is normally high. A low pulse is output after a read from the Data Register of port A.

This can be used, for example, to drive an analogue to digital converter (see Appendix F.5):

1 Initialise the PIA with port A Control Register bit 5 = 1, bit 4 = 0 and bit 3 = 1 for *Automatic Input Strobe* (bit 2 is also set to 1 to access the Data Register).
2 Read the port A Data Register which will send **Automatic Input Strobe** pulse to the A to D to start conversion; then wait for bit 7 of the Control Register to set.
3 The A to D finishes conversion, places the data on the data lines and strobes a *conversion finished* pulse on line CA1 of the PIA.
4 The program has been polling bit 7 of the Control Register which is set by the *conversion finished* pulse on CA1.
5 The program reads the character from the Data Register of port A which clear bit 7 and send the *Automatic Input Strobe* pulse to start the next conversion.

If the application did not require the conversion to start immediately after reading the value the *start conversion* signal could be sent using direct program control of CA2, i.e. by toggling bit 3 of the Control Register as in Program 20.2.

Problems for Chapter 20

1 Assuming that suitable hardware facilities exist, modify Program 20.3 to sense when the printer is ON/OFF line, out of paper, etc.

2 Connect up an A to D converter (Appendix F.5 - or emulate using the switches and lights board). Implement a program to read the A to D using Automatic Input Strobe.

3 Connect up a matrix keypad (see Appendix F.6). Write a subroutine which returns the value of key hit as a decimal digit (ignore the * and # keys). Test the subroutine with a suitable program.

4 Using a suitable motor/heater circuit (see Appendix F.7) implement the PWM (pulse width modulation) motor speed control described in Appendix F.7.1.

5 Using a suitable motor/heater circuit implement the simple temperature control using heater and motor described in Appendix F.7.6. Note that use can be made of the BSET and BCLR instructions to switch the motor and heater ON and OFF.

```
--------------------------------------------------------------------------------
 1                       * Program 20.3 - MC6821 PIA Parallel Printer Driver
 2                       *
 3                       * initialise PIA:  port B O/P with automatic STROBE
 4                       * initialise A0 to point to the output text array
 5                       * LOOP for ever
 6                       *     get next character
 7                       *     If character = 0
 8                       *        reset A0 and get first character
 9                       *     PRINT(character)
10                       *     WRITE(character) to screen
11                       *     WAIT for acknowledge pulse from printer to set CB1
12                       * END LOOP
13                       * STOP
14                       *
15        0005CEF1       PIA:      EQU      $5CEF1       PIA base address for FORCE
16        0005CEF3       ACTRL:    EQU      PIA+2        Control Register A
17        0005CEF5       BDATA:    EQU      PIA+4        Data Register B
18        0005CEF5       BDDR:     EQU      PIA+4        Data Direction Register B
19        0005CEF7       BCTRL:    EQU      PIA+6        Control Register B
20                       *
21                                 XREF     WRLINE,WRTEXT,WRCHAR,WRHEXB,DELAY
22 001000 13FC      0000 START:    MOVE.B   #0,ACTRL       clear the Control
             0005CEF3
23 001008 13FC      0000           MOVE.B   #0,BCTRL          Registers
             0005CEF7
24 001010 13FC      00FF           MOVE.B   #$0FF,BDDR     DDR port B for output &
             0005CEF5
25 001018 13FC      002C           MOVE.B   #$2C,BCTRL     automatic output STROBE
             0005CEF7
26 001020 1039 0005CEF5           MOVE.B   BDATA,D0       clear BCTRL bit 7 if set
27 001026 207C 0000104E SET_A0:   MOVEA.L  #TEXT,A0       point A0 to text
28                       *
29                       * LOOP for ever
30 00102C 1018           LOOP:     MOVE.B   (A0)+,D0       get next character
31 00102E 67F6                     BEQ.S    SET_A0         if 0 reset A0
32 001030 13C0 0005CEF5           MOVE.B   D0,BDATA       character to printer
33 001036 4EB9 00028028           JSR      WRCHAR         and to screen
34 00103C 0839      0007 POLL:     BTST.B   #7,BCTRL       ACKNLG received (bit 7) ?
             0005CEF7
35 001044 67F6                     BEQ.S    POLL           if not poll bit 7 again
36 001046 1039 0005CEF5           MOVE.B   BDATA,D0       yes, clear bit 7
37                       *         JSR      DELAY          delay to show handshaking
38                       *         DC.W     100               if required
39 00104C 60DE                     BRA.S    LOOP           LOOP for next character
40                       *
41                       * Text to be displayed to screen and printer
42 00104E 41 42 43 44 45TEXT:      DC.B     'ABCDEFGHIJKLMNOPQRSTUVWXYZ '
43 001069 31 32 33 34 35           DC.B     '1234567890 !"%^&*()_+{}[]:@;~#<>?'
44 00108A 0D 0A 00                 DC.B     $0D,$0A,0
45                                 END
--------------------------------------------------------------------------------
```

Program 20.3 MC6821 PIA driving a parallel printer using automatic handshaking

21

Introduction to the MC68230 PIT

The MC68230 PIT (Parallel Interface/Timer - Motorola 1983a) contains three 8-bit, bidirectional, parallel interface ports, called A, B and C with associated control lines (H1, H2, H3 and H4) plus a programmable timer (see Chapter 28.3). Each 8-bit parallel port has associated with it:

Data Lines. Eight bidirectional data lines. The port A data lines are called PA0 to PA7, port B data lines PB0 to PB7 and port C data lines PC0 to PC7.

Control Lines. There are four control or handshaking lines, H1 and H3 which are input lines and H2 and H4 which are programmable as input or output lines. In general H1 and H2 are associated with port A handshaking and H3 and H4 with port B handshaking. The active state of the lines can be high or low (specified by the program) and when used as input lines the instantaneous level can be read and/or can be used to set edge sensitive status bits (i.e. to sense transitions or changes in level) which can be polled or used to interrupt the program.

Data Register (8 bits). Contains output and/or input data to be transferred to/from the eight data lines.

Data Direction Register (8 bits). In unidirectional modes (see below) the bit pattern in this register determines which bits in the Data Register are used for input and which for output. If a bit in the Data Direction Register is:

0: the corresponding bit in the Data Register is used for input;

or 1: the corresponding bit in the Data Register is used for output.

Thus the bits in a Data Register (and the associated data lines) can be all input, or all output, or a mixture, as determined by the corresponding Data Direction Register bits.

The device is extremely versatile with ports A and B capable of being operated in a variety of modes/submodes and this book will only attempt an introduction to the device (see reference Motorola 1983a for full details). In summary the modes and submodes are:

Mode 0 Ports A and B operate as two independent unidirectional 8-bit parallel I/O ports. The direction of data for each bit (i.e. which pins are input and which are output) is set by the associated Data Direction Register which is loaded by the program (in the following description port A uses H1 and H2 and port B uses H3 and H4):

Submode 00 double buffered input: with H1(H3) and H2(H4) as handshake lines; generally used to connect a single input device, e.g. an analogue to digital converter (double buffering will be discussed fully in the next chapter).

Submode 01 double buffered output: with H1(H3) and H2(H4) as handshake lines; generally used to connect a single output device, e.g. a parallel printer.

Submode 1x bit I/O: is used for applications where a variety of devices are connected to a port; pins can be output or input as set by the Data Direction Register:

(a) if a pin is output the data written is single buffered

and (b) if input the instantaneous value of the pin is read.

The H1(H3) control line can be used as a status/interrupt input and H2(H4) as a status/interrupt input or a general purpose output.

Mode 1 Ports A and B are combined to form a single unidirectional 16-bit port which may be used for double buffered input or double buffered output.

Mode 2 Port A is used for simple bit I/O with no handshake pins. Port B is used for bidirectional 8-bit double buffered data transfers with the direction of transfer being determined at any instant by the H1 control line (as set by the external device), and H2, H3 and H4 used for other handshaking functions.

Mode 3 Ports A and B are combined to form a single 16-bit port used for bidirectional double buffered transfers with the direction set at any instant by H1 with H2, H3 and H4 are used for other handshaking functions.

Port C can be used as eight general purpose I/O lines (PC0 to PC7 - with **no** handshaking) with the direction of each pin (input or output) set by the port C Data Direction Register. It is important to note, however, that all pins of port C may not be available at all times because some pins have alternate functions (set under program control) thus:

PC0 and PC1	are always port C bits 0 and 1 respectively;
PC2/TIN	PC2 or as the external clock input for the timer (Timer IN)
PC3/TOUT	PC3 or the timer interrupt request output or timer output (Timer OUT)
PC4/DMAREQ	PC4 or the request to an external direct memory access controller
PC5/PIRQ	PC5 or the parallel I/O interrupt request output
PC6/PIACK	PC6 or the interrupt acknowledge to a parallel I/O interrupt request
PC7/TIACK	PC7 or the interrupt acknowledge to a timer interrupt request

Thus depending upon the application certain port C bits may not be available, e.g. if the timer is using an external clock this will be input on the combined PC2/TIN pin and port C bit 2 is not available (the program selects between PC2 and TIN when configuring the device).

The MC68000 uses memory mapped I/O to access the MC68230 PIT registers. Each PIT connected to the system has 23 byte sized addresses associated with it, 14 used by the parallel interface and 9 by the timer. Table 21.1 lists the PIT registers for the parallel interface, their associated memory address offsets and the register names as used in sample programs (it is assumed that alternate addresses are used to access the PIT registers).

The programs in this chapter define the register addresses by means of the **EQU** pseudo-operator, which are then accessed using MOVE and other instructions, with .B (byte) length operands. Readers implementing the programs may need to alter these addresses to suit the microcomputer used (refer to hardware manuals).

offset	MC68230 register	program name
00	Port General Control Register	PGCR
02	Port Service Request Register	PSRR
04	port A Data Direction Register	PADDR
06	port B Data Direction Register	PBDDR
08	port C Data Direction Register	PCDDR
0A	Port Interrupt Vector Register	PIVR
0C	port A Control Register	PACR
0E	port B Control Register	PBCR
10	port A Data Register	PADR
12	port B Data Register	PBDR
14	port A Alternate Register	PAAR
16	port B Alternate Register	PBAR
18	port C Data Register	PCDR
1A	Port Status Register	PSR

Table 21.1 Memory mapped register offsets for the MC68230 PIT
Base addresses are $130001 for the Bytronic board and $A00001 for the Kaycomp

21.1 MC68230 PIT register details (reference only)

This section presents details of the MC68230 parallel interface registers. The information is necessarily complex and only functions relevant to programs in this book will be described (for full details see reference Motorola 1983a). For each register the function of the bits is described followed by notes giving further explanation (note that an x in a bit setting means *don't care* and it can be 0 or 1). At this stage just scan through this section and then use it for reference when the sample programs are discussed.

Port A and B Data Registers (PADR & PBDR) ares used for moving data to and from the port A/B data pins and are dependent upon the mode and submode. The registers are readable and writable at all times and **are not** affected by RESET.

Port A and B alternate registers (PAAR & PBAR) (read only) the instantaneous port A/B pin levels are read.

Port C Data Register (PCDR) is used for moving data to and from the port C/alternate function pins. The exact pin functions are dependent upon the Data Direction Register and whatever alternate functions are enabled (the state of the pins can be read while used for alternate functions). The register is readable and writable at all times and **is not** affected by RESET.

Port A, B and C Data Direction Registers (PADDR, PBDDR & PCDDR)

In unidirectional modes the bit pattern in a Data Direction Register determines which bits in the associated Data Register are used for input and which for output. If a bit in the Data Direction Register is:

0: the corresponding bit in the Data Register is used for input;
or 1: the corresponding bit in the Data Register is used for output.

Notes:

1 All bits are reset to 0 on RESET and are readable and writable at all times.
2 The port A data direction register (PADDR) is ignored in mode 3.
3 The port B data direction register (PBDDR) is ignored in modes 2 and 3.
3 The operation of the port C data direction register (PCDDR) is independent of the PIT mode (Port C is general purpose bit I/O at all times and is not affected by the mode).

Port General Control Register (PGCR) sets the overall operating mode of the interface.

7	6	5	4	3	2	1	0
Port Mode Enable		H34 enable	H12 enable	H4 sense	H3 sense	H2 sense	H1 sense

bits **7 6** **Port Mode Enable** sets the overall operating mode of the interface (see note 2)
0 0 mode 0 (unidirectional 8-bit mode)
0 1 mode 1 (unidirectional 16-bit mode)
1 0 mode 2 (bidirectional 8-bit mode)
1 1 mode 3 (bidirectional 16-bit mode)

bits **5 & 4** **H34 & H12 enable** the H3/H4 and H1/H2 handshaking lines (see note 3):
0 disabled
1 enabled

bits **3, 2, 1 & 0** determine the voltage/logic level **sense of H4, H3, H2 & H1** (see note 4):
0 the associated pin is at the high voltage level (5 volts) when negated (reset to logical 0) and low voltage level (0 volt) when asserted (set to logical 1)
1 the associated pin is at the low voltage level (0 volt) when negated (reset to logical 0) and high voltage level (+5 volt) when asserted (set to logical 1)

Notes:

1 All bits are reset to 0 on RESET and are readable and writable at all times.
2 **Port Mode Enable (bits 7 & 6)** sets the operating mode of the device; **the H12 and H34 enable bits must be 0 when the mode is changed.** In general:
  (a) the PGCR is loaded to set the mode (with the H12 & H34 bits equal 0);
  (b) other registers are loaded setting the submode, etc.
  (c) the PGCR is reloaded with:
    (i) the same mode value as in (a)
    (ii) H12 and/or H34 set to enable the handshaking required
    (iii) the H4, H3, H2 and H1 sense bits set to enable the required active sense
3 **H34 (bit 5) and H12 (bit 4) Enable** enables/disables the handshaking during program execution (at any time as required by the application).
4 The **H4, H3, H2 and H1 sense** bits allow any combination of active high (+5 volt is logic 1, 0 volt is logic 0) or active low (0 volt is logic 1, +5 volt is logic 0) as required. When used as input status bits H1, H2, H3 and H4 are set on the asserted edge, i.e. a change from the negated (logical 0) to the asserted (logical 1) state. Thus the status bits can be set on 0 to +5 volt or +5 to 0 volt transitions.

Port Service Request Register (PSRR) is used to specify DMA and interrupt functions

7	6	5	4	3	2	1	0
*	SVCRQ select		interrupt PFS		port interrupt priority control		

bits **6 5** **SVCRQ (service request) Select** sets the function of the PC4/DMAREQ pin
- 0 x the PC4/DMAREQ pin is used for normal port C bit 4 I/O (see note 3)
- 0 1 the PC4/DMAREQ pin carries DMA requests associated with double buffered transfers controlled by H1. H1 interrupts are disabled.
- 1 1 the PC4/DMAREQ pin carries DMA requests associated with double buffered transfers controlled by H3. H3 interrupts are disabled.

bits **4 3** **Interrupt Pin Function Select** sets the PC5/PIRQ & PC6/PIACK pin functions
- 0 0 the PC5/PIRQ pin is used for normal port C bit 5 I/O (PC5) (see note 4)
  the PC6/PIACK pin is used for normal port C bit 6 I/O (PC6)
- 0 1 the PC5/PIRQ pin is used for the parallel interface interrupt request (PIRQ)
  the PC6/PIACK pin is used for normal port C bit 6 I/O (PC6)
  this is used for autovectored interrupts (see Chapter 23.7 and Program 24.3)
- 1 0 the PC5/PIRQ pin is used for normal port C bit 5 I/O (PC5)
  the PC6/PIACK pin is used for parallel interface interrupt acknowledge (PIACK)
- 1 1 the PC5/PIRQ pin is used for parallel interface interrupt request (PIRQ)
  the PC6/PIACK pin is used for parallel interface interrupt acknowledge (PIACK)
  this is used when the interface supplies an interrupt vector (see Chapter 23.7 and Program 24.2)

bits **2 1 0** **Port Interrupt Priority Control** sets the interrupt priority of the control pins
- 0 0 0 the priority is from H1 (highest), H2, H3 to H4 (lowest) (see note 5)
- 0 0 1 the priority is from H2 (highest), H1, H3 to H4 (lowest)
- 0 1 0 the priority is from H1 (highest), H2, H4 to H3 (lowest)
- 0 1 1 the priority is from H2 (highest), H1, H4 to H3 (lowest)
- 1 0 0 the priority is from H3 (highest), H4, H1 to H2 (lowest)
- 1 0 1 the priority is from H3 (highest), H4, H2 to H1 (lowest)
- 1 1 0 the priority is from H4 (highest), H3, H1 to H2 (lowest)
- 1 1 1 the priority is from H4 (highest), H3, H2 to H1 (lowest)

Notes:

1. All bits are reset to 0 on RESET and are readable and writable at all times.
2. bit 7 is unused, it always reads as 0 and writes are ignored
3. **SVCRQ Select (bits 6 & 5)** specifies whether interrupt or DMA requests are generated by activity on the H1 and H3 handshake pins (sample programs in this book will only use interrupts).
4. **Interrupt Pin Function Select (bits 4 & 3)** specifies whether the PC5/PIRQ and PC6/PIACK pins carry the port C pin 5 and 6 I/O function or are used for interrupt control. Autovectored interrupts will use PIRQ and vectored interrupts will use PIRQ and PIACK; see Chapter 23.7 and 24.2 (Programs 24.2 and 24.3).

5 **Port Interrupt Priority Control (bits 2, 1 and 0)** determines the priority among the interrupt sources, i.e. from H1, H2, H3 or H4 - see the Port Interrupt Vector Register (PIVR) below. Note that these bits should only be changed when the affected interrupt(s) are disabled or known to be inactive.

Port Interrupt Vector Register (PIVR) specifies the upper six bits of the four interrupt vectors

7	6	5	4	3	2	1	0
interrupt vector number						*	*

1 This is used with vectored interrupts, i.e. not autovectored (see Chapter 23.7 and 24.2).
2 On RESET the register is initialised to $0F, which if an interrupt occurs will cause the uninitialised interrupt vector (see Chapter 23.2). If the PIVR register is read at this stage (i.e. before being written to) the value $0F will be returned.
3 When writing, the upper six bits are loaded and the lower two bits are forced to 0. A read will then return this value.
4 When loaded the lower two bits are determined by the interrupt source, a H1 interrupt yields 00, H2 yields 01, H3 yields 10 and H4 yields 11. For example, if the Port Interrupt Vector Register (PIVR) had been loaded with 64:
    A H1 interrupt would yield an interrupt vector of 64 (address $100)
    A H2 interrupt would yield an interrupt vector of 65 (address $104)
    A H3 interrupt would yield an interrupt vector of 66 (address $108)
    A H4 interrupt would yield an interrupt vector of 67 (address $10C)

Port Status Register (PSR)

7	6	5	4	3	2	1	0
H4 level	H3 level	H2 level	H1 level	H4S status	H3S status	H2S status	H1S status

Bits **7 6 5 & 4** show the **instantaneous level** of the handshake pins (independent of the handshake pin sense bits in the Port General Control register)

bits **3 2 1 & 0** are the **status bits** for H4, H3, H2 and H1 respectively: a 1 being the active or asserted state. The interpretation depends upon the mode and submode.

The status bits (0 to 3) in this register are reset to 0 by the *direct method of resetting status*, in which, to reset a particular bit a 1 is written to it (0s are ignored), e.g.:

```
        MOVE.B    #2,PSR        reset H2S - H2 status bit
```

i.e. a 1 is written to a particular bit to reset it to 0.

If the status bits (in the Port Status Register) are going to be used it is good idea to reset them during initialisation in case they may have been set by a previous program, e.g.:

```
        MOVE.B    #$0F,PSR      clear status flags (if set)
```

Port A Control Register (PACR) specifies the port A submode, and the control of H2 & H1. The following only specifies the settings for Mode 0 Submode 1x; others will be introduced at a later stage.

7	6	5	4	3	2	1	0
Port A Submode		H2 Control			H2 Int Enable	H1 SVCRQ Enable	H1 Stat Ctrl

bits **7 6** **Port A submode** - specifies the submode (see note 2)
1x submode 1x (x can be 0 or 1 and is ignored)

bits **5 4 3** **H2 Control** - controls the function of the H2 pin (see note 3)
0 x x input pin - status only (see section 2.3 below)
1 x 0 output pin - always negated, i.e. logical 0 output (see section 2.4 below)
1 x 1 output pin - always asserted, i.e. logical 1 output (see section 2.4 below)

bit **2** **H2 Interrupt Enable** enables/disables H2 interrupts (see note 4)
0 H2 interrupt request disabled
1 H2 interrupt request enabled

bit **1** **H1 SVCRQ (service request) Enable** enables/disables H1 interrupt/DMA request
0 H1 interrupt and DMA request disabled
1 H1 interrupt or DMA request enabled (see note 5)

bit **0** **H1 Status Control** controls the operation of H1S status bit (see note 6)
x not used in this submode, i.e. its value is ignored

Notes:

1. All bits are reset to 0 on RESET and are readable and writable at all times.
2. **Port A Submode (Bits 7 & 6)** sets the submode and **the H12 enable bit in the PGCR must be 0 when it is changed.**
3. **H2 Control (bits 5, 4 & 3)** determines whether H2 is an input pin (see section 2.3 below) or an output pin (asserted or negated, see section 2.4 below)
4. **H2 Interrupt Enable (bit 2)** determines whether an interrupt will be generated when the H2S status bit in the Status Register is set (goes to logical 1; see below)
5. When **H1 SVCRQ Enable (bit 1)** is set the **SVCRQ Select** bits in the Port Service Request Register (PSRR) select between DMA requests and H1 interrupts, i.e. if interrupts are selected **H1 SVCRQ Enable** determines whether an interrupt will be generated when the H1S status bit in the Port Status Register is set (goes to logical 1)
6. **H1 Status Control (bit 0)** controls the operation of the H1S status bit (not used in submode 1x)

Port B Control Register (PBCR) specifies the port B submode, and control of H4 & H3.

7	6	5	4	3	2	1	0
Port B Submode		H4 Control			H4 Int Enable	H3 SVCRQ Enable	H3 Stat Ctrl

For mode 0 submode 1x the port B Control Register functions are similar to the port A with H3 and H4 (and H34) used (see reference Motorola 1983a for details of other modes).

21.2 Mode 0 Submode 1x - unidirectional bit input/output

Bit I/O is used for applications where a variety of devices are connected to a port where individual pins are input or output as set by the Data Direction Register.

21.2.1 Bit input and output with no handshaking

Program 21.1 (next page) uses the simple switches and lights circuit described in Appendix F.3 (eight switches attached to port A for input and eight LEDs attached to port B for output display). Note that *absolute addressing mode* is used to access the MC68230 I/O registers. The program sequence is:

line

18-26 `PIT: EQU $A00001` defines the absolute base address of the PIT memory mapped registers for the Kaycomp board; the PIT registers are then defined relative to this base (for another system change the base address, e.g. the PIT on the Bytronic 68000 board is at $130001)

29 `MOVE.B #0,PGCR` selects Mode 0 in Port General Control Register (all other bits are 0)

30 `MOVE.B #0,PSRR` clears the Port Service Request Register

31 `MOVE.B #$80,PACR` sets the port A Port Control Registers to submode 1x

32 `MOVE.B #$80,PBCR` sets the port B Port Control Registers to submode 1x

33 `MOVE.B #0,PADDR` sets all bits of port A for input (i.e. a 0 in the Data Direction Register sets the corresponding bit in the Data Register for input, and a 1 sets it for output)

34 `MOVE.B #$0FF,PBDDR` sets all bits of port B for output

37 `MOVE.B PADR,D0` data is read from port A (the switches) into D0

38 `MOVE.B D0,PBDR` the data is then output to port B (to light the LEDs)

39-42 the value is then displayed on the screen using WRHEXB

44-45 if the switch value was not 0 the program LOOPs back to line 37 otherwise

46 the program terminates

As the switches are changed the updated value is displayed on the LEDs and as a hexadecimal number on the display screen. In this case no handshaking is required to control the data transfer, i.e. it is a simple program loop which reads the switches and sets the LEDs. In other applications handshaking may be required to control the data flow between the transmitter and receiver (handshaking will be explained in detail later).

```
--------------------------------------------------------------------------------
 1                          * Program 21.1 - MC68230 PIT simple input/output
 2                          *
 3                          * Program which reads switch settings from port A
 4                          *  and outputs value to LEDs connected to port B.
 5                          *
 6                          * Program description in Pseudo-English:-
 7                          * initialise PIT: Bit I/O, A for input, B for output
 8                          * LOOP
 9                          *    READ(switch values from port A into D0)
10                          *    WRITE(D0 to port B to LEDs)
11                          *    WRITE('Port A Data Register is ',display D0)
12                          * UNTIL switch values = 0
13                          * STOP program
14                          *
15                                  XREF     WRLINE,WRTEXT,WRHEXB
16                          *
17                          * define MC68230 PIT registers for the Kaycomp board
18           00A00001       PIT:    EQU      $A00001   PIT base address for Kaycomp
19           00A00001       PGCR:   EQU      PIT+0     port general control register
20           00A00003       PSRR:   EQU      PIT+2     port service request register
21           00A00005       PADDR:  EQU      PIT+$04   data direction register A
22           00A00007       PBDDR:  EQU      PIT+$06   data direction register B
23           00A0000D       PACR:   EQU      PIT+$0C   control register port A
24           00A0000F       PBCR:   EQU      PIT+$0E   control register port B
25           00A00011       PADR:   EQU      PIT+$10   data register port A
26           00A00013       PBDR:   EQU      PIT+$12   data register port B
27                          *
28                          *initialise parallel interface, mode 0, A & B submode 1x
29 400400 13FC     0000 START:  MOVE.B   #0,PGCR       set mode 0 in PGCR
               00A00001
30 400408 13FC     0000         MOVE.B   #0,PSRR       clear the PSRR
               00A00003
31 400410 13FC     0080         MOVE.B   #$80,PACR     A submode 1x (bit I/O)
               00A0000D
32 400418 13FC     0080         MOVE.B   #$80,PBCR     B submode 1x (bit I/O)
                00A0000F
33 400420 13FC     0000         MOVE.B   #0,PADDR      port A all lines input
               00A00005
34 400428 13FC     00FF         MOVE.B   #$0FF,PBDDR   port B all lines output
               00A00007
35                          *
36                          * LOOP reading switches from A and lighting LEDs on B
37 400430 1039 00A00011 LOOP:   MOVE.B   PADR,D0       read the switches
38 400436 13C0 00A00013         MOVE.B   D0,PBDR       output to the LEDs
39 40043C 4EB9 0000802C         JSR      WRLINE        display value in hex
40 400442 4EB9 00008038         JSR      WRTEXT
41 400448 50 6F 72 74 20        DC.W     'Port A Data Register is $'
42 400462 4EB9 0000804A         JSR      WRHEXB
43                          * UNTIL switch values = 0
44 400468 4A00                  TST.B    D0            if <> 0 continue loop
45 40046A 66C4                  BNE.S    LOOP
46 40046C 4E4E                  TRAP     #14           terminate program
47                              END
--------------------------------------------------------------------------------
```

Program 21.1 To read switches and set LEDs via an MC68230 (mode 0 submode 1x)

Exercise 21.1 *(see Appendix G for sample answer)*

Enter Program 21.1, modify as required until operational. Alter the program to display a rotating bit on the LEDs (i.e. LEDs light in order bit 0 1 2 3 4 5 6 7 0 1 2 3 4 etc.). The bit should move approximately once every second (use subroutine DELAY in the library).

21.2.2 Handshaking lines H1 and H3 for input

When in Mode 0 the MC68230 PIT ports A and B each have an input control line, H1 and H3 respectively, which can be used for handshaking (Chapter 19.1 discussed parallel I/O handshaking techniques). Although the instantaneous level of the handshaking lines can be read via the Port Status Register (PSR) their main use is to set status bits (H1S, H2S, H3S and H4S) by detecting **transitions or changes** in the voltage levels on the lines.

Status bits in the Port Status Register (PSR bit 0 for H1S and Bit 2 for H3S) are set on an asserted edge (a change from a logical 0 to a logical 1 level), note:

1 The H12 (H34) enable bit must be set in the Port General Control Register (PGCR) to enable transitions to set H1S (H3S), i.e. if H12 (H34) is 0 H1S (H3S) is held at 0.
2 H1S (H3S) in the Port Status Register (PSR) is reset to 0 by the *direct method of resetting status*, in which, to reset a particular bit a 1 is written to it.
3 The voltages corresponding to logical 0 and 1 are specified by the Port General Control Register (PGCR), i.e. if an asserted edge is rising (0 to +5 volt) or falling (+5 to 0 volt).
4 The status bits in the Port Status Register (PSR) can be polled, or interrupts can be enabled by setting bit 1 in the Port A (for H1) and port B (for H3) Control Registers.

21.2.3 Using handshaking lines H2 and H4 for input

Setting bits 5, 4 and 3 to 0xx (x can be 0 or 1) of the port A and port B Control Registers sets the H2 and H4 lines, respectively, to be an edge sensitive input. These lines then operate in a similar way to H1 and H3 with the instantaneous levels and status bits being read from the Port Status Register. Again, H12 and/or H34 must be set in the Port General Control Register and bit 2 in the corresponding Port Control Register enables/disables interrupts.

21.2.4 Using handshaking lines H2 and H4 for direct output

The H2 and H4 handshaking lines may be used for output by setting bit 5, 4 and 3 of the Port A and Port B Port Control Register respectively:

bits **5 4 3**	**H2 control** (x can be 0 or 1 and is ignored)
1 x 0	output pin - always negated, i.e. logical 0 output
1 x 1	output pin - always asserted, i.e. logical 1 output

The voltage levels output corresponding to logical 0 and 1 are set by the H2 and H4 sense bits in the Port General Control Register (PGCR). Note that the H12 (H34) Enable bit in the Port General Control Register (PGCR) does not affect this output and H2S (H4S) in the Port Status Register (PSR) is always 0.

```
---------------------------------------------------------------------------------------
 1                           * Program 21.2 - MC68230 PIT Program
 2                           *  Using PIT to display a rotating LED pattern, etc.
 3                           *  Pattern reverses when switch attached to H1 is hit.
 4                           *
 5                           * initialise PIT: mode 0 bit I/O, port B output
 6                           * WRITE('hit H1 switch to reverse LED rotation')
 7                           * set test_pattern = 1 and initial direction = left
 8                           * LOOP FOR ever
 9                           *     LOOP
10                           *         output test_pattern to port B (to LEDs)
11                           *         IF direction = left
12                           *             rotate test_pattern one bit left
13                           *             blink LED attached to H2
14                           *         ELSE
15                           *             rotate test_pattern one bit right
16                           *             blink LED attached to H4
17                           *         delay for a time period
18                           *     UNTIL H1 switch is hit
19                           *     INVERT direction
20                           * END LOOP
21                           *
22          00130001         PIT:       EQU       $130001     PIT base address for Bytronic
23          00000000         PGCR:      EQU       0           port general control register
24          00000002         PSRR:      EQU       2           port service request register
25          00000006         PBDDR:     EQU       $06         data direction register B
26          0000000C         PACR:      EQU       $0C         control register A
27          0000000E         PBCR:      EQU       $0E         control register B
28          00000012         PBDR:      EQU       $12         data register B
29          0000001A         PSR:       EQU       $1A         port status register
30                           *
31                                      XREF      WRLINE,WRTEXT,WRHEXB,DELAY
32 001000 4EB9 00108038      START:     JSR       WRTEXT
33 001006 68 69 74 20 48                DC.W      'hit H1 switch to reverse LED rotation$'
34 00102C 207C 00130001                 MOVEA.L   #PIT,A0           set PIT base address
35 001032 117C 0000 0000                MOVE.B    #0,PGCR(A0)       PGCR set mode 0
36 001038 117C 0000 0002                MOVE.B    #0,PSRR(A0)       clear PSRR
37 00103E 117C 00A0 000C                MOVE.B    #$0A0,PACR(A0)    A submode 1x, H2 output
38 001044 117C 00A0 000E                MOVE.B    #$0A0,PBCR(A0)    B submode 1x, H4 output
39 00104A 117C 00FF 0006                MOVE.B    #$0FF,PBDDR(A0)   port B all lines output
40 001050 08E8 0004 0000                BSET.B    #4,PGCR(A0)       PGCR enable H12
41 001056 117C 000F 001A                MOVE.B    #$0F,PSR(A0)      clear flags (if set)
42 00105C 7201                          MOVEQ     #1,D1             initial test_pattern
43 00105E 4282                          CLR.L     D2                direction = left
44                            *
45                            * LOOP rotating left or right
46 001060 1141       0012     LOOP:     MOVE.B    D1,PBDR(A0)       test_pattern to LEDs
47 001064 4A02                          TST.B     D2                direction ?
48 001066 660A                          BNE.S     RIGHT
49 001068 E319                          ROL.B     #1,D1             rotate test_pattern left
50 00106A 0868 0003 000C                BCHG.B    #3,PACR(A0)       invert H2 to blink LED
51 001070 6008                          BRA.S     WAIT
52 001072 E219                RIGHT:    ROR.B     #1,D1            rotate test_pattern right
53 001074 0868 0003 000E                BCHG.B    #3,PBCR(A0)       invert H4 to blink LED
54 00107A 4EB9 0010809A       WAIT:     JSR       DELAY             wait for delay time
55 001080 0064                          DC.W      100               100 milliseconds
56 001082 0828 0000 001A                BTST.B    #0,PSR(A0)        has H1 switch been hit ?
57 001088 67D6                          BEQ.S     LOOP              if not rotate again
58                            *  UNTIL H1 switch is hit
59 00108A 117C 0001 001A                MOVE.B    #1,PSR(A0)        clear H1S status flag
60 001090 4602                          NOT.B     D2                and invert direction
61 001092 60CC                          BRA.S     LOOP
62                                      END
---------------------------------------------------------------------------------------
```

Program 21.2 Direct program control of the MC68230 PIT lines, H1, H2 and H4

Program 21.2

Program 21.2 (previous page) is an extension of Exercise 21.1 which makes use of a switch attached to the H1 handshaking line and LEDs attached to H2 and H4 (see sample circuit Appendix F.3). The program displays a rotating pattern on the LEDs attached to the data lines and blinks the LED attached to the H2 or H4 line. When the switch attached to H1 is pressed, the rotation of the pattern displayed on the LEDs is reversed and H2 and H4 change over. Note that *address register indirect with displacement addressing mode* is used to access the MC68230 I/O registers (Program 21.1 used *absolute addressing mode*; Chapter 18.4 discussed alternative addressing techniques). The sequence of events is:

22 `PIT: EQU $130001` defines the absolute base address of the PIT memory mapped registers for the Bytronic board is defined

23-29 the PIT registers are defined as offsets from the base; *address register indirect with displacement addressing mode* will then be used to access them

34 `MOVEA.L #PIT,A0` loads the base address of the PIT into A0

35 `MOVE.B #0,PGCR(A0)` sets Mode 0 in the Port General Control Register (PGCR) (all other bits are reset to 0)

36 `MOVE.B #0,PSRR(A0)` clears the Port Service Request Register

37-38 `MOVE.B #$0A0,PACR(A0)` sets up the port A (and B) Port Control Register:
(a) bits 6 and 7 are set to 10; to select submode 1x
(b) bits 5, 4 and 3 are set to 100; H2 & H4 are outputs always negated
(c) all other bits are 0 to disable interrupts

39 all bits port B are set for output (port A is not used in this program)

40 `BSET.B #4,PGCR(A0)` sets H12 Enable in the Port General Control Register (Mode 0 is still selected)

41 `MOVE.B #$0F,PSR(A0)` clears the status bits (H1S, H2S, H3S and H4S) in the Port Status Register (if set from a previous program) using the *direct method of resetting*, i.e. a 1 is written to a particular bit to reset it to 0

42 moves the initial bit pattern into D1 (to be rotated)

43 the initial rotation direction is set to left (D2=0)

46 `MOVE.B D1,PBDR(A0)` displays the bit pattern on the LEDs (via port B)

47-48 IF the direction is left
    49 the pattern is rotated left
    50 PACR bit 3 is inverted (H2 output line) to blink the yellow LED
  ELSE
    52 the pattern is rotated right
    53 PBCR bit 3 is inverted (H4 output line) to blink the green LED

54-55 DELAY for time period

56 `BTST.B #0,PSR(A0)` tests the H1S status (bit 0) in the Port Status Register

57 IF H1S was 0 (i.e. switch not pressed) loop back to line 46 to rotate again

59 IF H1S was set (switch hit) then reset it using the *direct method of resetting*, i.e.:
    `MOVE.B #1,PSR(A0)` to write a 1 to bit 0 of the Port Status Register
  note that if H1S was not reset the direction would invert on each loop

60 invert the rotation direction

61 loop back to line 46 to rotate again

The rotation reverses when the switch attached to H1 is pressed (+5 volt to 0 volt transition); the release has no effect. To amend the program so that the press of the switch is ignored, and the reverse of rotation occurs when the switch is released, the initialisation statement on line 40 should be changed so that (in addition to enabling H12) the H1 sense is asserted +5 volt and negated 0 volt; thus a 0 to 5 volt transition on H1 will set H1S:

```
        MOVE.B     #$11,PGCR(A0)     mode 0, H12 enable, H1 sense
```

Thus it is possible to handshake a parallel communications system directly under program control by toggling H2 (or H4) to generating output pulses (data ready STROBE in Fig. 19.1) and polling H1 (or H3) for the reply from the external device (data acknowledge ACKNLG).

Problems for Chapter 21

1 Connect up a keypad, e.g. as shown in Appendix F.6. Write a subroutine which returns the value of key hit as a decimal digit (ignore the * and # keys). Test the subroutine with a suitable program.

2 Using a suitable motor circuit (see Appendix F.7) implement the PWM (pulse width modulation) motor speed control described in Appendix F.7.2.

3 Using a suitable motor/heater circuit implement the simple temperature control described in Appendix F.7.6. Note that use can be made of the BSET and BCLR instructions to switch the motor and heater ON and OFF.

22

MC68230 PIT double buffered I/O

Chapter 21 introduced the MC68230 PIT and described Mode 0 Submode 1x bit input/output and direct program control of the handshaking lines, e.g. to handshake a parallel communications system by generating output pulses (see data ready STROBE in Fig. 19.2) by toggling H2 (or H4) and polling H1 (or H3) for the reply from the external device (see data acknowledge ACKNLG in Fig. 19.2).

This chapter introduces Mode 0 submode 10 double buffered output and submode 00 double input and describes more advanced handshaking techniques. Double buffering (explained below) has advantages when used for connecting devices of similar speeds, e.g. while the final output buffer is being transmitted to an external device the program can be filling the initial buffer with the next data so it is immediately available for transmission.

22.1 Mode 0 Submode 01 - unidirectional 8-bit double buffered output

Mode 0 Submode 01 double buffered output with H1(H3) and H2(H4) as handshake lines is generally used to connect a single output device, e.g. a parallel printer.

Port A Control Register (PACR) specifies the port A submode, and the control of H2 & H1. The following only specifies the settings for Mode 0 Submode 01.

7	6	5	4	3	2	1	0
Port A Submode		H2 Control			H2 Int Enable	H1 SVCRQ Enable	H1 Stat Ctrl

bits **7 6** **Port A submode** - specifies the submode
- 0 1 submode 01 double buffered output

bits **5 4 3** **H2 Control** (x can be 0 or 1 and is ignored)
- 0 x x input pin - status only
- 1 0 0 output pin - always negated, i.e. logical 0
- 1 0 1 output pin - always asserted, i.e. logical 1
- 1 1 0 output pin - interlocked output handshake protocol
- 1 1 1 output pin - pulsed output handshake protocol

bit **2** **H2 Interrupt Enable** enables/disables H2 interrupts (see Chapter 21)
- 0 H2 interrupt request disabled
- 1 H2 interrupt request enabled

bit **1** **H1 SVCRQ (service request) Enable** enables/disables H1 interrupt/DMA request
0 H1 interrupt and DMA request disabled
1 H1 interrupt or DMA request enabled (see Chapter 21)

bit **0** **H1 Status Control** (used to see if data can be loaded into the output buffers)
0 the H1 status bit is 1 when either the initial or final output latches can accept new data. It is 0 when both are full.
1 the H1 status bit is 1 when both of the port A output latches are empty. It is 0 when at least one is full.

```
write from CPU → —| initial data latch |—→—| final data latch |—— → data lines
```

Fig. 22.1 Programmer's model of the double buffered output data latches

Fig. 22.1 is a programmer's model of the double buffered output data latches. There are two data latches or buffers:

(a) the initial data latch,
and (b) the final data latch which drives the data lines (assuming the Data Direction Register (DDR) bits are set for output; see next paragraph).

Although the primary direction of this submode is output the bits of the Data Register and associated data lines can be input or output as specified by the Data Direction Register:

Output lines are double buffered (via the initial and final data latches) and controlled by the handshaking pins; see below.

Input lines are not buffered and are not affected by handshaking; data read is the instantaneous pin values. Data written to an input pin is passed to the initial and final data latches (as usual) but the output buffer which drives the lines is disabled.

From a programmer's viewpoint the double buffered output operates as follows:

1 Assuming both latches are empty the first data byte written to the Data Register by the program is passed directly to the final data latch and appears on the data lines (assuming the Data Direction Register (DDR) bits are set for output);
2 The next data byte written would be stored in the initial data latch.

Note, when both data latches are full:

(a) a read of the Data Register returns the value of the byte in the final data latch (except for input lines where the instantaneous pin value is read),
and (b) any further writes to the Data Register are ignored until the handshaking (see below) indicates that more data can be accepted by the external device (the H1S(H3S) status bit in the Port Status Register is used to check if there is room in the latches for more data).

The peripheral accepts data by asserting H1(H3) which causes the next data byte to be moved to the final data latch as soon as it is available, i.e.

(a) from the initial data latch if it contains a byte of data,
or (b) when the program writes the next data byte to the Data Register.

The H1S(H3S) status bit is used to see if there is room in the data latch for more data. This may be used in two ways depending upon the H1(H3) Status Control in the Port A or B Control Register:

(a) to check if **at least one** data latch is empty, i.e. that the next data byte can be written to the Data Register;

or (b) to check **if both** are empty, i.e. at the end of data transmission this will be used to ensure that all data has been physically transmitted before terminating the program (this can also be used to operate effectively single buffered, see section 4 below).

22.1.1 H2(H4) options with double buffered output

The function of H2(H4) is programmable via bits 5, 4 and 3 of the Port A (or B) Control Register; it may indicate to the peripheral that data has been moved into the port final output latch or some other function.

22.1.1.1 H2 is an input pin - status only H2 Control bits (5 4 3) are 0 x x

H2(H4) is an edge sensitive input pin independent of H1(H3) and the transfer of data. H2S(H4S) is set on the asserted edge of H2(H4) and reset by the *direct method of resetting* or when the H12 Enable bit in the PGCR is 0 (see Chapter 21.2.3).

22.1.1.2 H2 is an output pin - always negated H2 Control bits (5 4 3) are 1 0 0

H2(H4) is an output pin which is always negated (see Chapter 21.2.4). H2S(H4S) is always 0.

22.1.1.3 H2 is an output pin - always asserted H2 Control bits (5 4 3) are 1 0 1

H2(H4) is a general pin which is always asserted (see Chapter 21.2.4). H2S(H4S) is always 0.

22.1.1.4 H2 is an output pin in the interlocked output handshaking protocol

H2 Control bits (5 4 3) are 1 1 0

1. H2(H4) is asserted **two clock cycles** after data is transferred to the final output latch.
2. The data and H2(H4) remain stable until the next asserted edge of H1(H3) when H2(H4) is negated.
3. As soon as the next byte of data is available it is transferred to the final output latch and the process is repeated.
4. When H2(H4) is negated asserted transitions on H1(H3) have no effect.
5. The H2S(H4S) status bit is always 0.
6. When the H12(H34) enable bit in the PGCR is 0 H2(H4) is held negated and handshaking is suspended.

22.1.1.5 H2 is an output pin in the pulsed output handshaking protocol

H2 Control bits (5 4 3) are 1 1 1

1 H2(H4) is asserted **two clock cycles** after data is transferred to the final output latch.
2 H2(H4) remains asserted no longer than **four clock cycles** (i.e. a pulse is generated).
3 If an H1(H3) asserted edge occurs before termination of the pulse on H2(H4), the H2(H4) pulse is terminated.
3 The data remains stable until the next asserted edge of H1(H3).
4 As soon as the next data byte is available it is transferred to the final output latch and the process is repeated.
5 The H2S(H4S) status bit is always 0.
6 When the H12(H34) enable bit in the PGCR is 0 H2(H4) is held negated and handshaking is suspended.

Thus it is possible to handshake in various ways, e.g. by toggling H2(H4) under program control (as in Program 21.2) or using the interlocked or pulsed handshaking protocols.

22.2 Program to demonstrate double buffered output

Program 22.1 (next page) uses the switches and lights boards of Appendix F.3 to demonstrate PIT mode 0 submode 01 double buffered output:

1 values are written to the port B Data Register when the user terminal keyboard is hit;
2 the switch attached to the H3 line is used to signal acceptance of data and move the next data, if available, from the initial data latch to the final data latch.

Fig. 22.2 shows a sample run of the program. The sequence of Program 22.1 is:

line
30 `MOVE.B #0,PGCR(A0)` sets Mode 0 in the Port General Control Register
31 `MOVE.B #0,PSRR(A0)` clears the Port Service Request Register
32 `MOVE.B #$40,PBCR(A0)` sets submode 01 in the port B Port Control Register
33 `MOVE.B #$0FF,PBDDR(A0)` sets all bits of port B for output
34 `BSET.B #5,PGCR(A0)` sets H34 enable in the Port General Control Register (Mode 0 is still selected)
35 sets the initial data pattern (in D1) to 0
36-42 writes the initial value of the port B data register, i.e. the value left by the last program
44-45 checks to see if the keyboard has been hit, if not goto line 54
46 reads the character hit (which is not used)
47 increments the value of the data pattern
48 writes the data pattern to the port B Data Register (to the LEDs)
49-52 writes the value to the screen
53 loops back to line 38 to display the port B Data Register, etc.
54-55 checks if H3 is asserted (switch pressed), if not goto line 44
56-57 displays message indicating H3 switch has been pressed
58-59 waits for release of the H3 switch
60 loops back to line 38 to display the B Data Register, etc.

```
-------------------------------------------------------------------------------------
 1                           * Program 22.1, MC68230 PIT double buffered output to LEDs
 2                           *   waits for keyboard hit before writing data.
 3                           *
 4                           * initialise PIT mode 0 Port B submode 01 double buffered O/P
 5                           * data_pattern = 0
 6                           * write('Initial value of data register is ',B data register)
 7                           * LOOP FOR ever
 8                           * IF keyboard has been hit
 9                           *     read keyboard
10                           *     data_pattern = data_pattern + 1
11                           *     write data_pattern to port B data register
12                           *     write('KBD hit, write ',data_pattern)
13                           *     write('  data register is ',B data register)
14                           * IF H3 asserted
15                           *     write('H3 switch hit  ',data_pattern)
16                           *     LOOP   UNTIL H3 becomes negated
17                           *     write('  data register is ',B data register)
18                           * END LOOP
19                           *
20          00A00001         PIT:       EQU       $130001      PIT base address for Bytronic
21          00000000         PGCR:      EQU       0            port general control register
22          00000002         PSRR:      EQU       2            port service request register
23          00000006         PBDDR:     EQU       $06          data direction register B
24          0000000E         PBCR:      EQU       $0E          control register B
25          00000012         PBDR:      EQU       $12          data register B
26          0000001A         PSR:       EQU       $1A          port status register
27                           *
28                                      XREF      WRLINE,WRTEXT,WRHEXB,RDCHAR,RDCHECK
29 400400 207C 00130001 START:    MOVEA.L   #PIT,A0           base address of PIT
30 400406 117C 0000 0000          MOVE.B    #0,PGCR(A0)       PCGR set mode 0
31 40040C 117C 0000 0002          MOVE.B    #0,PSRR(A0)          clear PSRR
32 400412 117C 0040 000E          MOVE.B    #$40,PBCR(A0)     port B submode 01
33 400418 117C 00FF 0006          MOVE.B    #$0FF,PBDDR(A0)   port B all lines output
34 40041E 08E8 0005 0000          BSET.B    #5,PGCR(A0)       PGCR enable H34
35 400424 4281                    CLR.L     D1                initial data pattern
36 400426 4EB9 00008038           JSR       WRTEXT
37 40042C 49 6E 69 74 69          DC.W      'Initial value of $'
38 40043E 4EB9 00008038 B_DREG:   JSR       WRTEXT            display B data register
39 400444 20 20 64 61 74          DC.W      '  data register is $'
40 400458 1028      0012          MOVE.B    PBDR(A0),D0       read data register
41 40045C 4EB9 00008050           JSR       WRHEXB
42 400462 4EB9 0000802C           JSR       WRLINE
43                           * main program loop
44 400468 4EB9 00008080 LOOP:     JSR       RDCHECK           keyboard been hit ?
45 40046E 672C                    BEQ.S     TEST_H3           if not test H3 line
46 400470 4EB9 00008084           JSR       RDCHAR            yes, read keyboard
47 400476 5281                    ADDQ.L    #1,D1             increment data pattern
48 400478 1141      0012          MOVE.B    D1,PBDR(A0)       data pattern to LEDs
49 40047C 4EB9 00008038           JSR       WRTEXT            and to screen
50 400482 4B 42 44 20 68          DC.W      'KBD hit, write $'
51 400492 2001                    MOVE.L    D1,D0
52 400494 4EB9 00008050           JSR       WRHEXB
53 40049A 60A2                    BRA.S     B_DREG            display data register
54 40049C 0828 0006 001A TEST_H3: BTST.B    #6,PSR(A0)        is H3 asserted
55 4004A2 66C4                    BNE.S     LOOP              if not loop
56 4004A4 4EB9 00008038           JSR       WRTEXT            H3 is asserted
57 4004AA 48 33 20 73 77          DC.W      'H3 switch hit    $'
58 4004BC 0828 0006 001A NEGATE:  BTST.B    #6,PSR(A0)        wait until H3 negated
59 4004C2 67F8                    BEQ.S     NEGATE
60 4004C4 6000      FF78          BRA.L     B_DREG            display data register
61                                END
-------------------------------------------------------------------------------------
```

Program 22.1 Program to demonstrate PIT double buffered output

```
--------------------------------------------------------------------------------
 1: MC68000 monitor V1.04b, please enter command (<ESC> to abort, ? for help)
 2: M68> go/run program, from start address (<CR> for 00400400) ? $
 3: Initial value of   data register is 16        value left by last program
 4: KBD hit, write 01  data register is 01        01 written to final data latch
 5: KBD hit, write 02  data register is 01        02 written to initial data latch
 6: H3 switch hit      data register is 02        02 from initial to final data latch
 7: KBD hit, write 03  data register is 02        03 written to initial data latch
 8: H3 switch hit      data register is 03        03 from initial to final data latch
 9: KBD hit, write 04  data register is 03        04 written to initial data latch
10: H3 switch hit      data register is 04        04 from initial to final data latch
11: KBD hit, write 05  data register is 04        05 written to initial data latch
12: H3 switch hit      data register is 05        05 from initial to final data latch
13: KBD hit, write 06  data register is 05        06 written to initial data latch
14: KBD hit, write 07  data register is 05        write ignored, both latches full
15: KBD hit, write 08  data register is 05        write ignored, both latches full
16: KBD hit, write 09  data register is 05        write ignored, both latches full
17: KBD hit, write 0A  data register is 05        write ignored, both latches full
18: H3 switch hit      data register is 06        06 from initial to final data latch
19: KBD hit, write 0B  data register is 06        0B written to initial data latch
20: H3 switch hit      data register is 0B        0B from initial to final data latch
21: KBD hit, write 0C  data register is 0B        0C written to initial data latch
22: H3 switch hit      data register is 0C        0C from initial to final data latch
23: KBD hit, write 0D  data register is 0D        0D written to initial data latch
24: H3 switch hit      data register is 0D        0D from initial to final data latch
25: H3 switch hit      data register is 0D        no effect - initial latch is empty
26: H3 switch hit      data register is 0D        no effect - initial latch is empty
--------------------------------------------------------------------------------
```

Discussion - line number:

3 the initial read of the data register shows the value left by the last program
4 a write (value 01) goes directly to the final data latch and onto the data lines
5 a write (value 02) to the Data Register loads the initial data latch
6 the H3 switch hit loads the final data latch from the initial data latch (value 02)
7 a write (value 03) to the Data Register loads the initial data latch
8 the H3 switch hit loads the final data latch from the initial data latch (value 03)
9 a write (value 04) to the Data Register loads the initial data latch
10 the H3 switch hit loads the final data latch from the initial data latch (value 04)
11 a write (value 05) to the Data Register loads the initial data latch
12 the H3 switch hit loads the final data latch from the initial data latch (value 05)
13 a write (value 06) to the Data Register loads the initial data latch
14 a write (value 07) to the Data Register is ignored (i.e. initial latch is full)
15 a write (value 08) to the Data Register is ignored (i.e. initial latch is full)
16 a write (value 09) to the Data Register is ignored (i.e. initial latch is full)
17 a write (value 0A) to the Data Register is ignored (i.e. initial latch is full)
18 the H3 switch hit loads the final data latch from the initial data latch (value 06)
19 a write (value 0B) to the Data Register loads the initial data latch
20 the H3 switch hit loads the final data latch from the initial data latch (value 0B)
21 a write (value 0C) to the Data Register loads the initial data latch
22 the H3 switch hit loads the final data latch from the initial data latch (value 0C)
23 a write (value 0D) to the Data Register loads the initial data latch
24 the H3 switch hit loads the final data latch from the initial data latch (value 0D)
25 the H3 switch hit is ignored (the initial data latch is empty)
26 the H3 switch hit is ignored (the initial data latch is empty)

Fig. 22.2 Sample run of program 22.1 (line numbers and comments in **bold** added later)

22.3 Double buffered output with a Centronics compatible printer

A Centronics compatible parallel printer can be attached to the MC68230 (see Appendix F.4 and Motorola 1982):

1. the MC68230 GND (ground) to printer GND;
2. the eight data lines of port B to the printer input data lines;
3. STROBE (data ready from interface) handshaking line to H4;
4. ACKNLG (printer data acknowledge) handshaking line to H3.

In addition, signals from the printer which indicate printer ON/OFF line, out of paper, etc., can be connected to port A to enable a program to determine its current status (Motorola 1982).

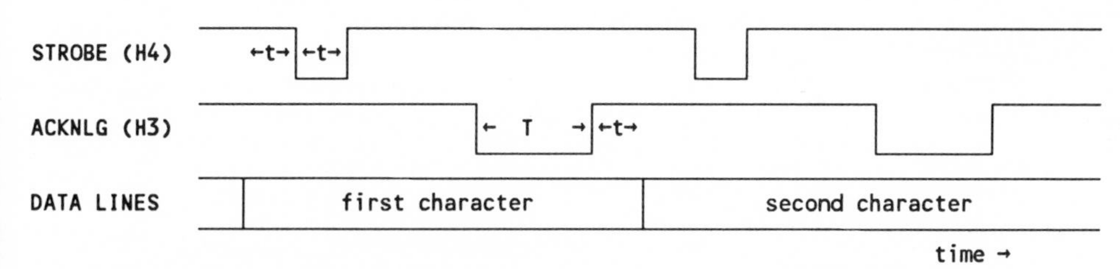

Fig. 22.3 Parallel printer timing diagram ($t>0.5$ & $T>1.5$ microseconds)

Fig 22.3 shows the timing diagram of an MC68230 operated using **double buffered pulsed output handshaking protocol**. The sequence of events when a program is ready to output a character string (bytes of data) to the parallel printer is:

1. The program writes the characters to the port B Data Register; the program waits for room in the output latches by polling the H3S status bit in the Port Status Register (two bytes can be written into the data latches in double buffered mode).
2. On a write to the port B Data Register, the *pulsed output handshaking protocol* sends a low pulse from H4 to the STROBE input of the printer (normally this line is high).
3. After the printer receives the STROBE pulse, it reads the character from the data lines. It may now buffer the character or print a line of data.
4. When the printer is ready to receive the next character, it sends the low ACKNLG pulse down the H3 line (this line is normally high), and then waits for the next STROBE pulse.
5. When the MC68230 receives the asserted transition on H3 it will transmit the next data byte if available in the initial latch or wait for the next byte to be loaded by the program.
5. The program has been waiting for room to appear in the output latches by polling the H3S status bit of the Port Status Register. As room becomes available it loads the next data byte.

```
-------------------------------------------------------------------------------
1                         * Program 22.2 - MC68230 PIT Parallel Printer Driver
2                         *
3                         * initialise PIT: mode 0 port B submode 01,
4                         *     H4 pulsed output handshaking protocol
5                         *     H3S status set when both latches are empty
6                         * initialise A0 to point to the output text array
7                         * LOOP for ever
8                         *     get next character
9                         *     If character = 0
10                        *        reset A0 and get first charcter
11                        *     PRINT(character)
12                        *     WRITE(character) to screen
13                        *     WAIT for acknowledge pulse from printer to set H3
14                        * END LOOP
15                        * STOP
16                        *
17         00130001       PIT:      EQU      $130001     PIT base address for Bytronic
18         00130001       PGCR:     EQU      PIT+0       port general control register
19         00130003       PSRR:     EQU      PIT+2       port service request register
20         00130007       PBDDR:    EQU      PIT+$06     data direction register B
21         0013000F       PBCR:     EQU      PIT+$0E     control register port B
22         00130013       PBDR:     EQU      PIT+$12     data register port B
23         0013001B       PSR:      EQU      PIT+$1A     port status register
24                        *
25                                  XREF     WRLINE,WRTEXT,WRCHAR,WRHEXB,DELAY
26 001000 13FC     0000 START:      MOVE.B   #0,PGCR        set mode 0
               00130001
27 001008 13FC     0000             MOVE.B   #0,PSRR        clear PSRR
               00130003
28                        * port B submode 01, H4 pulsed output handshakomg protocol
29                        *  H3S status set when both latches are empty
30 001010 13FC     0078             MOVE.B   #$78,PBCR      B mode 01, etc.
               0013000F
31 001018 13FC     00FF             MOVE.B   #$0FF,PBDDR    port B all lines output
               00130007
32 001020 13FC     0020             MOVE.B   #$20,PGCR      set mode 0, enable H34
               00130001
33 001028 13FC     000F             MOVE.B   #$0F,PSR       clear status flags (if set)
               0013001B
34 001030 207C 0000105A SET_A0:     MOVEA.L  #TEXT,A0       point A0 to text
35                        *
36                        * LOOP for ever
37 001036 1018            LOOP:     MOVE.B   (A0)+,D0       get next character
38 001038 67F6                      BEQ.S    SET_A0         if 0 reset A0
39 00103A 13C0 00130013             MOVE.B   D0,PBDR        character to printer
40 001040 4EB9 00108028             JSR      WRCHAR         and to screen
41 001046 0839     0002 POLL:       BTST.B   #2,PSR         ACKNLG pulse received ?
               0013001B
42 00104E 67F6                      BEQ.S    POLL           if not poll H3 again
43 001050 08F9     0002             BSET.B   #2,PSR         clear H3S status flag
               0013001B
44                        *         JSR      DELAY          delay to show handshaking
45                        *         DC.W     100                if required
46 001058 60DC                      BRA.S    LOOP           LOOP for next character
47                        *
48                        * Text to be displayed to screen and printer
49 00105A 41 42 43 44 45TEXT:       DC.B     'ABCDEFGHIJKLMNOPQRSTUVWXYZ '
50 001075 31 32 33 34 35            DC.B     '1234567890 !"%^&*()_+{}[]:@;~#<>?'
51 001096 0D 0A 00                  DC.B     $0D,$0A,0
52                                  END
-------------------------------------------------------------------------------
```

Program 22.2 Driving a parallel printer with handshaking using an MC68230 PIT

Program 22.2 (previous page) uses **pulsed output handshaking protocol** to print text on a parallel printer (an oscilloscope can be used to observe the handshaking). The sequence is:

26 the Port General Control Register is initialised to Mode 0
27 the Port Service Request Register is cleared
30 `MOVE.B #$78,PBCR` sets up the port B Port Control Register:
(a) bits 6 and 7 are set to 01; to select submode 01
(b) bits 5, 4 and 3 are set to 111; H4 pulsed output handshaking protocol
(c) bit 0 reset to 0; H3S set when **either** latch is empty (i.e. double buffered)
(d) bits 1 and 2 are 0 to disable interrupts
31 all bits of port B are set for output
32 H34 enable is set in the Port General Control Register (mode 0)
33 clear status bits in Port Status Register (if set by previous program)
34 moves the long word address of the start of the text into A0.

The program loop then transmits the text character by character:

37 the next character is moved into D0
38 if the character was 0 goto line 34 to reset A0 to the start of the text
39 the character is moved into port B Data Register for transmission to the printer; also the *pulsed output handshaking protocol* sends a low pulse down H4 to inform the printer that data is available
40 the character is displayed on the screen
41-42 polling loop; the H3S status bit is polled to see if the next character can be loaded into the port B data register
43 the H3S status bit is reset by the *direct method of resetting*
46 the program loops back to line 37 to get the next character, etc.

When executing Program 22.2 using a printer with a buffer the screen will display a block of characters (transmitted until the buffer is full) followed by small blocks of characters (as the printer empties the buffer by printing characters and it is refilled). Switching the printer OFF LINE will also stop transmission of characters.

The handshaking of Program 22.2 can be seen visually by using the switches and lights board of Appendix F.3 with the data output and STROBE signal shown on the LEDs and a press switch serves as the ACKNLG signal. Line 30 should be modified to use *interlocked output handshaking protocol* with H3S being set when both latches are empty:

```
        MOVE.B    #$71,PBCR         interlocked handshaking
```

The comments should also be removed from lines 44 and 45. The sequence is then:

1 when the data is sent to port B it appears on the eight LEDs;
2 the green LED attached to H4 goes out as the interlocked output handshaking protocol transmits the asserted edge (a high to low transition occurs);
3 the switch attached to H3 is hit to emulate the data acknowledge from the printer;
4 when the PIT senses the transition the H3S status bit will be set and the green LED attached to H4 will light (as the interlocked output handshaking protocol is reset);
4 after the DELAY time the next character appears on the LEDs and screen and the green LED goes out as the interlocked output handshaking protocol transmits the next asserted edge.

The reason for using *interlocked output handshaking protocol* with H3S being set when both latches are empty is to be able to see the H4 line go high (during the DELAY time). Otherwise the time between the end of one handshaking pulse (on H4) and the start of the next would be so short that the LED would never appear to light.

22.4 Practical problems with handshake timing using an MC68230

The MC68230 PIT (unlike the MC6821 PIA) is a modern chip designed for attaching to fast peripherals. Problems can be encountered when driving some low cost Centronics compatible printers with long (in relative terms) pulse widths, etc.

For example, problems were encountered with Program 22.2 when driving **some** Centronics compatible printers from a Kaycomp board equipped with an 8MHz crystal in that the program locked up after transmitting a few characters:

1 some printers require the **minimum** width of the STROBE pulse to be 0.5 microseconds (as shown in Fig. 22.3);
2 the *pulsed output handshaking protocol* generates a pulse no longer than four clock cycles wide, i.e. with an 8MHz clock this is a **maximum** of 0.5 microseconds.

Thus the STROBE pulse width was outside the timing constraints of some printers so after a few characters the handshaking was lost and the program locked up. It was found that fitting a 4MHz crystal and/or using *interlocked output handshaking protocol* (to lengthen the STROBE pulse) overcame the problem.

Another problem encountered with some printers is that MC68230 double buffered output would not work. If the H3 Status Control in the Control Register is 0 so that H3S was set when at least one latch was empty (as in Program 22.2) the program locked up after transmitting a few characters, i.e.:

1 the byte in the final latch has been transmitted and the H4 STROBE pulse sent;
2 the program has loaded the next byte into the initial latch;
3 on the next asserted edge of H3 the next byte is moved from the initial latch into the final latch and appears on the data lines;
4 two clock cycles later (0.25 microseconds with an 8 MHz clock and 0.5 microseconds with a 4MHz clock) the next H4 STROBE pulse is transmitted.

Some printers require a **minimum** of 0.5 microseconds to elapse between the data being applied to the data lines and the start of the STROBE pulse thus making the above timing very critical (even with a 4MHz clock). This problem can be overcome by operating with the H3 Status Control in the Control Register set to 1 so that H3S status was set when **both** output latches are empty, i.e. the program was essentially running single buffered. Thus line 30 of Program 22.2 becomes:

```
            MOVE.B     #$79,PBCR      single buffered mode
or          MOVE.B     #$71,PBCR      with interlocked handshaking
```

The latter being single buffered with interlocked handshaking (to lengthen the STROBE pulse). If problems are encountered when interfacing equipment using pulsed or interlocked handshaking modes external circuitry may have to be installed to lengthen pulses, or the handshake lines controlled directly by the program (as in Program 21.2).

22.5 Mode 0 Submode 00 - unidirectional 8-bit double buffered input

Mode 0 Submode 00 double buffered input with H1(H3) and H2(H4) as handshake lines; generally used to connect a single input device, e.g. an analogue to digital converter.

Port A Control Register (PACR) specifies the port A submode, and the control of H2 & H1. The following only specifies the settings for Mode 0 Submode 00.

7	6	5	4	3	2	1	0
Port A Submode		H2 Control			H2 Int Enable	H1 SVCRQ Enable	H1 Stat Ctrl

bits **7 6** **Port A submode** - specifies the submode
0 0 submode 00 double buffered input

bits **5 4 3** **H2 Control** (x can be 0 or 1 and is ignored)
0 x x input pin - status only
1 0 0 output pin - always negated, i.e. logical 0
1 0 1 output pin - always asserted, i.e. logical 1
1 1 0 output pin - interlocked input handshake protocol
1 1 1 output pin - pulsed input handshake protocol

bit **2** **H2 Interrupt Enable** enables/disables H2 interrupts (see Chapter 21)
0 H2 interrupt request disabled
1 H2 interrupt request enabled

bit **1** **H1 SVCRQ (service request) Enable** enables/disables H1 interrupt/DMA request
0 H1 interrupt and DMA request disabled
1 H1 interrupt or DMA request enabled (see Chapter 21)

bit **0** **H1 Status Control**
x not used in this submode

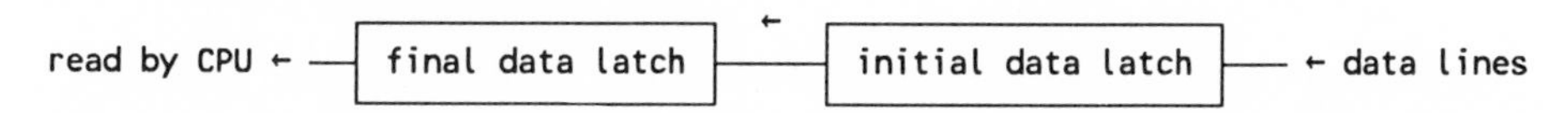

Fig. 22.4 Programmer's model of the MC68230 double buffered input data latches

Fig. 22.4 is a programmers model of the double buffered input data latches. This allows a program to be reading a data byte from the Data Register (from the final data latch) while the external device strobes the next data byte into the initial data latch.

Although the primary direction of this submode is input the bits of the Data Register and associated data lines can be input or output as specified by the Data Direction Register:

Input lines are double buffered and controlled by the handshaking pins; see below.
Output lines are single buffered (via a data latch) and are not affected by handshaking.

From a programmers viewpoint double buffered input operates as follows:

IF there is data in the final data latch
    the contents of the final data latch are read by the program
    IF there is data in the initial data latch
        the initial data latch is loaded into the final data latch
ELSE
    the instantaneous input pin values are read

The peripheral loads data into the latches by asserting H1(H3), i.e.

(a) if both latches are empty directly into the final data latch,
or (b) into the initial data latch if it is empty,
or (c) if both are full the H1(H3) edge is ignored.

The H1S(H3S) status bit is set if there is data in the latches which has not been read by the program.

22.5.1 H2(H4) options with double buffered input

The function of H2(H4) is programmable via bits 5, 4 and 3 of the Port A or B Control Register; it may indicate to the peripheral that there is room in the latches for more data or some other function.

22.5.1.1 H2(H4) is an input pin - status only. H2 Control bits (5 4 3) are 0 x x

H2(H4) is an edge sensitive input pin independent of H1(H3) and the transfer of data. H2S(H4S) is set on the asserted edge of H2(H4) and reset by the *direct method of resetting* or when the H12 Enable bit in the PGCR is 0 (see Chapter 21.2.3).

22.5.1.2 H2(H4) is an output pin - always negated. H2 Control bits (5 4 3) are 1 0 0

H2(H4) is a negated output pin (see Chapter 21.2.4). H2S(H4S) is 0.

22.5.1.3 H2(H4) is an output pin - always asserted. H2 Control bits (5 4 3) are 1 0 1

H2(H4) is asserted output pin (see Chapter 21.2.4). H2S(H4S) is 0.

22.5.1.4 H2(H4) is an output pin in the interlocked input handshaking protocol

H2 Control bits (5 4 3) are 1 1 0

1 H2(H4) is asserted when the input latches can accept new data.
2 H2(H4) is negated following an asserted edge of H1(H3).
3 H2(H4) is asserted again when the input latches can accept new data.
4 H2(H4) is held negated if both latches are full.
5 If H2(H4) is negated transitions on H1(H3) are ignored.
6 The H2S(H4S) status bit is always 0
7 When the H12(H34) enable bit in the PGCR is 0 H2(H4) is held negated and handshaking is suspended.

i.e. any time H2(H4) is asserted new data can be loaded by asserting H1(H3).

22.5.1.5 H2(H4) is an output pin in the pulsed input handshaking protocol

H2 Control bits (5 4 3) are 1 1 1

1. H2(H4) is asserted for no longer than **four clock cycles** (i.e. a pulse is generated) when the input latches can accept new data.
2. H2(H4) is negated following an asserted edge of H1(H3).
3. H2(H4) is asserted again (for no longer than **four clock cycles**) when the input latches can accept new data.
4. If an H1(H3) asserted edge occurs before termination of the pulse on H2(H4), the H2(H4) pulse is terminated.
5. H2(H4) is held negated if both latches are full.
6. If H2(H4) is negated transitions on H1(H3) are ignored.
7. The H2S(H4S) status bit is always 0
8. When the H12(H34) enable bit in the PGCR is 0 H2(H4) is held negated and handshaking is suspended.

22.6 Mode 1 - unidirectional 16-bit double buffered input/output

In mode 1 ports A and B are combined to form a single unidirectional 16-bit port which may be used for double buffered input or output with handshaking controlled by H3. H1 and H2 are used in the data transfer; H1 can be used as an input and H2 as an input or a general purpose output. The Port B Control Register selects the submode of mode 1.

22.6.1 Mode 1 submode x0 16-bit Double Buffered Input

This is similar to mode 0 submode 00 (8-bit double buffered input, section 5 above). Ports A and B are combined into a single 16-bit port controlled by the handshaking lines H3 and H4. Although the primary direction is input the Data Direction bits set which pins are input or output (single buffered). The peripheral loads data into the latches by asserting H3 (see section 5) and H4 can be programmed for any of the options in section 5.1. To operate correctly port A contains the most significant byte of data and is read first. Port B, which contains the least significant byte, is read last and controls the handshaking.

22.6.2 Mode 1 submode x1 16-bit Double Buffered Output

This is similar to mode 0 submode 01 (8-bit double buffered output; see section 1 above). Ports A and B are combined into a single 16-bit port controlled by the handshaking lines H3 and H4. Although the primary direction is output the Data Direction bits set which pins are input or output (inputs read the instantaneous pin value). To operate correctly the most significant byte should be written to port A followed by the least significant byte to port B. The peripheral accepts data by asserting H3 and H4 can be programmed for any of the handshake options described in section 1.1 above.

Problem for Chapter 22

1. Write a program to verify that double buffered input operates as described in section 5 above (e.g. similar to Program 22.1 for double buffered output).
2. Connect two microcomputers with PITs and implement data transfer using double buffered input and output.

23

Exceptions

Chapter 18 discussed the principles of input/output programming and the advantages of using interrupts. In the case of the MC68000 microprocessor, interrupts are but one category of events called exceptions. Other exceptions are caused by TRAP instructions and various error conditions.

23.1 Processor operating modes

The MC68000 microprocessor can operate either in Supervisor Mode (the most privileged operating mode allowing the execution of any instruction) or User Mode. Normally, the operating system or monitor executes in Supervisor Mode with other system programs (i.e. the editors, compilers, assemblers, etc.) and the user programs, execute in User Mode. When the processor is running in User Mode, instructions which interfere with the overall operation of the computer cannot be used (e.g. in a multi-user environment, a user program would not be allowed to execute the RESET instruction, which would reset all the I/O device interfaces). If an attempt is made to execute such a **privileged** instruction, an exception occurs which transfers control to the operating system where an error message would be displayed.

The S bit in the processor Status Register (SR) determines the processor's current operating mode:

bit	15	14	13	12	11	10	9	8	7	6	5	4	3	2	1	0
	T		S			I	I	I				X	N	Z	V	C

The low-order byte of the SR is the CCR (Condition Code Register), which contains the X, N, Z, V and C bits. The high-order byte of the SR contains the following bits:

T **Trace Bit**, if set, the processor is in trace mode, and after the execution of each instruction an exception occurs which transfers control to the operating system or monitor. Debugging information (PC value, data and address register contents, etc.) is displayed and then program execution is resumed.

S **Supervisor Bit**, if set, the processor is in Supervisor Mode.

I **Interrupt Mask**, three bits which define the interrupt level mask (see section 7 below).

The value of the Status Register can be set explicitly by the use of the following **word** length **privileged** instructions:

```
MOVE.W    #$2000,SR     set Supervisor Mode clear other bits
ANDI.W    #$0DFFF,SR    set User Mode
ORI.W     #$2000,SR     set Supervisor Mode
EORI.W    #$2000,SR     switch processor modes
```

A full list of the source addressing modes for the above can be obtained from Appendix B. MOVE to SR can be used to set all the bits of the SR and ANDI, ORI or EORI can clear, set or invert particular bits. After one of these instructions has cleared the S bit the processor executes the following instructions in User Mode. For example, an operating system would switch the processor into User Mode immediately before starting a user program. Once operating in User Mode the processor cannot be switched back into supervisor mode except by the occurrence of an exception. The contents of the SR can be read by the following instruction (**non-privileged** on the MC68000 and **privileged** on the MC68010/20/30/40):

```
        MOVE.W    SR,D0          move contents of SR into D0
```

The MC68000 chip has three processor status output lines which indicate the current operating mode of the processor and other information. External circuits can examine these signals to ensure that certain operations can only be performed while the processor is in Supervisor Mode. User programs executing on a multi-user machine must not interfere with other programs or the operating system, therefore:

1. Critical instructions which affect the overall operation of the processor are **privileged** and may not be executed in User Mode, i.e. STOP, RESET, RTE, MOVE to/from USP, MOVE to SR, ANDI to SR, ORI to SR, or EORI to SR.
2. User programs would be restricted to accessing memory which contained only their own code and data, i.e. not allowed access to memory used by the operating system, other user programs, exception vector table, I/O device registers, etc.

When the processor makes a memory read/write request the address decoding circuits would examine the processor status lines (see Wilcox 1987 for an example circuit) and if these indicated Supervisor Mode the access would continue. Otherwise, if the processor was in User Mode, checks would be carried out to ensure that the address was within the user program address space, and, if not, a bus error exception would be generated (see section 6 below). This exception would switch the processor into Supervisor Mode where the operating system would display an error message. Note that an MC68010/20/30/40 running a virtual memory environment would use bus error exceptions when generating page faults (Tanenbaum 1987).

Low-level development systems (e.g. the single board microcomputers used to test the sample programs in this book) normally run in Supervisor Mode all the time. This enables the implementation of programs which access memory, I/O registers, etc., as required.

23.1.1 The SSP and USP stack pointers

Address register A7 is used by the MC68000 processor as the general system stack pointer. Within the MC68000 there are **two A7 address registers**, the **SSP** (System Stack Pointer) which is used when the processor is in Supervisor Mode, and the **USP** (User Stack Pointer) which is used when it is in User Mode. When executing instructions that operate upon the stack, the processor uses the appropriate version of A7 corresponding to the current mode. This means that user programs running in User Mode have their own stack pointer (A7) and stack area, and are prevented from interfering with the SSP and system stack area (which would possibly damage the operating system or other programs). To start a user program the operating system (running in Supervisor Mode) loads the USP with the address of the stack area (in the user RAM data space), switches the processor into User Mode (using an MOVE/ANDI/ORI/EORI to SR instruction) and then jumps to the start of the user's program. The following **privileged** instructions access the USP:

```
        MOVE.L    A0,USP       move A0 to user stack pointer
        MOVE.L    USP,A1       read the value of the USP
```

The operands of the MOVE to/from USP instructions are of length long word, and the other operand must be an **address register**. These instructions allow a program running in Supervisor Mode to read and write the User Mode A7 (i.e. in Supervisor Mode moves to and from A7 using normal instructions would access the SSP). If the monitor or operating system does not set up the USP (e.g. in a small single board system), the user program must do so:

```
START:    MOVE.L    #STACK,A7    set up the stack
          .....
* RAM data area
          DS.W      1000         stack area of 1000 words
STACK:
```

The MC68000 stack works from high to low addresses, therefore the stack area is defined to be in memory before the label STACK. This address is moved into A7 (which will be the SSP or USP depending upon the processor mode when the program starts execution).

Some single user operating systems, which run user programs in User Mode, provide a system call (via a TRAP instruction, see Chapters 9.4 and 16.7 and section 5 below) to switch user programs from User Mode into Supervisor Mode (e.g. to enable the writing of special I/O routines outside the operating system). Care must be taken when using such a call as it then becomes possible to crash the operating system and it is important for the user program to switch back into User Mode (by clearing the S bit in the SR) as soon as possible. In particular, remember that when the processor mode is changed A7 points to a **different stack area**.

23.2 Exception types, priorities and vector table

Exceptions can be generated internally within the processor or by external circuits sending signals to the processor.

23.2.1 Internally generated exceptions

Exceptions generated within the processor fall into three types (a) internally detected errors, (b) instruction traps and (c) the TRACE function.

Internally detected errors which cause exceptions are:

Address error. The MC68000 attempts to access a word or long word operand, or an instruction, at an odd address.

Privilege violation. An attempt has been made to execute a privileged instruction when running in User Mode (i.e. STOP, RESET, RTE, MOVE to/from USP, MOVE to SR, ANDI to SR, ORI to SR, or EORI to SR).

Illegal instruction code. An instruction has been fetched whose bit pattern does not correspond to one of the processors machine code instructions, e.g. executing `MOVE.W CCR,D0` on a 68000 would generate an illegal instruction (this can be used to see if the processor is a 68000 or 68010).

Unimplemented instruction. If bits 12 to 15 of an instruction are 1010 or 1111, these correspond to the two unimplemented instruction operation-codes. This allow special instructions (e.g. floating point arithmetic) to be emulated in software.

In the case of the **MC68020/30/40** an instruction with bits 12 to 15 equal to 1111 is implemented as a co-processor instruction. The processor checks to see if the co-processor (e.g. a 68881 floating point co-processor) is fitted :

(a) if not a 1111 unimplemented instruction code exception is generated,

or (b) if fitted the instruction is passed to the co-processor and the 68020/30/40 acts as a slave.

Instruction traps are caused by the execution of special trap instructions in the program:

TRAP #n is the standard trap instruction, where n (range 0 to 15) specifies the trap number. Programs use it to call operating system functions (Chapter 9.4 & 16.7).

TRAPV will cause an exception if the overflow (V) bit is set in the condition code register. This can be used to check for arithmetic errors in user programs.

CHK will cause an exception if the contents of a specified destination **data register** are less than zero or greater than a source operand (specified using any addressing mode except address register direct). This instruction is useful to ensure that an array index value is within the array bounds, e.g. when using address register indirect addressing modes to access arrays or records. Consider an array of ten byte values; if D0 contains the index (range 0 to 9), the bounds test would be:

```
        CHK         #9,D0       check array index bounds
```

If the value in D0 was less than 0 or greater than 9, an exception would be generated, and the operating system would display a *value range error* message.

DIVS and DIVU. If an attempt is made to divide an operand by 0, the instruction is aborted, an exception generated and an error message displayed.

The TRACE function. If the T bit in the Status Register is set (it can only be set when in Supervisor Mode) an exception will be generated after the execution of each instruction. This can be used to trace the execution path of a program.

23.2.2 Externally generated exceptions

External exceptions are caused by external circuits sending signals to the processor.

RESET. The system reset signal has been generated by external logic (e.g. Reset button) and applied as an input to the processor RESET pin (this pin is also used as an output by the RESET instruction to reset I/O devices). In general, to restart or reboot the monitor or operating system, a press switch is provided which generates this signal.

Bus error. Occurs when circuits external to the processor assert the BERR signal, for example, if a program has attempted to access memory which does not exist, or has attempted to access privileged memory when the processor is in User Mode.

Interrupt request. An interrupt request has been received from an I/O device interface.

Exercise 23.1 *(see Appendix G for sample answer)*

Write a program which will generate an address error and observe any response from the monitor or operating system.

23.2.3 Exception priorities

Certain exceptions have more importance than others (i.e. so that error conditions can be dealt with as soon as possible). The MC68000 exceptions are placed into three groups, as shown by Table 23.1, with group 0 having the highest priority and group 2 the lowest:

1 The higher priority exceptions (in group 0 and 1) indicate major trouble and instruction execution is aborted within two clock cycles.
2 The lower priority exceptions in group 1 and 2 are none fatal and the instruction is completed.

Group	Exception Cause	Processing Response
0	Reset Bus error Address error	Abort current bus cycle (within 2 clock cycles) then process exception
1	Illegal instruction Unimplemented instruction Privilege violation	Finish current bus cycle then process exception
	Trace Interrupt request	Complete current instruction then process exception
2	TRAP, TRAPV, CHK DIVS, DIVU	Instruction execution causes exception processing

Table 23.1 MC68000 exception priority grouping

Interrupts have a further seven interrupt priority levels, to enable I/O devices to be assigned an order of importance (see section 7 below).

23.2.4 The exception vector table

When an exception occurs, some mechanism is required to transfer control to an exception processing routine (which is similar in structure to a subroutine). The MC68000 uses an exception vector table in primary memory which contains the addresses of the exception processing routines. When an exception occurs, the corresponding entry (the routine address) is obtained from the vector table and placed in the program counter (i.e. transferring control to the service routine in a similar way to a subroutine call). The MC68000 exception vector table contains 256 entries (long word values) starting at memory address 0 (i.e. using the first 1024 bytes in memory). The entries in the exception vector table are either:

(a) predefined by Motorola to be used for a particular exception (e.g. entry number 6, at address $18, is used by the CHK instruction);

or (b) supplied by the I/O interface when an interrupt occurs, either directly (vectored) or by a calculation based on the interrupt priority level (autovectored interrupt).

Vector	Address	Exception
0	$000000	RESET: initial SSP (in ROM on RESET)
1	$000004	RESET: initial PC (in ROM on RESET)
2	$000008	Bus error
3	$00000C	Address error
4	$000010	Illegal instruction
5	$000014	DIVS and DIVU: division by Zero
6	$000018	CHK: out of bounds error
7	$00001C	TRAPV: overflow bit V is set
8	$000020	Privilege violation
9	$000024	Trace: T bit in SR is set
10	$000028	Unimplemented instruction: 1010
11	$00002C	Unimplemented instruction: 1111
12	$000030	reserved for future expansion
13	$000034	reserved for future expansion
14	$000038	Format error (MC68010/20/30/40)
15	$00003C	Uninitialised interrupt vector
16	$000040	reserved for future expansion
..	......	"
to 23	$00005C	"
24	$000060	Spurious interrupt, i.e. unknown cause
25	$000064	Level 1 interrupt auto-vector
26	$000068	Level 2 interrupt auto-vector
27	$00006C	Level 3 interrupt auto-vector
28	$000070	Level 4 interrupt auto-vector
29	$000074	Level 5 interrupt auto-vector
30	$000078	Level 6 interrupt auto-vector
31	$00007C	Level 7 interrupt auto-vector
32	$000080	TRAP #0 instruction
33	$000084	TRAP #1 instruction
..	......	
to 47	$0000BC	TRAP #15 instruction
48	$0000C0	reserved for future expansion
..	......	"
to 63	$0000FC	"
64	$000100	User interrupt vectors for I/O devices
..	......	"
to 255	$0003FC	"

Table 23.2 Exception vector table of the MC68000

Table 23.2 lists the exception vector table entries of the MC68000. The first 64 vector entries are predefined with the other 192 available for I/O device interrupts (information on I/O device registers, interrupt vectors, etc., can be found in the microcomputer manual). The RESET entries (0 and 1) must be in ROM (Read Only Memory) when the power is switched on (or reset button pressed) to enable RESET exception processing (see section 4 below). Other table entries may be in ROM or RAM (Read/Write Memory) (the latter will have to be initialised by the operating system or monitor after RESET processing). In some microcomputers the vector table memory area may only be accessed when the processor is in Supervisor Mode (to prevent user programs interfering with the table).

The MC68010 has additional vector entries, and saves more information onto the stack when an exception occurs.

What exactly happens when an exception occurs depends upon the cause of exception but the following is a general outline:

1 Current program execution is halted (precisely when depends upon the exception priority - see Table 23.1).
2 The processor is switched into Supervisor Mode.
3 Information is saved on the Supervisor Stack, e.g. current PC and Status Register value (full details in sections 4 to 7 below).
4 Depending upon the source of the exception the vector is:
   (a) predefined (e.g. address error is number 3);
   (b) calculated from the interrupt priority (autovectored interrupt);
   or (c) obtained from the I/O device (vectored interrupts).
5 The exception routine address (long word) is obtained from the corresponding vector table entry and moved into the PC.
6 Execution then continues from the start of the exception service routine.
7 The exception service routine *processes* the exception, e.g. displays an error message or carries out some I/O task.

If the exception was fatal (e.g. an address error) control returns to the monitor or operating system. Otherwise the exception service routine is terminated by a return from exception (**RTE**) instruction which resumes the program interrupted by the exception.

23.2.4.1 The Vector Base Register (MC68010)

In the MC68000 the exception vector table is always at the bottom of memory and cannot be moved. The MC68010/20/30/40 have a **Vector Base Register** which allows multiple exception vector tables to be positioned in memory as required. The Vector Base Register is accessed using the **privileged** Move to/from Control Register MOVEC instruction:

```
        MOVEC       D1,VBR      load Vector Bases Register
        MOVEC       VBR,A0      read Vector Base Register
```

The other operand is a data or address register and the length is 32-bits. When an exception occurs the 32-bit address in the Vector Base Register (truncated to 24-bits on the MC68010) is used as the start address of the exception vector table. On RESET the Vector Base Register is initialised to 0 and exceptions then function as in the MC68000 (i.e. the RESET vector table entries must be in ROM at addresses 0 to 7) until the Vector Base Register is loaded.

23.3 Programming an exception processing routine

Although the full detail of what happens in the hardware when the processor gets an exception is complex (see sections 4 to 7 below for an outline) a programmer's view can be much simpler. Before a particular exception can be processed the address of the service routine is moved into the corresponding exception vector table entry. For example, Program 16.4 showed the setting up a `TRAP #0` processing routine in which the address of the routine (label `TRP_0`) was moved into the `TRAP #0` table entry (number 32 at address $80). If the vector table had been in ROM a JMP would have been set up (see section 4.3 below).

Programming an I/O device to use interrupts is more complex in that the device interface has to be set up and interrupts enabled in the processor Status Register:

(a) The I/O device is initialised, e.g. see sample programs in Chapters 20, 21 and 22.
(b) The address of the interrupt service routine is moved into the appropriate exception vector table entry (the device vector may be fixed by hardware or programmable - see section 7 below).
(c) The device interrupts are enabled via the device control register.
then (d) $2000 is moved into the MC68000 Status Register to enable interrupts of any priority level (see section 7 below).

When an interrupt occurs the service routine will be called, the interrupt processed (e.g. data read), and the routine terminated with an RTE (Return from Exception) instruction.

The above explanation assumed that user programs are able to set up entries in the exception vector table. Some machines may need to be switched from User Mode to Supervisor Mode to do this (see software manuals). Multi-user systems generally run user programs in user mode where they are not allowed to access I/O registers and the exception vector table (so that a single user cannot crash the whole machine, which is easy to do by making errors in vector table processing).

The exception processing routines are written in a similar way to program subroutines, keeping the following guidelines in mind:

1 TRAP instructions enable communication between user programs and the operating system (see Chapters 9.4 and 16.7). Data is generally passed via registers or on the stack (remember User and Supervisor modes use different stacks).
2 An interrupt may occur at any time when the processor may be executing the user program, library routines, operating system functions, etc. Therefore information should only be passed between a main program and an interrupt service routine via named global variables in memory (i.e. **not** in registers).
3 Apart from note 1 above no assumptions can be made about register contents. In particular, registers used within the exception service routine should be saved on entry and restored on exit (the CCR is automatically saved and restored).
4 On entry to an interrupt service routine, remove the cause of the interrupt as soon as possible by clearing interrupt flags or status bits, e.g. to allow the device to accept the next byte of data.
5 Make the routine as short as possible; only carry out tasks which must be done within the routine, do anything else in the main program (a common fault is to write information to the screen from within an interrupt service routine which can upset I/O device timing).

23.4 Vector table addressing and RESET processing

As mentioned in the previous section at least part of the exception vector table must be in ROM to enable processing of the RESET exception. However, if the operating system, monitor or user programs (if allowed) are to set up exception vector table entries at run time (e.g. to process interrupts from an I/O device) the relevant parts of the exception vector have to be in RAM, or some other means provided to access these facilities. Consider:

1 A washing machine controller would probably have its exception vector table in ROM, i.e. there will never be any need to reprogram any of the entries (to upgrade the program the whole ROM will be replaced).
2 The operating system of an end user workstation will need to be able to reprogram exception vector table entries as different I/O drivers are loaded or even different operating systems are used. Hence the table will need to be in RAM.

The following subsection describes techniques which enable the programming of exceptions at run time (see Wilcox 1987 for further discussion and sample circuits).

23.4.1 Separate decoding of RESET entries

The RESET vector table entries (addresses 0 to 7) are permanently decoded to address the first eight bytes of ROM (the remainder of which is mapped elsewhere in the memory map) and the RAM effectively starts at address 8. For example, from Fig. 4.3 it would appear that the RAM on the Force board occupies addresses $000000 to $01FFFF and the ROM addresses $020000 to $02FFFF. This is not quite true, the first eight bytes of ROM, $020000 to $020007, are also mapped to addresses 0 to 7 and the first eight bytes of RAM are unused.

23.4.2 Memory bank switching on RESET

An alternative to mapping the first eight bytes of ROM permanently to addressees 0 to 7 is to map them only when needed, i.e. when they are being read by the processor during RESET exception processing. For example, on power up or when the RESET button is pressed the Bytronic 68000 board responds as follows:

1 the RAM addressing is disabled;
2 the ROM is mapped from address 0;
3 this situation is maintained during four *address strobes* while the processor reads the first four words (eight bytes) of ROM to obtain the RESET exception vector table entries;
4 after four address strobes the memory map of Fig. 4.3 is enabled:
    (a) the RAM is mapped to start at address 0, i.e. $000000 to $0FFFFF
    and (b) the ROM is mapped to appear at its *normal* place in the memory map, i.e. $100000 to $11FFFF.

The initial value of the PC (in addresses 4 to 7 on RESET) would be the address of the start of the monitor or bootstrap loader within the ROM when it is decoded to its *normal* position (i.e. starting at $100000 in stage 4 above).

23.4.3 ROM at the bottom of the memory map

Both the above techniques require some address decoding and manipulation logic which, although only a few chips, would significantly increase the price of a low cost system. For example, the Kaycomp 68000 single board computer (Coats 1985/86) was designed for the home computer market and low end educational use. To achieve its low price the decoding and addressing logic was reduced to a minimum. The ROM is permanently mapped from address $000000 to 00FFFF and the RAM from $400000 to 40FFFF. However, the M68 monitor (and KAYBUG) does provide facilities on the Kaycomp to allow programs to implement exception processing routines. Particular exception vector table entries hold pointers to six byte blocks of RAM within the system monitor work area. When a program wishes to use a particular exception a JMP instruction, using *long absolute* addressing mode, is inserted into the corresponding six byte block followed by the absolute address of the exception processing routine. When an exception occurs the sequence is:

1 the vector is obtained (e.g. from the I/O device);
2 the contents of the vector table are loaded into the PC, i.e. the address of the JMP;
3 the processor executes the JMP instruction;
4 the processor starts executing the exception processing routine.

The vector table entries available on the KAYCOMP and their JMP addresses are:

```
Autovector Level 1  $400016     Trap #0  $400040     Vector 64  $400082
Autovector Level 2  $40001C     Trap #1  $400046     Vector 65  $400088
Autovector Level 3  $400022     Trap #2  $40004C     Vector 66  $40008E
Autovector Level 4  $400028     Trap #3  $400052     Vector 67  $400094
Autovector Level 5  $40002E     Trap #4  $400058     Vector 68  $40009A
Autovector Level 6  $400034     Trap #5  $40005E     Vector 69  $4000A0
Autovector Level 7  $40003A     Trap #6  $400064     Vector 70  $4000A6
                                Trap #7  $40006A     Vector 71  $4000AC
                                Trap #8  $400070
                                Trap #9  $400076
                                Trap #10 $40007C
```

Although this allows the use of the more commonly used vectors it is still very restrictive, e.g. a program is unable to process its own address errors, etc. (in practice it would be fairly easy to reprogram the ROMs to extend the tables).

23.4.4 RESET exception processing

Table 23.2 shows that the RESET exception vector has two vector table entries at positions 0 and 1 (memory addresses 0 and 4):

Vector	Address	Contents
0	0	The initial value of the Supervisor Stack Pointer SSP
1	4	Initial PC, start address of RESET exception processing routine

RESET is used to start the operating system or monitor and nothing can be assumed about the state of the processor and RAM memory. Therefore these two vector table entries and the associated exception processing routine **must be in ROM on RESET**, otherwise there is no means of starting the monitor or bootstrapping the operating system. The RESET exception processing routine will usually check out the computer hardware (carry out memory tests, etc.), and then start the ROM based monitor, or bootstrap the operating system off disk. After an external RESET the current machine cycle is aborted, then:

1 The S-bit in the Status Register is set to 1 to place the processor in Supervisor Mode.
2 The T-bit is reset to 0 to disable any Trace.
3 The three interrupt mask bits in the SR are set to 1, which sets the initial interrupt priority level to 7 (i.e. no interrupts except level 7 are allowed; see section 7 below).
4 The SSP (Supervisor Stack Pointer) is loaded with exception vector table entry 0 (memory addresses 0 to 3).
5 The PC is loaded with exception vector table entry 1 (memory addresses 4 to 7).
6 Instruction execution then starts at the address in the PC (which will be the system check-out and monitor start-up routine or bootstrap).

The vector table entries to process the other Motorola defined exceptions (bus error, address error, divide by 0, etc.) will either be in ROM or set up in RAM during RESET exception processing (possibly by the operating system after bootstrapping). In addition any table entries required to process TRAP functions and interrupt driven I/O will be set up at this time.

23.5 Internal exceptions

Exceptions due to TRAP, Trace, TRAPV, DIVS, DIVU, CHK, illegal or unimplemented instruction or a privilege violation are processed as follows:

1 The SR (Status Register) is saved to an internal register.
2 The S-Bit in the SR is set to 1.
3 The T-bit in the SR is reset to 0.
4 The PC is pushed onto the (Supervisor) stack .
5 The saved SR is pushed onto the stack.
6 The PC is loaded from the appropriate vector table entry.
7 Instruction execution resumes at the new PC.

Thus, after switching into Supervisor Mode the PC and Status Register are saved onto the Stack and control transferred to the exception service routine which will:

(a) TRAP: perform some monitor or operating system function (see Chapter 9.4) and terminate with an RTE (resuming interrupted program),
(b) TRACE: display program trace information and terminate with an RTE,
or (c) display an error message and return control to monitor or operating system, i.e. TRAPV, DIVS, DIVU, CHK, illegal or unimplemented instruction or privilege violation.

23.5.1 MC68010 internal exceptions

In addition to the PC and SR the MC68010/20/30/40 push an additional format/offset word onto the stack prior to the PC (i.e. before step 4 above). This format/offset word is composed of two parts:

bits 0 to 11 vector offset: offset from the start of the exception vector table, i.e. vector number * 4

bits 11 to 15 format code: indicates the number of words pushed - in this case 0000 for 4 words, i.e. format/offset (1 word) plus PC (two words) plus SR (1 word)

23.6 Bus and address error exception processing

In response to an internal address error or external bus error the current instruction cycle is aborted and:

1 The SR (Status Register) is saved to an internal register.
2 The S-bit in the SR is set to 1 to switch into Supervisor Mode.
3 The T-bit is reset to 0 to disable Trace.
4 The PC and the saved value of the SR are pushed onto the Stack (Supervisor Stack).
5 The contents of the IR (Instruction Register) are pushed onto the Stack. This would be the first word of the instruction being executed when the error occurred.
6 The 32-bit address which caused the exception is pushed onto the Stack.
7 A word that provides information on the type of processor cycle in progress is pushed onto the stack (details below):
   bits 0, 1 & 2: processor function code outputs
   bit 3: is 0 - instruction or group 2 exception in progress
   or 1 - group 0 or 1 exception processing
   bit 4: is 0 - write cycle aborted, or 1 - read cycle aborted.
8 The PC is loaded from the vector table entry.
9 Instruction execution resumes at the new PC.

The information pushed onto the stack enables the monitor or operating system to determine the source of the error. A message would then indicate the PC, instruction and memory address being accessed at the time of the error. The word pushed at step 8 above, which provides information on the type of cycle, can be used to determine (a) if a normal program instruction or exception processing was in progress, and (b) if a read or write cycle was in progress. If the error occurred during the processing of a previous bus error, address error or RESET exception something is very wrong, probably a hardware fault, and the MC68000 will enter the HALT state and remain there. The value of the PC could be within an instruction, depending upon when the error occurred during instruction execution. Such an error is usually fatal and it is impossible to resume program execution (the MC68010 saves additional information onto the stack to enable recovery from bus errors when running a virtual memory environment - see next subsection).

23.6.1 MC68010 Bus and Address error

When an MC68010/20/30/40 bus or address error exception occurs:

1 26 words of information (see reference Motorola 1989) representing the internal state of the processor are pushed onto the Supervisor Stack. The information enables the handling of page faults in a virtual memory environment (Tanenbaum 1987).
2 A format/offset word is pushed onto the stack:
   (a) bits 0 to 11 vector offset: offset from the start of the exception vector table, i.e. vector number * 4
   (b) bits 11 to 15 format code: indicates the number of words pushed - in this case 1000 for 29 words.
4 The PC is pushed onto the Stack.
5 The saved value of the SR is pushed onto the Stack.

Exercise 23.2 *(see Appendix G for sample answer)*

Write a program which will generate an address error, then when exception processing starts generate another address error (this will cause the processor to halt; note the Force and Bytronic boards have a halt LED to show this condition.

23.7 Interrupt exception processing

Interrupts have eight levels of priority, from the lowest 0, to the highest 7. Bits 8, 9 and 10 of the Status Register specify the current value of the interrupt mask. An I/O device requests an interrupt by encoding an interrupt request level on three processor interrupt input lines (IPL0, IPL1 & IPL2). The processor compares this interrupt level with the value of the interrupt mask in the SR:

(a) If the request level is less than or equal to the interrupt mask the interrupt request will be ignored (it could be processed later if the mask is changed).

or (b) If the request level is higher than the interrupt mask, the interrupt request will be serviced (Priority level 7 is a special case in that it is always serviced, and cannot be inhibited by the interrupt mask, i.e. making it a *non-maskable* interrupt facility for critical applications).

The interrupt servicing sequence is:

1 The SR is saved into an internal register.
2 The S-bit in the SR is set to 1.
3 The T-bit in the SR is reset to 0.
4 The interrupt mask bits in the SR are changed to the level of the interrupt request, i.e. allowing higher priority devices to interrupt the servicing of this interrupt, while lower or equal priority devices are ignored until this one has been serviced.
5 The processor then performs an interrupt acknowledgment cycle:
    (a) the processor informs the I/O interface that the interrupt will now be serviced;
  then (b) the interface responds with:
      (i) the **vectored response** and supplies the vector number,
    or (ii) the **autovectored response** and the processor calculates the vector number from the priority level (i.e. vector numbers 25 to 31.
6 **MC68010/20/30/40 only:** the format/offset word is pushed onto the stack (see section 5.1 above)
7 The PC is pushed onto the stack (Supervisor).
8 The saved value of the SR is pushed onto the stack.

Then, if the I/O device interface gave the vectored response and **supplied the vector number**:

9 The PC is loaded with four bytes from the exception vector table as defined by the vector number (e.g. if the vector number supplied was 64 decimal the four bytes starting at address 100 hexadecimal will be loaded into the PC).
10 Instruction execution resumes at the new PC.

Otherwise, if the I/O interface gave the **autovectored** response:

9 The PC is loaded with four bytes from the autovector entry in the exception vector table which corresponds to the interrupt request level from the interface (e.g. if the request level was 3 the four bytes starting at address 6C hexadecimal will be loaded into the PC).
10 Instruction execution resumes at the new PC.

The interrupt processing routine would then be executed and, after servicing the request, would be terminated with an RTE instruction (see next section). The structure of the stack and the setting of the interrupt priority mask in the SR to the level of the interrupting device enables interrupt servicing in order of priority. In general, high speed devices (e.g. a hard disk) would be assigned a higher priority than lower speed devices (e.g. a serial interface connected to a user's terminal).

The vectored and autovectored responses from the interface to the granting of the interrupt request, allow interfaces of different complexity to be connected to the bus. Device interfaces that can supply the exception vector byte are more complex than those which use the autovectored response. In particular, the autovectored response may be used by device chips designed for the MC6800 family of 8-bit processors. In a simple system with few devices autovectored response may be used for all I/O requests. When using autovectored interrupts, if two or more devices have the same priority, the first job of the interrupt service routine is to find out which device interface requested the interrupt. This can be done by examining the status bits in the status registers of the interfaces.

If, during the interrupt acknowledge cycle (when the processor is granting the interrupt request) no device responds, external circuits assert the bus error signal. In this case the exception is not treated as a bus error, but as a *spurious interrupt* and table entry 24 is loaded into the PC. This would generally indicate a design or hardware fault.

A number of MC68000 family peripheral interface circuits allow the interrupt vector table entry value to be loaded into a vector register (within the interface) by the program (see the Port Interrupt Vector Register (PIVR) of the MC68230 PIT in Chapter 21.1). If the vector register has not been set up in this way, and the device interrupts the processor, the interface will provide vector 15 ($0F), the uninitialised interrupt vector.

23.8 The Return from Exception (RTE) instruction

After exceptions such as RESET, address error, privilege violation, etc. (which are fatal run time errors on the 68000), it is impossible to restart the interrupted program. Other exceptions can be processed (e.g. process the I/O device interrupt or display an error message) and then program execution can be resumed. In such cases the last instruction of an exception processing routine should be the RTE instruction that will resume execution of the program that was interrupted by the exception. The action of RTE is:

1 The top word of the stack is pushed into the Status Register.
2 The next long word from the top of the stack is pushed into the PC.

Thus RTE restores the original SR and PC (which were saved onto the supervisor stack during exception processing). Hence the processor mode, interrupt mask, etc., are reset to that before the exception occurred, allowing user programs to resume in User Mode.

23.8.1 MC68010 RTE

The RTE instruction on the MC68010/20/30/40 processors is more complex in that the amount of information pushed by exceptions varies, i.e. the majority of exceptions push four words but bus and address errors push twenty nine words. The action of RTE on MC68010/20/30/40 processors is:

1 The top word of the stack is pushed into the Status Register.
2 The next long word from the top of the stack is pushed into the PC.
3 The next word read from the stack is the format/offset word.

Bits 12-15 of the format/offset word should be either 0000 (4 words stacked) or 1000 (29 words stacked), anything else will cause a *format error* exception (see Table 23.2 vector entry 14). If the stack frame was four words the RTE is complete and execution continues at the new program counter address. If the stack frame was twenty nine words the next twenty five are read off the stack and processed (see reference Motorola 1989 for full details).

23.9 The STOP instruction

The **privileged instruction STOP**, has an immediate operand:

```
        STOP        #xxx
```

When the instruction is executed the immediate operand is moved into the Status Register (SR), the PC is advanced to point to the next instruction, and then the processor stops fetching and executing instructions. Instruction execution will resume when a reset, trace, or interrupt exception occurs (the interrupt priority must be higher than the value of the interrupt mask for the interrupt to have an effect). This instruction allows a program to set the processor mode and interrupt priority mask, and then suspend execution until an exception occurs (e.g. if the program is waiting for input from an I/O device).

23.10 Saving registers during exception processing

When an exception request is granted any registers used by the service routine must be saved on entry and restored on exit (e.g. by using MOVEM in a similar way to subroutines). There is no need to save the condition codes as this is done automatically by the exception processing.

Problem for Chapter 23

Using the hardware manual for your microcomputer determine what priorities and exception vector table entries are used by the I/O device interfaces connected to the system.

24

Interrupt I/O with the MC6821 PIA & MC68230 PIT

Program 20.2 (for the MC6821 PIA) and 21.2 (for the MC68230 PIT) polled status bits to check if a switch had been pressed (to reverse the rotation of the LEDs). Polled input/output programming is suitable for simple applications but not for complex programs with time critical I/O. By using interrupts an I/O device can immediately gain the attention of the processor by interrupting the current program and transferring control to an interrupt service routine which deals with the I/O request. After the I/O request has been processed the interrupted program is resumed. The interrupt service routine can be considered as a subroutine which is called automatically on interrupt, and, on execution of the return from exception (RTE) instruction, control is returned to the interrupted program.

24.1 Interrupt I/O with the MC6821 PIA (Motorola 1985a)

The microcomputer hardware manual will indicate which exception vectors are used by the PIA and if vectored or autovectored interrupts are used. The Force microcomputer (Force 1984) used to test the sample programs in this section used the autovectored interrupt response (section 2 below presents an example of vectored response), with port A having a priority of 2 (vector number 26 at address $68) and port B a priority of 3 (vector 27 at $6C). The sample programs may therefore have to be modified to run on other machines.

Program 24.1 (next page) is a modification of Program 20.2. The program uses the sample circuit of Appendix F.3 to display a rotating pattern on the LEDs attached to the port B data lines and blinks the LED attached to the CA2 or CB2 line. When the switch attached to CA1 is pressed, the rotation of the pattern displayed on the LEDs is reversed and CA2 or CB2 change over. Program 20.2 polled bit 7 the Port A Control Register to detect a switch hit. Program 24.1 enables interrupts on the CA1 line and detects a switch hit via an interrupt service routine, note:

1. The PIA is initialised.
2. The exception vector table entry is loaded with the address of the service routine.
3. CA1 interrupts are enabled in the PIA Port A Control register
4. The 68000 Status Register is set to enable system wide interrupts.
5. The interrupt service routine can make no assumptions about the state of processor registers, therefore:
   (a) The main program accesses the PIA using *address register with displacement addressing mode* with A0 holding the base address.
   (b) The interrupt service routine cannot assume that A0 is intact and must either reload A0 (or another register) or use *absolute addressing mode* (which is used in the sample program).
   (c) Information is passed between the main program and the interrupt service routine via the *global variable DIRECTION*, i.e. **NOT** a processor register.

```
---------------------------------------------------------------------------------
 1                             * Program 24.1 - MC6821 PIA Program interrupts
 2                             *   Program using a PIA to display a rotating LED pattern.
 3                             *    Pattern reverses when switch attached to CA1 is hit.
 4                             *
 5                             * PIA set port B for output, CA2 & CB2 as output lines
 6                             * set up interrupt routine address in exception table
 7                             * PIA enable interrupts on CA1 (switch)
 8                             * WRITE('hit CA1 switch to reverse LED rotation')
 9                             * set test_pattern = 1
10                             * set up MC68000 Status Register to enable interrupts
11                             * LOOP FOR ever
12                             *     output test_pattern to port B (to LEDs)
13                             *     IF direction = left
14                             *          rotate test_pattern one bit left
15                             *          blink LED attached to CA2
16                             *     ELSE
17                             *          rotate test_pattern one bit right
18                             *          blink LED attached to CB2
19                             *     delay for a time period
20                             * END LOOP
21                             *
22          0005CEF1           PIA:      EQU      $5CEF1      PIA base address Force board
23          00000000           ADATA:    EQU      0           Data Register A
24          00000002           ACTRL:    EQU      2           Control Register A
25          00000004           BDATA:    EQU      4           Data Register B
26          00000004           BDDR:     EQU      4           Data Direction Register B
27          00000006           BCTRL:    EQU      6           Control Register B
28          00000068           AVECT:    EQU      $68         Port A interrupt vector
29                             *
30                                       XREF     WRLINE,WRTEXT,WRHEXB,DELAY
31 001000 4EB9 00028038       START:    JSR      WRTEXT
32 001006 68 69 74 20 43                DC.W     'hit CA1 switch to reverse LED rotation$'
33 00102E 207C 0005CEF1                 MOVEA.L  #PIA,A0          base address of PIA
34 001034 117C 0000 0002                MOVE.B   #0,ACTRL(A0)     clear A Control Reg
35 00103A 117C 0000 0006                MOVE.B   #0,BCTRL(A0)     clear B Control Reg
36 001040 117C 00FF 0004                MOVE.B   #$0FF,BDDR(A0)   port B all bits output
37 001046 117C 0034 0002                MOVE.B   #$34,ACTRL(A0)   access data registers &
38 00104C 117C 0034 0006                MOVE.B   #$34,BCTRL(A0)   CA2 & CB2 for output
39 001052 1628      0000                MOVE.B   ADATA(A0),D3     clear ACTRL bit 7 if set
40 001056 21FC 00001092                 MOVE.L   #INT_SRV,AVECT   set up exception table
                 0068
41 00105E 08E8 0000 0002                BSET.B   #0,ACTRL(A0)     enable CA1 interrupts
42 001064 7201                          MOVEQ    #1,D1            initial test_pattern
43 001066 46FC      2000                MOVE.W   #$2000,SR        enable interrupts in SR
44                             *
45                             * LOOP rotating left or right
46 00106A 1141      0004       LOOP:     MOVE.B   D1,BDATA(A0)  output test_pattern to LEDs
47 00106E 4A39 000010A4                 TST.B    DIRECTION     direction ?
48 001074 660A                          BNE.S    RIGHT
49 001076 E319                          ROL.B    #1,D1         rotate test_pattern left
50 001078 0868 0003 0002                BCHG.B   #3,ACTRL(A0)  invert CA2 to blink LED
51 00107E 6008                          BRA.S    WAIT
52 001080 E219                RIGHT:    ROR.B    #1,D1         rotate test_pattern right
53 001082 0868 0003 0006                BCHG.B   #3,BCTRL(A0)  invert CB2 to blink LED
54 001088 4EB9 0002809A       WAIT:     JSR      DELAY         wait for a delay time
55 00108E 0064                          DC.W     100           100 milliseconds
56 001090 60D8                          BRA.S    LOOP
```

**** continued on next page ****

Program 24.1a MC6821 PIA interrupts, main program (Force board)

```
------------------------------------------------------------------------------------
57                           *
58                           * PIA interrupt service routine
59                           * save registers
60                           * read A Data Register to clear bit 7 of Control Register
61                           * invert DIRECTION
62                           * restore registers
63                           * RETURN from exception
64                           *
65                           INT_SRV:
66 001092 2F00                         MOVE.L   D0,-(A7)
67 001094 1039 0005CEF1                MOVE.B   PIA+ADATA,D0  read A data to clear bit 7
68 00109A 4639 000010A4                NOT.B    DIRECTION     and invert direction
69 0010A0 201F                         MOVE.L   (A7)+,D0
70 0010A2 4E73                         RTE
71                           *
72                           * Global data area
73 0010A4 00                 DIRECTION: DC.B    0             rotation direction = left
74                                     END
------------------------------------------------------------------------------------
```

Program 24.1b MC6821 PIA interrupts - interrupt service routine (Force board)

The sequence of events in Program 24.1 is:

33 loads the base address of the PIA into A0

34 clears the Port A Control Register

35 clears the port B Control Register

36 all bits of port B are set for output

37-38 `MOVE.B #$34,ACTRL(A0)` sets up the port A (and B) Control Register:
- (a) bit 0 is cleared to 0 disabling CA1/CB1 interrupts
- (b) bit 1 is cleared to 0, so that bit 7 will be set on a high to low transition of CA1
- (c) bit 2 is set to 1 to address the Data Register.
- (d) bit 5 is set to 1 to make CA2 & CB2 output lines
- (e) bit 4 is set to 1 to output the value of bit 3 on CA2 & CB2
- (f) bit 3 is cleared to 0 to output a 0 initially on CA2 & CB2

39 read the port A Data Register to clear Bit 7 of the Control Register (if set)

40 `MOVE.L #INT_SRV,AVECT` moves the address (long word) of the interrupt service routine (INT_SRV) into the exception vector address (AVECT)

41 `BSET.B #0,ACTRL(A0)` sets bit 0 of the port A Control Register to enable interrupts on the CA1 input line (from the switches)

42 moves the initial bit pattern into D1 (to be rotated)

43 `MOVE.W #$2000,SR` enables interrupts in the MC68000 Status Register

46 the bit pattern is moved to the port B Data Register (to display on the LEDs)

47-48 IF the direction is left
- 49 the pattern is rotated left
- 50 the CA2 output line is inverted to blink the yellow LED

ELSE
- 52 the pattern is rotated right
- 53 the CB2 output line is inverted to blink the green LED

54-55 DELAY for time period

56 loop back to line 46 to rotate again

When an interrupt occurs the service routine INT_SRV is called:

66 save data register D0
67 `MOVE.B PIA+ADATA,D0` reads the port A Data Register to clear bit 7 of the Control Register (note the use of *absolute addressing mode*)
68 inverts DIRECTION
69 restores data register D0
70 RTE (Return from Exception) instruction resumes the interrupted program

If, in line 67, the Data Register was not read to clear Control Register bit 7 the PIA would interrupt immediately after the RTE instruction. If a program using interrupts appears to 'lock up' check that interrupt service routines are clearing the cause of any interrupts.

24.2 Interrupt I/O with the MC68230 PIT (Motorola 1983a)

Reference to the microcomputer hardware manual will indicate which exception vectors are used by the PIT and if vectored or autovectored interrupts are used. This section will use the Bytronic and Kaycomp 68000 boards with sample programs:

The Bytronic board uses vectored interrupts with the exception vector table in RAM.

The Kaycomp Board is a low cost system:

1 The exception vector table is in ROM with interrupt service routines accessed via JMP instructions in RAM (see Chapter 23.4.3).
2 MC68000 microcomputers generally have a decoder which translates the seven possible interrupt priority levels to signals to drive the MC68000 IPL0, IPL1 and IPL2 interrupt control lines (Wilcox 1987, Clements 1987). The Kaycomp does not have such circuits and the interrupt request signals from I/O devices are connected directly to IPL0, IPL1 and IPL2. The Kaycomp board used to test the sample programs in this section was wired as follows:
   (a) The MC68681 DUART interrupt request signal (IRQ) was connected to IPL1 and used vectored interrupts.
   (b) The MC68230 PIT timer interrupt request (TOUT) was connected directly to IPL0 and used level 1 autovectored interrupts.
   (c) The MC68230 PIT parallel interface interrupt request (PIRQ) was connected to IPL2 and used level 4 autovectored interrupts.

   This configuration causes problems when both PIT interrupts are enabled and a level 5 autovectored interrupt occurs (i.e. the timer and parallel interface interrupt at the same time). In such a case the software would have to take account of the level 5 interrupt.

Programs 24.2 (Bytronic board - next page) and 24.3 (Kaycomp board - end of chapter) are modifications of Program 21.2. The programs use the circuit of Appendix F.3 to display a rotating pattern on the port B data lines and blink the LED attached to the H2 or H4 line. When the switch attached to H1 is pressed, the rotation of the pattern displayed on the LEDs is reversed and H2 or H4 change over. Programs 24.2 and 24.3 use interrupts to detect a switch hit (Program 21.2 polled the H1S status bit in the Port Status Register).

```
--------------------------------------------------------------------------------
 1                        * Program 24.2 - MC68230 PIT interrupts BYTRONIC board
 2                        *  Program using a PIT to display a rotating LED pattern.
 3                        *   Pattern reverses when switch attached to H1 is hit.
 4                        *
 5                        * initialise PIT: mode 0 bit I/O, port B output
 6                        * set up Port Service Request Reg
 7                        *  - PC5/PIRQ pin carries PIRQ & PC6/PIACK carries PIACK
 8                        * set up interrupt service routine address in vector table
 9                        * set up A Control Reg to enable interrupts on H1 (switch)
10                        * WRITE('hit H1 switch to reverse LED rotation')
11                        * set test_pattern = 1
12                        * set up MC68000 Status Register to enable interrupts
13                        * LOOP FOR ever
14                        *     output test_pattern to port B (to LEDs)
15                        *     IF direction = left
16                        *          rotate test_pattern one bit left
17                        *          blink LED attached to H2
18                        *     ELSE
19                        *         rotate test_pattern one bit right
20                        *         blink LED attached to H4
21                        *     delay for a time period
22                        * END LOOP
23                        *
24        00130001        PIT:      EQU       $130001    PIT base address Bytronic board
25        00000000        PGCR:     EQU       0          port general control register
26        00000002        PSRR:     EQU       2          port service request register
27        00000006        PBDDR:    EQU       $06        data direction register B
28        0000000A        PIVR:     EQU       $0A        port interrupt vector reg
29        0000000C        PACR:     EQU       $0C        control register A
30        0000000E        PBCR:     EQU       $0E        control register B
31        00000012        PBDR:     EQU       $12        data register B
32        0000001A        PSR:      EQU       $1A        port status register
33        00000044        P_VECT:   EQU       68         use vector 68 for PIT
34                        *
35                                  XREF      WRLINE,WRTEXT,WRHEXB,DELAY
36 001000 4EB9 00108038   START:    JSR       WRTEXT
37 001006 68 69 74 20 48            DC.W      'hit H1 switch to reverse LED rotation$'
38 00102C 207C 00130001             MOVEA.L   #PIT,A0           set PIT base address
39 001032 117C 0000 0000            MOVE.B    #0,PGCR(A0)       PGCR set mode 0
40                        * set up PSRR - PC5/PIRQ & PC6/PIACK  carry PIRQ & PIACK
41 001038 117C 0018 0002            MOVE.B    #$18,PSRR(A0)     set up PSRR
42 00103E 117C 00A0 000C            MOVE.B    #$0A0,PACR(A0)    A submode 1x, H2 output
43 001044 117C 00A0 000E            MOVE.B    #$0A0,PBCR(A0)    B submode 1x, H4 output
44 00104A 117C 00FF 0006            MOVE.B    #$0FF,PBDDR(A0)   port B all lines output
45 001050 117C 000F 001A            MOVE.B    #$0F,PSR(A0)      reset flags (if set)
46 001056 117C 0044 000A            MOVE.B    #P_VECT,PIVR(A0)  set up vector register
47 00105C 21FC 0000109E             MOVE.L    #P_INT,P_VECT*4   service routine address
                   0110
48 001064 08E8 0001 000C            BSET.B    #1,PACR(A0)       enable H1 interrupts
49 00106A 08E8 0004 0000            BSET.B    #4,PGCR(A0)       PGCR enable H12
50 001070 7201                      MOVEQ     #1,D1             initial test_pattern
51 001072 46FC      2000            MOVE.W    #$2000,SR         enable system interrupts

                         **** continued on next page ****
--------------------------------------------------------------------------------
```

Program 24.2a MC68230 PIT interrupts, main program (Bytronic board)

```
-------------------------------------------------------------------------------
52                           *
53                           * LOOP rotating left or right
54 001076 1141      0012     LOOP:      MOVE.B    D1,PBDR(A0)      test_pattern to LEDs
55 00107A 4A39 000010AE                 TST.B     DIRECTION        direction ?
56 001080 660A                          BNE.S     RIGHT
57 001082 E319                          ROL.B     #1,D1            rotate test_pattern left
58 001084 0868 0003 000C                BCHG.B    #3,PACR(A0)      invert H2 to blink LED
59 00108A 6008                          BRA.S     WAIT
60 00108C E219               RIGHT:     ROR.B     #1,D1            test_pattern right
61 00108E 0868 0003 000E                BCHG.B    #3,PBCR(A0)      invert H4 to blink LED
62 001094 4EB9 0010809A      WAIT:      JSR       DELAY            wait for delay time
63 00109A 0064                          DC.W      100              100 milliseconds
64 00109C 60D8                          BRA.S     LOOP             rotate again
65                           *
66                           * PIT parallel interface interrupt service routine
67                           * it is assumed that only H1 interrupts are enabled
68                           *
69                           * wait for switch bounce to finish (if required)
70                           * reset H1S bit in PSR using direct method of resetting
71                           * invert DIRECTION
72                           * Return from Exception (resume main program)
73                           *
74                           P_INT:
75                           *          JSR       DELAY              wait for switch bounce
76                           *          DC.W      500
77 00109E 13FC      0001                MOVE.B    #1,PIT+PSR         reset H1S status bit
                0013001B
78 0010A6 4639 000010AE                 NOT.B     DIRECTION          reverse direction
79 0010AC 4E73                          RTE
80                           *
81                           * Global data area
82 0010AE 00                 DIRECTION: DC.B      0          rotation direction = left
83                                      END
-------------------------------------------------------------------------------
```

Program 24.2b MC68230 PIT main program and interrupt routine (Bytronic board)

In general, the outline program sequence when using interrupts with the MC68230 PIT is:

1 The required mode is set in the Port General Control Register.
2 The interrupt mechanism (vectored or autovectored) is set in the Port Service Request Register (plus any priorities, etc.)
3 The submode is selected in the Port A and B Control Registers.
4 The Data Direction Registers are set up.
5 Either:
   (a) Bytronic board: the Port Interrupt Vector Register (PIVR) is loaded with the vector number and the corresponding exception vector table entry is loaded with the address of the service routine,
   or (b) Kaycomp board: a JMP instruction (with the address of the interrupt service routine) is set up in RAM.
6 Enable interrupts in the Port A and/or Port B Control Registers.
7 Enable H12 and/or H34 handshaking in the Port General Control Register (mode as 1).
8 The 68000 Status Register is set to enable system wide interrupts.

Note that the interrupt service routine can make no assumptions about the state of processor registers, therefore in Programs 24.2 and 24.3:

(a) The main program accesses the PIT using *address register with displacement addressing mode* and A0 contains the base address.
(b) The interrupt service routine cannot assume that A0 is intact and must either reload A0 (or another register) or use *absolute addressing mode.*
(c) Information is passed between the main program and the interrupt service routine via the *global variable DIRECTION*, i.e. **NOT** a processor register.

In Program 24.2 (for the Bytronic board) the sequence of events is:

line
38 `MOVEA.L #PIT,A0`, loads the base address of the PIT into A0
39 the Port General Control Register (PGCR) is set to Mode 0 (all other bits to 0)
40 `MOVE.B #$18,PSRR(A0)` sets the Port Service Request Register so that the PC5/PIRQ pin carries PIRQ parallel interface interrupt request and PC6/PIACK carries PIACK parallel interface interrupt acknowledge (vectored interrupts)
42-43 `MOVE.B #$0A0,PACR(A0)` sets up the port A (and B) Port Control Register:
(a) bits 6 and 7 are set to 10; to select submode 1x
(b) bits 5, 4 and 3 are set to 100; H2 (& H4) are output lines always negated
(c) all other bits are 0 to disable interrupts
44 all bits of port B are set for output
45 resets the status flags in the Port Status Register (if set from a previous program)
46 `MOVE.B #P_VECT,PIVR(A0)` moves exception vector number (to be used by the parallel interface) into the Port Interrupt Vector Register (PIVR)
47 `MOVE.L #P_INT,P_VECT*4` moves the address (long word) of the interrupt service routine (P_INT) into the exception vector table (address P_VECT*4)
48 `BSET.B #1,PACR(A0)` enables H1 interrupts in the Port A Control Register
49 H12 Enable is set in the Port General Control Register (Mode 0 is still selected)
50 moves the initial bit pattern into D1 (to be rotated)
51 `MOVE.W #$2000,SR` enables interrupts in the MC68000 Status Register

54 the bit pattern is moved to the port B Data Register (to display on the LEDs)
55-56 IF the direction is left
57 the pattern is rotated left
58 PACR bit 3 is inverted (H2 output line) to blink the yellow LED
ELSE
60 the pattern is rotated right
61 PBCR bit 3 is inverted (H4 output line) to blink the green LED
62-63 DELAY for time period
64 loop back to line 54 to rotate again

When an interrupt occurs the service routine P_INT is called:

75-76 a delay to remove switch bounce (not required in this case)
77 reset the H1S bit in the Port Status Register by the *direct method of resetting*
78 invert DIRECTION
79 RTE (Return from Exception) instruction resumes the interrupted program

Notes:

1 If line 77 did not reset the H1S bit in the Port Status Register the PIT would interrupt immediately after the RTE instruction. If a program using interrupts appears to 'lock up' check that interrupt service routines are clearing the causes of interrupts.

2 If line 41 did not set up the Port Service Request Register (PSRR) so that the PC6/PIACK pin carried PIACK a *spurious interrupt would be generated (number 24).*

3 If line 46, `MOVE.B #P_VECT,PIVR(A0)`, did not set up the exception vector number for the parallel interface the *uninitialised interrupt vector (number 15)* would be generated when an interrupt occurred.

In Program 24.2 (for the Bytronic board) the address of the interrupt service routine could be loaded directly into the exception vector table in RAM. The Kaycomp board has its exception vector table in ROM with JMP instructions in RAM used to access interrupt service routines (see Chapter 23.4.3). The sequence of events in Program 24.3 (next page), a modified version of Program 24.2 for the Kaycomp board, is:

41 `MOVEA.L #PIT,A0` loads the base address of the PIT into A0
42 the Port General Control Register (PGCR) is set to Mode 0 (all other bits to 0)
43 `MOVE.B #$08,PSRR(A0)` set up the Port Service Request Register (PSRR) so that the PC5/PIRQ pin carries PIRQ parallel interface interrupt request (autovectored interrupts)
45-46 `MOVE.B #$0A0,PACR(A0)` sets up the port A (and B) Port Control Register:
(a) bits 6 and 7 are set to 10; to select submode 1x
(b) bits 5, 4 and 3 are set to 100; H2 & H4 are output lines always negated
(c) all other bits are 0 to disable interrupts
47 all bits of port B are set for output
48 resets the status flags in the Port Status Register (if set from a previous program)
49 `MOVE.W JUMP,P_VECT` moves a JMP instruction into the RAM memory location corresponding to a autovectored level 4 interrupt, i.e. address $4000028
50 `MOVE.L JUMP+2,P_VECT+2` moves the absolute address of the interrupt service routine into memory following the JMP, i.e. address $400028 + 2
51 H1 interrupts are enabled in the Port A Control Register (PACR)
52 H12 Enable is set in the Port General Control Register (Mode 0 is still selected)
53 moves the initial bit pattern into D1 (to be rotated)
54 `MOVE.W #$2000,SR` enables interrupts in the MC68000 Status Register

57 the bit pattern is displayed on the LEDs (via port B)
58-59 IF the direction is left
    60 the pattern is rotated left
    61 PACR bit 3 is inverted (H2 output line) to blink the yellow LED
ELSE
    63 the pattern is rotated right
    64 PBCR bit 3 is inverted (H4 output line) to blink the green LED
65-66 DELAY for time period
67 loop back to line 57 to rotate again
68 JMP instruction to the interrupt service routine (used in lines 49 & 50)

When an interrupt occurs the service routine P_INT is called:

79-80 a delay to remove switch bounce (not required in this case)
81 reset the H1S bit in the Port Status Register by the *direct method of resetting*
82 invert DIRECTION
83 RTE (Return from Exception) instruction resumes the interrupted program

The major differences between Program 24.2 and 24.3 are:

1 The Port Service Request Register (PSRR) is set up for vectored interrupts in Program 24.2 and autovectored interrupts in Program 24.3.
2 The exception vector number to be used by the parallel interface is loaded into the Port Interrupt vector Register (PIVR) in Program 24.2 (not required in Program 24.3).
3 The address of the interrupt service routine is loaded into the exception vector table (in RAM) in Program 24.2 and a JMP instruction is set up in Program 24.3.

Problem for Chapter 24

Extend the PWM motor speed control program implemented for the Problem of Chapter 20 or 21 to sense the motor revolutions using interrupts (see Appendix F.7.1, F.7.2 & F.7.3).

```
---------------------------------------------------------------------------------
 1                      * Program 24.3 - MC68230 PIT interrupts on the Kaycomp board
 2                      *   Program using a PIT to display a rotating LED pattern.
 3                      *     Pattern reverses when switch attached to H1is hit.
 4                      *
 5                      * It is assumed that the MC68230 PC5/PIRQ pin 35 is linked
 6                      *   to the MC68000 IPL2 pin 23, i.e. the  parallel interface
 7                      *   will generate a level 4 autovectored request
 8                      *
 9                      * initialise PIT: mode 0 bit I/O, port B output
10                      * set up Port Service Request Reg - PC5/PIRQ pin carries PIRQ
11                      * set up JMP instruction & nterrupt service routine address
12                      * set up A Control Reg to enable interrupts on H1 (switch)
13                      * WRITE('hit H1 switch to reverse LED rotation')
14                      * set test_pattern = 1
15                      * set up MC68000 Status Register to enable interrupts
16                      * LOOP FOR ever
17                      *     output test_pattern to port B (to LEDs)
18                      *     IF direction = left
19                      *          rotate test_pattern one bit left
20                      *          blink LED attached to H2
21                      *     ELSE
22                      *         rotate test_pattern one bit right
23                      *         blink LED attached to H4
24                      *     delay for a time period
25                      * END LOOP
26                      *
27       00A00001       PIT:      EQU       $A00001   PIT base address Kaycomp board
28       00000000       PGCR:     EQU       0         port general control register
29       00000002       PSRR:     EQU       2         port service request register
30       00000006       PBDDR:    EQU       $06       data direction register B
31       0000000C       PACR:     EQU       $0C       control register A
32       0000000E       PBCR:     EQU       $0E       control register B
33       00000012       PBDR:     EQU       $12       data register B
34       0000001A       PSR:      EQU       $1A       port status register
35                      * interrupt vector JMP address for KAYCOMP board
36       00400028       P_VECT    EQU       $400028   autovectored interrupt level 4
```

**** continued on next page ****

Program 24.3a MC68230 PIT interrupts, program heading (Kaycomp board)

```
------------------------------------------------------------------------------------
37                           *
38                                     XREF       WRLINE,WRTEXT,WRHEXB,DELAY
39 400400 4EB9 00008038  START:        JSR        WRTEXT
40 400406 68 69 74 20 48               DC.W       'hit H1 switch to reverse LED rotation$'
41 40042C 207C 00A00001                MOVEA.L    #PIT,A0            set PIT base address
42 400432 117C 0000 0000               MOVE.B     #0,PGCR(A0)        PGCR set mode 0
43                           * set up Port Service Request Reg - PC5/PIRQ pin carries PIRQ
44 400438 117C 0008 0002               MOVE.B     #$08,PSRR(A0)      set up PSRR
45 40043E 117C 00A0 000C               MOVE.B     #$0A0,PACR(A0)     A submode 1x, H2 output
46 400444 117C 00A0 000E               MOVE.B     #$0A0,PBCR(A0)     B submode 1x, H4 output
47 40044A 117C 00FF 0006               MOVE.B     #$0FF,PBDDR(A0)    port B all lines output
48 400450 117C 000F 001A               MOVE.B     #$0F,PSR(A0)       reset flags (if set)
49 400456 33F9 004004A4                MOVE.W     JUMP,P_VECT        copy JMP instruction &
               00400028
50 400460 23F9 004004A6                MOVE.L     JUMP+2,P_VECT+2    service routine address
               0040002A
51 40046A 08E8 0001 000C               BSET.B     #1,PACR(A0)        enable H1 interrupts
52 400470 08E8 0004 0000               BSET.B     #4,PGCR(A0)        PGCR enable H12
53 400476 7201                         MOVEQ      #1,D1              initial test_pattern
54 400478 46FC      2000               MOVE.W     #$2000,SR          enable system interrupts
55                           *
56                           * LOOP rotating left or right
57 40047C 1141      0012     LOOP:     MOVE.B     D1,PBDR(A0)        test_pattern to LEDs
58 400480 4A39 004004BA                TST.B      DIRECTION          direction ?
59 400486 660A                         BNE.S      RIGHT
60 400488 E319                         ROL.B      #1,D1              rotate test_pattern left
61 40048A 0868 0003 000C               BCHG.B     #3,PACR(A0)        invert H2 to blink LED
62 400490 6008                         BRA.S      WAIT
63 400492 E219               RIGHT:    ROR.B      #1,D1              test_pattern right
64 400494 0868 0003 000E               BCHG.B     #3,PBCR(A0)        invert H4 to blink LED
65 40049A 4EB9 0000809A      WAIT:     JSR        DELAY              wait for delay time
66 4004A0 0064                         DC.W       100                100 milliseconds
67 4004A2 60D8                         BRA.S      LOOP               rotate again
68 4004A4 4EF9 004004AA      JUMP:     JMP        P_INT              a JMP instruction
69                           *
70                           * parallel interface interrupt service routine
71                           * it is assumed that only H1 interrupts are enabled
72                           *
73                           * wait for switch bounce to finish (if required)
74                           * reset H1S bit in PSR using direct method of resetting
75                           * invert DIRECTION
76                           * Return from Exception (resume main program)
77                           *
78                           P_INT:
79                           *         JSR        DELAY               wait for switch bounce
80                           *         DC.W       500
81 4004AA 13FC      0001               MOVE.B     #1,PIT+PSR          reset H1S status bit
                 00A0001B
82 4004B2 4639 004004BA                NOT.B      DIRECTION           reverse direction
83 4004B8 4E73                         RTE
84                           *
85                           * Global data area
86 4004BA 00                 DIRECTION: DC.B      0            rotation direction = left
87                                     END
------------------------------------------------------------------------------------
```

Program 24.3b MC68230 PIT interrupts, main program & interrupt routine (Kaycomp)

25

The MC6850 ACIA

The MC6850 ACIA (Asynchronous Communications Interface Adaptor, see Motorola 1985b) is a UART (Universal Asynchronous Receiver Transmitter) designed for use with the Motorola range of microprocessors. The ACIA may be used for connecting computer systems to a printers, terminals, other computers, etc. using serial communications techniques (Chapter 19.2 introduced serial communications). Associated with the ACIA are a number of lines:

TxD and RxD are the serial transmit and receive data lines.

$\overline{\text{RTS}}$ (Request To Send - output): allows the device to control a peripheral or modem:

(a) when used for **transmit handshaking** $\overline{\text{RTS}}$ would be asserted to indicated that the ACIA is ready to transmit data;

or (b) when used for **receive handshaking** $\overline{\text{RTS}}$ would be asserted to indicate that the ACIA can accept data.

$\overline{\text{CTS}}$ (Clear To Send - input): is used for **transmit handshaking**, i.e. is asserted by the external device in response to $\overline{\text{RTS}}$ to indicate that transmission may proceed.

$\overline{\text{DCD}}$ (Data Carrier Detect - input): is used for **receive handshaking**, i.e. a modem would assert this signal to indicate that the remote connection is complete and valid signals being received; a low to high transition would then indicate loss of connection.

The TxD and RxD lines enable the transmission and receiving of serial data on two independent communications channels simultaneously (i.e. full duplex). Generally $\overline{\text{CTS}}$ is used for transmit handshaking and $\overline{\text{RTS}}$ and $\overline{\text{DCD}}$ for receive handshaking (or other functions depending upon the application). $\overline{\text{RTS}}$, $\overline{\text{CTS}}$ and $\overline{\text{DCD}}$ are all active low (i.e. if the signal is asserted the line is at logical 0, +6 volt, and if negated at logical 1, -6 volt).

The sample programs in this chapter use polled and interrupt driven I/O to a terminal or host computer running a terminal emulator program with $\overline{\text{RTS}}$ and $\overline{\text{CTS}}$ used as receive and transmit handshaking lines respectively.

IBM PC			ACIA
pin	Signal name	connections	Signal name
1	ground	————————	ground
2	TxD data	→ ———————— →	RxD data
3	RxD data	← ———————— ←	TxD data
4	$\overline{\text{RTS}}$ output	→ ———————— →	$\overline{\text{CTS}}$ TxD handshake
5	$\overline{\text{CTS}}$ input	← ———————— ←	$\overline{\text{RTS}}$ RxD handshake
6	$\overline{\text{DSR}}$ input	← ——— (linked to 5)	
8	$\overline{\text{DCD}}$ input	← ——— (linked to 5)	
7	data common	————————	data common

Fig. 25.1 Typical connection of an ACIA to an IBM/PC compatible host computer

Fig. 25.1 shows a typical connection of an IBM/PC compatible host computer (which would be running a terminal emulator program) to an MC68000 via an MC6850 ACIA (pin numbers are for an RS232C 25-way D type connector):

(a) the IBM/PC host can control the transmission of characters from the 68000 board by negating and asserting its $\overline{RTS}$ output signal ($\overline{CTS}$ input to the ACIA);

(b) the 68000 board can control transmission of characters from the host by negating and asserting the ACIA $\overline{RTS}$ output signal ($\overline{CTS}/\overline{DSR}/\overline{DCD}$ input to host).

and (c) the ACIA $\overline{DCD}$ input should be connected to a low (logical 0).

In many cases the handshaking lines are not used (i.e. the START bit indicates the arrival of serial data, unlike a parallel interface where handshaking has to be used to indicate when valid data is on the lines). In such applications $\overline{CTS}$ and $\overline{DCD}$ should be connected to a logical 0 signal otherwise the ACIA will not work correctly (the $\overline{RTS}$ output line is set low or is ignored by the external device). Examination of the microcomputer hardware manuals should determine which handshaking signals are used and for what purpose.

The MC6850 ACIA serial interface contains four registers, two read only and two write only, which are accessed via a pair of memory locations (see Fig. 18.3).

Data Register:

Transmit: on a memory write, data is placed in the ACIA transmit data register and is then transmitted down the serial line.

Receive: on a memory read, the ACIA receive data register is read (this contains the last data received from the serial line).

Control/Status Register:

Control: on a memory write data is written to the ACIA control register.

Status: on a memory read the contents of the ACIA status register are read.

The register pairs are accessed via a pair of memory addresses. The sample programs in this chapter used for Force 68000 board (Force 1984) which used the following addresses:

Transmit/Receive Data Register: $50042
Control/Status Register: $50040

Note that:

1 the transmit (write only) and receive (read only) data registers share memory address $50042,

2 the control register (write only) and status register (read only) share memory address $50040.

Thus it is not possible to:

(a) read the transmit data register or the control register;

or (b) write to the receive data register or the status register.

Instructions which perform memory-to-memory operations cannot be used. If the program requires a note of the control register settings a copy should be made into a local variable each time it is set.

The format of the ACIA **Status Register** (read only) is:

7	6	5	4	3	2	1	0
IRQ	PE error	OVRN error	FE error	$\overline{\text{CTS}}$	$\overline{\text{DCD}}$	XMIT empty	RCV full

bit **0** **RCV Full** indicates the state of the receiver buffer
- 0 the receive data register is empty - a high signal on $\overline{\text{DCD}}$ will also cause this bit to be low (e.g. the data carrier signal has been lost); if not used $\overline{\text{DCD}}$ should be connected to logical 0
- 1 the receive data register contains data which can be read by the program (the IRQ bit is also set)

bit **1** **XMIT empty** indicates the state of the transmitter buffer
- 0 the transmit data register contains data being transmitted - a high signal on $\overline{\text{CTS}}$ will also cause this bit to be low even if the transmit data register is empty, i.e. the modem or printer is busy and cannot accept data
- 1 the transmit data register is empty and the ACIA is ready to transmit the next data byte (the IRQ bit is also set)

bit **2** **$\overline{\text{DCD}}$ bit** indicates the state of the $\overline{\text{DCD}}$ (Data Carrier Detect) line
- 0 the $\overline{\text{DCD}}$ line is at logical 0, i.e. the connection is complete and the modem is receiving the data carrier
- 1 the $\overline{\text{DCD}}$ line is at logical 1, i.e. the modem has lost the data carrier - the IRQ bit is also set and the RCV Full Bit is held low; if not used $\overline{\text{DCD}}$ should be connected to logical 0

bit **3** **$\overline{\text{CTS}}$ bit** indicates the state of the $\overline{\text{CTS}}$ (Clear To Send) line
- 0 the $\overline{\text{CTS}}$ line is 0, i.e. the external device can accept data
- 1 the $\overline{\text{CTS}}$ line is at logical 1 - the XMIT Empty Bit is also held at 0 (i.e. the modem or printer is busy and cannot accept data); if not used the $\overline{\text{CTS}}$ line should be connected to logical 0

bit **4** **FE (Framing Error)**
- 0 no error
- 1 received data was not properly framed by the required START and STOP bits, i.e. a STOP bit was 0 (when it should be a 1); this may be due to noise on the line corrupting the data signals or incorrect setting of the data characteristics (baud rate, number of data bits, number of stop bits or parity)

bit **5** **OVRN (Receiver Overrun)**
- 0 no error
- 1 indicates that data has been lost - it is set if more data is received while the receiver data register already contains data (i.e. the program has not read the last character received)

bit **6** **PE (Parity Error)**
- 0 no error
- 1 a parity error has been detected in received data, e.g. due to a noisy communications line

bit **7** **IRQ** interrupt request
- 0 no interrupt request pending
- 1 when there is an interrupt request pending (interrupts must be enabled via the Control Register for the interrupt to be passed to the processor) - IRQ is cleared by reading the receive data register or writing to the transmit data register

The format of the **Control Register** (write only) is:

7	6	5	4	3	2	1	0
RCV ctrl	XMIT ctrl		word size select			counter divide value	

bit **7** **RCV ctrl** receiver interrupt control
- 0 receiver interrupts disabled
- 1 receiver interrupts enabled, an interrupt request will be issued when:
  - (a) the receiver data register is full, i.e. a character has been received and the **RCV Full** bit in the Status Register is set,
  - or (b) on a low to high transition of the $\overline{\text{DCD}}$ input, i.e. the modem has lost the data carrier.

bits **6 5** **XMIT ctrl** transmit interrupt and $\overline{\text{RTS}}$ (Request To Send) line control
- 0 0 set $\overline{\text{RTS}}$ low, disable transmitter interrupt
- 0 1 set $\overline{\text{RTS}}$ low, enable transmitter interrupt
- 1 0 set $\overline{\text{RTS}}$ high, disable transmitter interrupt
- 1 1 set $\overline{\text{RTS}}$ low, disable transmitter interrupt and transmit a BREAK signal (logical 0)

Note:

1. $\overline{\text{RTS}}$ (Request to Send) can be used as the transmit or receive handshaking output with bits 5 and 6 controlling the $\overline{\text{RTS}}$ line and enabling/disabling the transmitter interrupt.
2. BREAK is a special signal where the transmit line is held at logical 0 (+6 volt). This is used to abort programs on some computer systems.
3. If the transmitter interrupt is enabled, an interrupt request will be sent to the processor when the **XMIT empty** bit sets, i.e. transmit data register empty and ready for the next byte.

bits **4 3 2** **word size select** sets the number of data bits, parity and number of stop bits
- 0 0 0 7 data bits + even parity + 2 stop bits
- 0 0 1 7 data bits + odd parity + 2 stop bits
- 0 1 0 7 data bits + even parity + 1 stop bits
- 0 1 1 7 data bits + odd parity + 1 stop bits
- 1 0 0 8 data bits + 2 stop bits
- 1 0 1 8 data bits + 1 stop bits
- 1 1 0 8 data bits + even parity + 1 stop bits
- 1 1 1 8 data bits + odd parity + 1 stop bits

bits **1 0 counter divide value** controls Master Reset (see note 2 below) and the baud rate:
- 0 0 divide clock by 1
- 0 1 divide clock by 16
- 1 0 divide clock by 64
- 1 1 Master Reset

Note:

1. An external clock is used to set the baud rate. The divisor subdivides the clock to set the actual baud rate (refer to the microcomputer manual for the clock speed).
2. Master Reset: before the ACIA is used bits 0 and 1 of the control register should be set to 1 to reset the device. This clears any byte already in the receiver data register.
3. Data bit 0 is transmitted first followed by bit 1, bit 2, etc.
4. The parity is transmitted after the data bits.
5. If 7-bit data bits plus parity mode is being used:
   (a) on transmit bit 7 is ignored;
   and (b) on receive bit 7 is set to 0.

Receive (data register full) and transmit (data register empty) interrupts can be enabled separately using bits 7 and 5 respectively (bit 6 cleared). If an interrupt occurs the interrupt priority mask in the MC68000 Status Register must be set to the appropriate level for it to cause an exception. If both receive and transmit interrupts are enabled, the cause of an interrupt can be found by examination of the ACIA status register. When using transmit interrupts bit 5 will have to be cleared by the program after all data has been sent.

25.1 Using the MC6850 ACIA for polled I/O

To transmit (or receive) characters using the ACIA with polled I/O techniques is simply a matter of checking the appropriate bits in the ACIA status register. For example, Program 25.1 (below) outputs a character string to a serial line which is linked to a terminal or host PC computer running a terminal emulator (it is assumed that the $\overline{CTS}$ and $\overline{RTS}$ handshaking lines are connected correctly). The program loops transmitting the character string TEXT. On each loop a check is made to see if a character has been received, if so, the character is read and replaces a character in the string TEXT.

```
--------------------------------------------------------------------------------
1                    * Program 25.1 - MC6850 ACIA polled input/output
2                    *  (a) continuously write text string to display screen
3                    *  (b) on keyboard hit replace characters in string
4                    *
5                    * initialise ACIA
6                    * set write/read pointers to array TEXT
7                    * LOOP for ever
8                    *     get next character from array TEXT
9                    *     IF at end of array TEXT
10                   *         reset pointer to start of TEXT & get first character
11                   *     WRITE character to serial line
12                   *     IF character in receiver buffer
13                   *        READ character from serial line
14                   *        move character into array TEXT
15                   * ENDLOOP
--------------------------------------------------------------------------------
```

Program 25.1a MC6850 ACIA polled input/output - program description

```
--------------------------------------------------------------------------------
16                          *
17                          * Define ACIA memory mapped addresses for FORCE 68000 board
18          00050040        ACIA:      EQU      $50040        base address of ACIA
19          00050040        CONTROL:   EQU      ACIA          ACIA Control Register
20          00050040        STATUS:    EQU      ACIA          ACIA Status Register
21          00050042        DATA:      EQU      ACIA+2        ACIA Data Register
22                                     XREF     DELAY
23 001000 4EB9 0002809A START:     JSR      DELAY            wait for ports to clear
24 001006 0064                     DC.W     100
25                          * set up ACIA Control Register with master RESET, then
26                          * divide by 16, 8 data bits, no parity, 1 stop bit
27 001008 13FC      0003           MOVE.B   #$3,CONTROL      master RESET to ACIA
              00050040
28 001010 13FC      0015           MOVE.B   #$15,CONTROL     set up Control Register
              00050040
29 001018 227C 0000106E            MOVEA.L  #TEXT,A1         set input text pointer
30 00101E 207C 0000106E SET_A0:    MOVEA.L  #TEXT,A0         set output text pointer
31 001024 1018          LOOP:      MOVE.B   (A0)+,D0         get next Tx character
32 001026 67F6                     BEQ.S    SET_A0           if 0 reset pointer
33 001028 4EB9 0000105A            JSR      Tx_CHAR          write character
34 00102E 6100      0020           BSR      Rx_CHECK         keyboard hit ?
35 001032 67F0                     BEQ.S    LOOP
36 001034 6100      0006           BSR      Rx_CHAR          yes, read character
37 001038 12C0                     MOVE.B   D0,(A1)+         put in text table
38 00103A 60E8                     BRA.S    LOOP
39                          *
40                          * subroutine to read character from serial line into D0
41                          *     LOOP
42                          *     UNTIL receiver ready
43                          *     READ character from ACIA
44                          Rx_CHAR:
45 00103C 0839      0000           BTST     #0,STATUS        character in Rx buffer ?
              00050040
46 001044 6700      FFF6           BEQ      Rx_CHAR          no, wait
47 001048 1039 00050042            MOVE.B   DATA,D0          yes, read character
48 00104E 4E75                     RTS
49                          *
50                          * subroutine clear Z bit if character is in receiver buffer
51                          *     TEST receiver ready bit in ACIA status register
52                          Rx_CHECK:
53 001050 0839      0000           BTST     #0,STATUS        character in Rx buffer ?
              00050040
54 001058 4E75                     RTS
55                          *
56                          * subroutine to write character in  D0 to serial line
57                          *     LOOP
58                          *     UNTIL transmitter ready
59                          *     WRITE character to ACIA
60                          Tx_CHAR:
61 00105A 0839      0001           BTST.B   #1,STATUS        transmitter ready ?
              00050040
62 001062 6700      FFF6           BEQ      Tx_CHAR          no, wait
63 001066 13C0 00050042            MOVE.B   D0,DATA          yes, transmit character
64 00106C 4E75                     RTS
65                          * TEXT data area
66 00106E 30 31 32 33   TEXT:      DC.B     '01234567890 abcdefghijklmnopqrstuvwxyz '
67 0010A0 0A 0D 00                 DC.B     10,13,0
68                                 END
--------------------------------------------------------------------------------
```

Program 25.1   MC6850 ACIA polled input/output - program code

The sequence of events in Program 25.1 is:

18-21	define ACIA base address for the Force board and then the I/O registers
23-24	delay to complete any outstanding I/O to port (e.g. from the monitor)
27	Master Reset to reset the ACIA (by setting bits 0 and 1 of the Control Register)
28	set up the clock divisor to 16, data bits to 8, no parity and 1 STOP bit
29-30	initialise input and output pointers to TEXT buffer
31-32	get next output character; if 0 goto line 30 to reset the output pointer
33	call subroutine Tx_CHAR to write the character to the display screen
34-35	call Rx_CHECK to see if a character as been received, if not goto line 31
36	call subroutine Rx_CHAR to read the character received
37	replace the next character in the TEXT buffer with the character read

The subroutine Rx_CHAR reads a character from the serial line:

45-46	loop until the **RCV full** bit in the Status Register is set, i.e. character received
47	read the character from the receiver Data Register into D0

The subroutine Rx_CHECK clears the Z bit if a character has been received:

53	test the **RCV full** bit in the Status Register

The subroutine Tx_CHAR transmits a character:

61-62	loops until the **XMIT empty** bit in the Status Register is set, i.e. ready to transmit
63	moves the character to the transmitter Data Register

25.2 Using the MC6850 ACIA for interrupt driven I/O

In Program 25.1 the receiver full bit (RCV full) in the Status Register was polled on every loop of the main program. In Program 25.2 (next page) the main program loop transmits characters and the receiver is interrupt driven. The program sequence is:

28	set the clock divisor to 16, data bits to 8, no parity, 1 STOP bit and enables receiver interrupts (i.e. the **RCV ctrl** bit is set)
29	load the exception vector table with the address of the interrupt service routine (the ACIA used autovector level 5)
30	enable system wide interrupts via MC68000 Status Register
31-35	transmit the string TEXT using polled I/O by calling subroutine Tx_CHAR

Receiver interrupts have been enabled (the RCV ctrl bit in the Control Register is set in line 28) and when an interrupt occurs the service routine is called:

52	get the input text pointer (from global variable TEXT_PNT) into A0
53	read the Receiver Buffer (clears the RCV Full bit and interrupt)
54	save the input text pointer to global variable TEXT_PNT

Problem for Chapter 25

Modify Program 25.2 to use transmitter and receiver interrupts, note:

1. Both receive and transmit interrupts will have to be enabled in the Control Register.
2. When an interrupt occurs the service routine will have to test the Status Register to determine whether the receiver or transmitter was the cause.

```
-----------------------------------------------------------------------------
 1                           * Program 25.2 - MC6850 ACIA interrupt input/output
 2                           *  (a) continuously write text string to display screen
 3                           *  (b) on keyboard hit replace characters in string
 4                           *
 5                           * initialise ACIA with receiver interrupts enabled
 6                           * set MC68000 Status Register to enable interrupts
 7                           * set write pointer to array TEXT
 8                           * LOOP for ever
 9                           *     get next character from array TEXT
10                           *     IF at end of array TEXT
11                           *         reset pointer to start of TEXT & get first character
12                           *     WRITE character to serial line
13                           * ENDLOOP
14                           *
15                           * Define ACIA memory mapped addresses for FORCE 68000 board
16        00050040           ACIA:      EQU       $50040         base address of ACIA
17        00050040           CONTROL:   EQU       ACIA           ACIA Control Register
18        00050040           STATUS:    EQU       ACIA           ACIA Status Register
19        00050042           DATA:      EQU       ACIA+2         ACIA Data Register
20        00000074           VECTOR:    EQU       $74            ACIA vector address
21                                      XREF      DELAY
22 001000 4EB9 0002809A START:      JSR       DELAY            wait for ports to clear
23 001006 0064                          DC.W      100
24                           * set up ACIA Control Register with master RESET, then
25                           * divide by 16, 8 data bits, no parity, 1 stop bit
26                           *  and receiver interrupts enabled
27 001008 13FC      0003                MOVE.B    #$3,CONTROL      master RESET to ACIA
                00050040
28 001010 13FC      0095                MOVE.B    #$95,CONTROL     set up Control Register
                00050040
29 001018 21FC 0000104A                MOVE.L    #A_INT,VECTOR    set up exception vector
                    0074
30 001020 46FC      2000                MOVE.W    #$2000,SR        enable system interrupts
31 001024 207C 0000106A SET_A0:     MOVEA.L   #TEXT,A0         set output text pointer
32 00102A 1018          LOOP:       MOVE.B    (A0)+,D0         get next Tx character
33 00102C 67F6                          BEQ.S     SET_A0           if 0 reset pointer
34 00102E 4EB9 00001036                JSR       Tx_CHAR          write character
35 001034 60F4                          BRA.S     LOOP
36                           *
37                           * subroutine to write character in  D0 to port B serial line
38                           *     LOOP
39                           *     UNTIL transmitter ready
40                           *     WRITE character to ACIA
41                           Tx_CHAR:
42 001036 0839      0001                BTST.B    #1,STATUS        transmitter ready ?
                00050040
43 00103E 6700      FFF6                BEQ       Tx_CHAR          no, wait
44 001042 13C0 00050042                MOVE.B    D0,DATA          yes, transmit character
45 001048 4E75                          RTS
46                           *
47                           * ACIA interrupt service routine, port A receiver only
48                           *    GET pointer to input TEXT buffer
49                           *    READ character from serial line into TEXT buffer
50                           *    SAVE pointer to input TEXT buffer
51 00104A 48E7      0080 A_INT:      MOVEM.L   A0,-(A7)
52 00104E 2079 00001066                MOVEA.L   TEXT_PNT,A0        get TEXT pointer
53 001054 10F9 00050042                MOVE.B    DATA,(A0)+         read character
54 00105A 23C8 00001066                MOVE.L    A0,TEXT_PNT        save TEXT pointer
55 001060 4CDF      0100                MOVEM.L   (A7)+,A0
56 001064 4E73                          RTE
57                           *
58 001066 0000106A          TEXT_PNT: DC.L      TEXT               input text pointer
59                           * data area
60 00106A 30 31 32 33       TEXT:       DC.B      '01234567890 abcdefghijklmnopqrstuvwxyz '
61 00109C 0A 0D 00                      DC.B      10,13,0
62                                      END
-----------------------------------------------------------------------------
```

Program 25.2 MC6850 ACIA interrupt driven input/output

26

The MC68681 DUART

The MC68681 DUART (Dual Asynchronous Receiver/Transmitter) is a member of the MC68000 family of peripheral interfaces and contains two independent full-duplex asynchronous receiver/transmitter channels called A and B respectively (the receivers are quadruple buffered and the transmitters double buffered). In addition the device contains:

1 a 6-bit parallel input port (input lines are called IP0 to IP5);
2 an 8-bit parallel output port (output lines are called OP0 to OP7);
3 a 16-bit programmable counter/timer.

The MC68681 DUART is very versatile (Motorola 1985c, Motorola 1984a, Motorola 1987) and this chapter will only introduce some of its functions and its use in simple polled and interrupt I/O applications (see Motorola 1985c for full details).

Associated with the serial ports are a number of signals including (Chapter 19.2 introduced serial communications):

TxDA/RxDA and TxDB/RxDB are the serial transmit and receive data lines for ports A and B (the receivers are quadruple buffered and the transmitters are double buffered).

$\overline{\text{RTSA}}$ and $\overline{\text{RTSB}}$ (Request To Send - output). The parallel output pins OP0 and OP1 can be used as the Request to Send handshaking output signals for ports A and B respectively. When selected for this function (by the program) the signals are automatically negated and asserted by the receiver or transmitter. For example, when used for **receive handshaking** $\overline{\text{RTS}}$ would be asserted to indicate that the DUART can accept data (this output line would be attached to the $\overline{\text{CTS}}$ input of the external device).

$\overline{\text{CTSA}}$ and $\overline{\text{CTSB}}$ (Clear To Send - input). The parallel input pins IP0 and IP1 can be used as the Clear to Send handshaking input signals for ports A and B respectively. When selected for this function (by the program) the transmitter checks the status of this signal before attempting to transmit (this input line would be connected to the $\overline{\text{RTS}}$ output of the external device).

The TxD and RxD signals enable the transmission and receiving of serial data on two independent communications channels simultaneously (i.e. full duplex). $\overline{\text{RTS}}$ and $\overline{\text{CTS}}$ are generally used for receive and transmit handshaking respectively and are active low (i.e. if the signal is asserted the line is at logical 0, +6 volt, and if negated at logical 1, -6 volt). The other parallel I/O lines can be used for other RS232C functions ($\overline{\text{DCD}}$, $\overline{\text{DSR}}$, etc.) or as general purpose I/O lines.

The sample programs in this chapter use polled and interrupt driven I/O to a terminal or host computer running a terminal emulator program with $\overline{\text{RTS}}$ and $\overline{\text{CTS}}$ used as receive and transmit handshaking lines respectively.

Fig. 26.1 shows a typical connection of an IBM/PC compatible host computer (which would be running a terminal emulator program) to an MC68000 via an MC68681 DUART (pin numbers are for an RS232C 25-way D type connector):

(a) the IBM/PC host can control the transmission of characters from the 68000 board by negating and asserting its $\overline{RTS}$ output signal ($\overline{CTS}$ input to the DUART);

and (b) the 68000 system can control transmission of characters from the host by negating and asserting the DUART $\overline{RTS}$ output signal ($\overline{CTS}/\overline{DSR}/\overline{DCD}$ input to host).

IBM PC			DUART
pin	Signal name	connections	Signal name
1	ground	———————	ground
2	TxD data	→ ——————— →	RxD data
3	RxD data	← ——————— ←	TxD data
4	$\overline{RTS}$ output	→ ——————— →	$\overline{CTS}$ TxD handshake
5	$\overline{CTS}$ input	← ——————— ←	$\overline{RTS}$ RxD handshake
6	$\overline{DSR}$ input	← ———┘ (linked to pin 5)	
8	$\overline{DCD}$ input	← ———┘ (linked to pin 5)	
7	data common	———————	data common

Fig. 26.1 Typical connected to attach a DUART to an IBM/PC compatible host computer

In many cases the handshaking lines are not used (i.e. the START bit indicates the arrival of serial data, unlike a parallel interface where handshaking has to be used to indicate when valid data is on the lines) and IP0/OP0 and IP1/OP1 can be used as general purpose parallel I/O pins under program control.

26.1 MC68681 DUART Registers

The MC68681 DUART contains 26 registers (see Motorola 1985c for full details). The information is necessarily complex and only functions relevant to the polled and interrupt driven I/O programs presented in this chapter will be described.

offset	MC68681 register	program name
$00	Mode Register port A	D_MRA
$02	Status Register port A	D_SRA
$02	Clock Select Register port A	D_CSA
$04	Command Register port A	D_CRA
$06	Receiver Buffer port A	D_RBA
$06	Transmitter Buffer port A	D_TBA
$08	Auxiliary Control Register	D_ACR
$0A	Interrupt Mask register	D_IMR
$10	Mode Register port B	D_MRB
$12	Status Register port B	D_SRB
$12	Clock Select Register port B	D_CSB
$14	Command Register port B	D_CRB
$16	Receiver Buffer port B	D_RBB
$16	Transmitter Buffer port B	D_TBB
$18	Interrupt Vector Register	D_IVR

Table 26.1 Memory mapped register offsets for the MC68681 DUART
Base addresses are $120001 for the Bytronic board and $800001 for the Kaycomp

Table 26.1 lists the MC68681 DUART registers which will be used in this chapter, their associated memory address offsets and the register names used in sample programs. To simplify address decoding it is assumed that alternate (odd) addresses are used to access the DUART registers.

Associated with each port of the MC68681 are two Mode Registers which are accessed via a single address (D_MRA for port A and D_MRB for port B in Table 26.1). The required Mode Register is selected using an internal pointer which is set up using the following procedure:

1 A RESET (hardware or instruction) or a *reset pointer* command to the corresponding Control Register (see below) will reset the internal pointer to Mode Register 1.
2 The required control bits are written to Mode Register 1.
3 The write to Mode Register 1 automatically moves the pointer to Mode Register 2.
4 The required control bits are written to Mode Register 2.
5 Further accesses (read or write) will access Mode Register 2.

A RESET (hardware reset or RESET instruction) carries out the following:

(a) sets the Mode Registers pointer to Mode Register 1;
(b) clears the Status Registers, the Interrupt Mask Register, the Interrupt Status Register, the Output Port Register and the Output Port Configuration Register;
(c) sets the Interrupt Vector Register to $0F (the uninitialised vector);
(d) sets the Auxiliary Command Register to timer mode and clears IP0 to IP3 interrupt enables;
and (e) disables the receivers and transmitters, clearing all buffers.

Mode Register 1 (MR1A) for port A (MR1B for port B is similar)

7	6	5	4	3	2	1	0
RxRTS control	RxIRQ select	error mode	parity mode		parity type	bits per character	

bit 7 **RxRTS control** enables receiver $\overline{RTS}$ handshaking
0 OP0 (OP1) is a parallel output pin.
1 OP0 is used as the port A (OP1 for port B) receiver $\overline{RTS}$ (Request to Send) handshaking output pin; when the receiver buffers are full $\overline{RTS}$ is negated and when space is available it is asserted. This can be used to control the data flow via the $\overline{CTS}$ (Clear to Send) input of the external device.

bit 6 **RxIRQ select** sets when a receiver interrupt is generated (must also be enabled)
0 A receiver interrupt is generated when the RxRDY (receiver ready status bit) in the Status Register indicates that a character has been received.
1 A receiver interrupt is generated when the receiver input buffers are full.

bit 5 **error mode** sets the operating mode of the error bits FE, PE and RCVed BREAK
0 Character mode. The error bits apply to the first character in the buffers.
1 Block mode. The error bits are set by the accumulated errors of all characters in the receiver buffers.

bit **4 3 Parity mode** enables/disables parity bit generation (see **Parity type** below)
0 0 With parity mode: a parity bit is added on transmit and checked on receive
0 1 Force parity mode: a parity bit is added on transmit and checked on receive
1 0 no parity
1 1 Multi-drop mode: see Motorola 1985c

bit **2 Parity type** determines the type of parity generated (if enabled)
0 With parity mode: an even parity bit is generated
Force parity mode: a low bit is generated
1 With parity mode: an odd parity bit is generated
Force parity mode: a high bit is generated

bit **1 0 Bits per character** sets the number of data bits used in transmit and receive
0 0 5 bits per character
0 1 6 bits per character
1 0 7 bits per character
1 1 8 bits per character

Mode Register 2 (MR2A) for port A (MR2B for port B is similar)

7	6	5	4	3	2	1	0
Channel mode		TxRTS control	TxCTS control	stop bit length			

bit **7 6 Channel mode** selects normal operation of the channel or test modes
0 0 normal operation: transmitter and receiver operate independently
0 1 Automatic echo: data received is automatically retransmitted (see Motorola 1985c)
1 0 Local loopback: used for testing (see Motorola 1985c)
1 1 Remote loopback: used for testing (see Motorola 1985c)

bit **5 TxRTS Control** enables transmitter $\overline{\text{RTS}}$ handshaking
0 disabled
1 enabled: see Motorola 1985c

bit **4 TxCTS Control** enables transmitter $\overline{\text{CTS}}$ handshaking
0 disabled: IP0 (IP1) is a parallel input pin
1 enabled: input pin IP0 (IP1) is used as the transmit $\overline{\text{CTS}}$ handshaking input. It is checked when the transmitter is ready to send a character; if asserted (low) the character is transmitted otherwise, if negated (high), the transmitter waits. This can be used to control the data flow via the $\overline{\text{RTS}}$ (Request To Send) output of the external device.

bit **3 2 1 0 Stop bit length** determines the number of stop bits following the character
0 1 1 1 1 stop bit will be transmitted (bits per character are 6, 7 or 8)
1 1 1 1 2 stop bits will be transmitted (bits per character are 6, 7 or 8)
? ? ? ? for other combinations see Motorola 1985c

Auxiliary Control Register (ACR - write only) enables the counter/timer and interrupts from the parallel input port. See Motorola 1985C for full details; set to 0 for now.

Clock Select Registers (CSRA and CSRB - write only) are used to set up the baud rate.

7	6	5	4	3	2	1	0
Receiver clock select				Transmitter clock select			

In each case four bits are used to select the baud rate from a wide range of values. The following table presents a few (see Motorola 1985c for full details).

bit		
bit	**7 6 5 4**	**receiver clock select**
bit	**3 2 1 0**	**transmitter clock select**
	0 1 0 1	600 baud
	0 1 1 0	1200 baud
	1 0 0 0	2400 baud
	1 0 0 1	4800 baud
	1 0 1 1	9600 baud

Channel Command Registers (CRA and CRB - write only) control the I/O channels.

7	6	5	4	3	2	1	0
*	Miscellaneous commands			Transmitter commands		Receiver commands	

bit **6 5 4** **Miscellaneous Commands**

0 0 0	No command
0 0 1	Reset Mode Register pointer to Mode Register 1
0 1 0	Reset Receiver: clears receiver status bits and receiver buffers
0 1 1	Reset Transmitter: clears transmitter status bits and buffers
1 0 0	Reset Error Status: clears the error flag bits in the **Status Register**
1 0 1	Reset Channel's Break-Change Interrupt: see Motorola 1985c
1 1 0	Start Break: transmits BREAK; transmit line goes to *SPACE* (low)
1 1 1	Stop Break: terminates BREAK; transmit line goes to *MARK* (high)

bit **3 2** **Transmitter Commands**

0 0	No action, stays in present mode
0 1	Transmitter Enabled: enables transmitter and asserts transmit-ready status bit
1 0	Transmitter Disabled: completes transmission of current character then disables the transmitter
1 1	Not used

bit **1 0** **Receiver Commands**

0 0	No action, stays in present mode
0 1	Receiver Enabled: enables receiver, i.e. looks for next *START* bit
1 0	Receiver Disabled: disables receiver
1 1	Not used

Receive Buffers (RBA and RBB - read only) 8-bit registers used to transfer data from the input serial lines to the processor.

Transmitter Buffers (TBA and TBB - write only) 8-bit registers used to transfer data from the processor to the output serial lines.

Status Registers (SRA and SRB - read only) provide status information on the channels.

7	6	5	4	3	2	1	0
RCVed BREAK	FE error	PE error	OVRN error	Tx empty	Tx RDY	FFULL	Rx RDY

This register is cleared on RESET.

bit **7** **RCVed BREAK** indicates the receive line is held at logical 0, i.e. *BREAK*
- 0 no *BREAK*
- 1 Received *BREAK*, i.e. a zero character without a STOP bit (one zero character is placed in the receiver buffers; the receiver then waits for *BREAK* to terminate)

bit **6** **FE (Framing Error)**
- 0 no error
- 1 received data was not properly framed by the required START and STOP bits, i.e. a STOP bit was 0 (when it should be a 1) due to noise on the line or incorrect baud rate, number of data bits, number of stop bits or parity

bit **5** **PE (Parity Error)**
- 0 no error
- 1 a parity error has been detected in received data, e.g. due to a noisy communications line or incorrect setting of data length or parity

bit **4** **OVRN (Receiver Overrun)**
- 0 no error
- 1 indicates that data has been lost - it is set if data is received when the receiver data buffers are full (i.e. the program has not read the last characters received)

bit **3** **Tx Empty** indicates the state of the double buffered transmitter
- 0 the transmit holding register or shift register contain data (to be transmitted) **OR** the transmitter is disabled
- 1 both the transmit holding register and shift register are empty

bit **2** **Tx RDY** indicates that the transmitter is ready to accept the next character
- 0 the transmit holding register contains data (to be transmitted) **OR** the transmitter is disabled
- 1 the transmit holding register is empty and ready for the next character (this bit is set when the transmitter is enabled and when a character is transferred from the transmitter holding register to the shift register)

bit **1** **FFULL** indicates the state of receiver data buffers
- 0 space is available in receiver data buffers for more character(s)
- 1 all receiver data buffers are full

bit **0** **Rx RDY** indicates that data in the receiver buffer(s) is ready to be read
- 0 the receiver data buffers are empty
- 1 the receiver data buffer(s) contain data which can be read by the program

Interrupt Mask Register (IMR - write only) enables/disables particular interrupts.

7	6	5	4	3	2	1	0
Input Port Change	Delta Break port B	RxRDY/ FFULL port B	TxRDY port B	Counter Timer Ready	Delta Break port A	RxRDY/ FFULL port A	TxRDY port A

Cleared on RESET. When set each bit enables an interrupt source, Delta=*change of state*:

bit **7** **Input port change:** see Motorola 1985c
bit **6** **Delta Break port B:** enable interrupts on port B *BREAK* start and finish
bit **5** **RxRDY/FFULL port B:** enable interrupts on port B RxRDY and/or FFULL set
bit **4** **TxRDY port B:** enable interrupts on port B TxRDY status set
bit **3** **Counter Timer Ready:** see Motorola 1985c
bit **2** **Delta Break port A:** enable interrupts on port A *BREAK* start and finish
bit **1** **RxRDY/FFULL port A:** enable interrupts on port A RxRDY and/or FFULL set
bit **0** **TxRDY port A:** enable interrupts on port A TxRDY status set

Interrupt Vector Register (IVR - read/write) specifies the interrupt vector number

7	6	5	4	3	2	1	0
interrupt vector number							

1 The MC68681 DUART supports vectored interrupts only, i.e. not autovectored interrupts (see Chapter 23.7).
2 Initialised on RESET to $0F (the uninitialised interrupt vector, see Chapter 23).

```
-----------------------------------------------------------------------------
1                          * Program 26.1 - MC68681 DUART polled input/output
2                          *  (a) continuously write text string to display screen
3                          *  (b) on keyboard hit replace characters in string
4                          *
5                          * initialise DUART
6                          * set write/read pointers to array TEXT
7                          * LOOP for ever
8                          * get next character from array TEXT
9                          * IF at end of array TEXT
10                         *     reset pointer to start of TEXT & get first character
11                         * WRITE character to serial line
12                         * IF character in receiver buffer
13                         *    READ character from serial line
14                         *    move character into array TEXT
15                         * ENDLOOP
16                         *
17        00800001         DUART:    EQU $800001         DUART base address
18        00800001         D_MRA:    EQU DUART           Mode Register 1/2 port A
19        00800003         D_SRA:    EQU DUART+$2        Status Register port A
20        00800003         D_CSA:    EQU DUART+$2        Clock Select Register port A
21        00800005         D_CRA:    EQU DUART+$4        Command Register port A
22        00800007         D_RBA:    EQU DUART+$6        Receive Register port A
23        00800007         D_TBA:    EQU DUART+$6        Transmit Register port A
24        00800009         D_ACR:    EQU DUART+$8        Auxiliary Control Reg
25        0080000B         D_IMR:    EQU DUART+$A        Interrupt Mask register
-----------------------------------------------------------------------------
```

Program 26.1 MC68681 DUART polled input/output (Kaycomp board)

```
------------------------------------------------------------------------------------------
26                                   XREF      DELAY
27 400400 4EB9 00008094 START:       JSR       DELAY             wait for ports to clear
28 400406 0064                       DC.W      100
29                      ;            RESET                       reset system
30 400408 13FC     0000              MOVE.B    #0,D_IMR          clear interrupt enables
                00080000B
31 400410 13FC     0000              MOVE.B    #0,D_ACR          timer & interrupts off
                00800009
32 400418 13FC     001A              MOVE.B    #$1A,D_CRA        MR to MR1, disable Rx/Tx
                00800005
33 400420 13FC     0020              MOVE.B    #$20,D_CRA        reset receiver
                00800005
34 400428 13FC     0030              MOVE.B    #$30,D_CRA        reset transmitter
                00800005
35                      * set up MR1A to Rx RTS handshaking, 8 data bits, no parity
36 400430 13FC     0093              MOVE.B    #$80+$13,D_MRA    set up MR1A
                00800001
37                      * set up MR2A to Tx CTS handshaking & 1 stop bit
38 400438 13FC     0017              MOVE.B    #$17,D_MRA        set up MR2A
                00800001
39 400440 13FC     00BB              MOVE.B    #$0BB,D_CSA       Rx and Tx at 9600 baud
                00800003
40 400448 13FC     0005              MOVE.B    #5,D_CRA          enable transmit & receive
                00800005
41                      * DUART initialised
42 400450 227C 004004A6              MOVEA.L   #TEXT,A1          set input text pointer
43 400456 207C 004004A6 SET_A0:      MOVEA.L   #TEXT,A0          set output text pointer
44 40045C 1018          LOOP:        MOVE.B    (A0)+,D0          get next Tx character
45 40045E 67F6                       BEQ.S     SET_A0            if 0 reset pointer
46 400460 4EB9 00400492              JSR       Tx_CHAR           write character
47 400466 6100     0020              BSR       Rx_CHECK          keyboard hit ?
48 40046A 67F0                       BEQ.S     LOOP
49 40046C 6100     0006              BSR       Rx_CHAR           yes, read character
50 400470 12C0                       MOVE.B    D0,(A1)+          put in text table
51 400472 60E8                       BRA.S     LOOP
52                      ;
53                      ; subroutine to read character from serial line into D0
54                      Rx_CHAR:
55 400474 0839     0000              BTST      #0,D_SRA          character in Rx buffer ?
                00800003
56 40047C 6700     FFF6              BEQ       Rx_CHAR           no, wait
57 400480 1039 00800007              MOVE.B    D_RBA,D0          yes, read character
58 400486 4E75                       RTS
59                      ;
60                      ; subroutine clear Z bit if character is in receiver buffer
61                      Rx_CHECK:
62 400488 0839     0000              BTST      #0,D_SRA          character in Rx buffer ?
                00800003
63 400490 4E75                       RTS
64                      ;
65                      ; subroutine to write character in  D0 to serial line
66                      Tx_CHAR:
67 400492 0839     0002              BTST      #2,D_SRA          transmitter ready ?
                00800003
68 40049A 6700     FFF6              BEQ       Tx_CHAR           no, wait
69 40049E 13C0 00800007              MOVE.B    D0,D_TBA          yes, transmit character
70 4004A4 4E75                       RTS
71                      * data area
72 4004A6 30 31 32 33   TEXT:        DC.B      '01234567890 abcdefghijklmnopqrstuvwxyz '
73 4004D8 0A 0D 00                   DC.B      10,13,0
74                                   END
------------------------------------------------------------------------------------------
```

Program 26.1 MC68681 DUART polled input/output (Kaycomp board)

26.2 Using the MC68681 DUART for polled I/O

To transmit (or receive) characters using the DUART with polled I/O techniques is simply a matter of checking the appropriate bits in the port Status Register. For example, Program 26.1 (previous page) loops transmitting the character string TEXT to the port A serial line which is linked to a terminal or host PC computer running a terminal emulator. On each loop a check is made to see if a character has been received, if so, the character is read and replaces a character in the string TEXT. It is assumed that the $\overline{\texttt{CTS}}$ and $\overline{\texttt{RTS}}$ handshaking lines are connected correctly (if handshaking is not required the appropriate bits should not be set in the Mode Registers. The sequence of events in Program 26.1 is:

line

17-25 define DUART base address for the Kaycomp board and then the I/O registers.

27-28 delay to complete any outstanding I/O to port (e.g. from the monitor)

29 RESET may be used to initialise the DUART otherwise lines 30 to 34 are required (in many circumstances the RESET instruction may not be used because it will initialise other I/O interfaces which may be in the middle of data transfers)

30 clear the Interrupt Mask Register to disable all interrupts

31 clear the Auxiliary Control Register to disable IP0 to IP3 interrupts

32 `MOVE.B #$1A,D_CRA` resets the Mode Register pointer to MR1 and disables the transmitter and receiver

33-34 resets the receiver and transmitter (clearing buffers, status bits, etc.)

36 `MOVE.B #$80+$13,D_MRA` sets Mode Register 1 RxRTS control to enable Rx $\overline{\texttt{RTS}}$ handshaking, no parity and 8 data bits

38 `MOVE.B #$17,D_MRA` sets Mode Register 2 to enable Tx $\overline{\texttt{CTS}}$ handshaking and 1 stop bit

39 `MOVE.B #$0BB,D_CSA` sets both receiver and transmitter to 9600 baud

40 `MOVE.B #5,D_CRA` enables the receiver and transmitter

42-43 initialise input and output pointers to TEXT buffer

44-45 get the next output character; if 0 GOTO line 43 to reset the output pointer to the start of TEXT buffer

46 call subroutine Tx_CHAR to write the character to display screen

47-48 call subroutine Rx_CHECK to see if a character has been received; if not GOTO line 44

49 call subroutine Rx_CHAR to read the character received

50 replace the next character in the TEXT buffer with the character read

The subroutine Rx_CHAR reads a character from the serial line:

55-56 loop until **RxRDY** bit is set in the Status Register , i.e. character received

57 read the character from receiver buffer

The subroutine Rx_CHECK clears the Z bit if a character has been received:

62 test the **RxRDY** bit in the Status Register

The subroutine Tx_CHAR transmits a character:

67-68 loops until **TxRDY** is set in the Status Register, i.e. ready to transmit

69 moves the character to transmitter buffer

```
--------------------------------------------------------------------------------
 1                         * Program 26.2 - MC68681 DUART interrupt input/output
 2                         *  (a) continuously write text string to display screen
 3                         *  (b) on keyboard hit replace characters in string
 4                         *
 5                         * initialise DUART with port B receiver interrupts enabled
 6                         * set MC68000 Status Register to enable interrupts
 7                         * set write pointer to array TEXT
 8                         * LOOP for ever
 9                         *     get next character from array TEXT
10                         *     IF at end of array TEXT
11                         *         reset pointer to start of TEXT & get first character
12                         *     WRITE character to serial line
13                         * ENDLOOP
14                         *
15        00120001         DUART:    EQU      $120001          Bytronic DUART address
16        00120009         D_ACR:    EQU      DUART+$8         Auxiliary Control Reg
17        0012000B         D_IMR:    EQU      DUART+$A         Interrupt Mask register
18        00120011         D_MRB:    EQU      DUART+$10        Mode Register port B
19        00120013         D_SRB:    EQU      DUART+$12        Status register B
20        00120013         D_CSB:    EQU      DUART+$12        clock select register B
21        00120015         D_CRB:    EQU      DUART+$14        Command register B
22        00120017         D_RBB:    EQU      DUART+$16        receive register B
23        00120017         D_TBB:    EQU      DUART+$16        transmit register B
24        00120019         D_IVR:    EQU      DUART+$18        Interrupt Vector Register
25        00000040         D_VECT:   EQU      64               Exception Vector number
26                                   XREF     DELAY
27 001000 4EB9 0010809A START:       JSR      DELAY            wait for ports to clear
28 001006 0064                       DC.W     100
29 001008 4E70                       RESET
30                         * set up MR1B to Rx RTS handshaking, 8 data bits, no parity
31 00100A 13FC     0093              MOVE.B   #$80+$13,D_MRB   set up MR1B
               00120011
32                         * set up MR2B to Tx CTS handshaking & 1 stop bit
33 001012 13FC     0017              MOVE.B   #$17,D_MRB       set up MR2B
               00120011
34 00101A 13FC     00BB              MOVE.B   #$0BB,D_CSB      Rx and Tx at 9600 baud
               00120013
35 001022 13FC     0040              MOVE.B   #D_VECT,D_IVR    set up interrupt number
               00120019
36 00102A 21FC 0000106C              MOVE.L   #D_INT,D_VECT*4  set up exception vector
                   0100
37 001032 13FC     0020              MOVE.B   #$20,D_IMR       port B Rx interrupts on
               0012000B
38 00103A 13FC     0005              MOVE.B   #5,D_CRB         enable transmit & receive
               00120015
39 001042 46FC     2000              MOVE.W   #$2000,SR        enable system interrupts
40 001046 207C 0000108C SET_A0:      MOVEA.L  #TEXT,A0         set output text pointer
41 00104C 1018          LOOP:        MOVE.B   (A0)+,D0         get next Tx character
42 00104E 67F6                       BEQ.S    SET_A0           if 0 reset pointer
43 001050 4EB9 00001058              JSR      Tx_CHAR          write character
44 001056 60F4                       BRA.S    LOOP
45                         *
46                         * subroutine to write character in  D0 to port B serial line
47                         *     LOOP .. UNTIL transmitter ready
48                         *     WRITE character to DUART
49                         Tx_CHAR:
50 001058 0839     0002              BTST     #2,D_SRB         transmitter ready ?
               00120013
51 001060 6700     FFF6              BEQ      Tx_CHAR          no, wait
52 001064 13C0 00120017              MOVE.B   D0,D_TBB         yes, transmit character
53 00106A 4E75                       RTS

                           ***** continued over page *****
--------------------------------------------------------------------------------
```

Program 26.2a MC68681 DUART interrupt driven I/O (Bytronic board)

```
--------------------------------------------------------------------------------
54                          *
55                          * DUART interrupt service routine, port B receiver only
56                          *     GET pointer to input TEXT buffer
57                          *     READ character from serial line into TEXT buffer
58                          *     SAVE pointer to input TEXT buffer
59 00106C 48E7       0080 D_INT:     MOVEM.L   A0,-(A7)
60 001070 2079 00001088              MOVEA.L   TEXT_PNT,A0        get TEXT pointer
61 001076 10F9 00120017              MOVE.B    D_RBB,(A0)+        read character
62 00107C 23C8 00001088              MOVE.L    A0,TEXT_PNT        save TEXT pointer
63 001082 4CDF       0100            MOVEM.L   (A7)+,A0
64 001086 4E73                       RTE
65                          * data area
66 001088 0000108C          TEXT_PNT: DC.L      TEXT               input text pointer
67 00108C 30 31 32 33 34TEXT:        DC.B      '01234567890 abcdefghijklmnopqrstuvwxyz'
68 0010BE 0A 0D 00                   DC.B      10,13,0
69                                   END
--------------------------------------------------------------------------------
```

Program 26.2b MC68681 DUART interrupt driven I/O (Bytronic board)

26.3 Using the MC68681 DUART for interrupt driven I/O

In Program 26.1 the receiver RxRDY bit was polled on every loop of the main program. In Program 26.2 the main program loop (previous page) transmits characters with the receiver interrupt driven (this page). The program sequence is (port B is used in this case):

29 RESET the DUART; or use DUART commands as in Program 26.1 lines 30 to 34

31-34 initialise DUART (handshaking, baud rate, etc.) as in Program 26.1

35 `MOVE.B #D_VECT,D_IVR` loads the interrupt vector number (D_VECT=64) into the Interrupt Vector Register

36 `MOVE.L #D_INT,D_VECT*4` loads the corresponding entry in the exception vector table with the address (long word) of the interrupt service routine

37 `MOVE.B #$20,D_IMR` enables port B RxRDY interrupts in the Interrupt Mask Register

38 enable the transmitter and receiver

39 enable system wide interrupts via the MC68000 Status Register

40-44 transmit the string TEXT using polled I/O by calling subroutine Tx_CHAR

Port B receiver (RxRDY) interrupts were enabled in line 37 and when an interrupt occurs the interrupt service routine D_INT is called:

60 get the input text pointer (from global variable TEXT_PNT) into A0

61 read the character from the Receiver Buffer replacing the next character in the string TEXT (this also clears RxRDY in the Status Register and the interrupt)

62 save the input text pointer to global variable TEXT_PNT

Problem for Chapter 26

Modify Program 26.2 to use transmitter and receiver interrupts, note:

1 RxRDY and TxRDY interrupts will have to be enabled in the Interrupt Mask Register.

2 When an interrupt occurs the service routine will have to test the port Status Register to check if the receiver or transmitter was the cause.

27

Macros and Conditional Assembly

The macro and conditional assembly facilities vary from assembler to assembler. This chapter introduces the principles and then presents sample programs using a native Motorola assembler (Motorola 1983b).

27.1 Macros

Programming operations frequently require the coding of a repeated pattern of instructions or data items. If such instruction sequences form a logical module they may be coded as a subroutine but in some cases this is not possible. A macro allows a sequence of instructions and/or data definitions to be defined, which may then be **replicated** in the program code as required. Macro definition and use depends upon the assembler being used, but an example of macro definition is (for a typical MC68000 native assembler):

```
label       MACRO                   start of macro definition
            .....                   macro body
            ENDM                    end of macro definition
```

The label is the name by which the macro will be 'called' or invoked from within the program. When the assembler finds a macro definition it does not produce any code but stores the definition for later use (ENDM is the macro terminator). When the macro is invoked (by specifying the macro name in the operation-code field) the macro code is **replicated** at that point in the program and then assembled as normal program code. The macro is 'called' as follows:

```
            label     parameter list
```

The label which was used to name the macro is placed in the operation-code field. Following this, in the operand field, may be an optional list of parameters separated by commas. In the case of the Motorola native assembler used for the sample programs in this chapter the parameters are identified within the macro body by the argument designations \1 to \9 and \A to \Z. When the macro is invoked the first parameter from the parameter list replaces the \1, the second \2, etc. The substitution is performed as a **literal string substitution** (i.e. the text of the first parameter is substituted wherever the corresponding argument \1 is found).

Program 27.1a (next page) shows two examples of macro use, the first with a single parameter and the second with six parameters. The first macro (called READ) moves a byte from memory address $50000 into a destination operand specified by the first (and only) macro parameter (e.g. this macro could read the contents of an I/O register). The parameter is assumed to be a data register and is long word cleared before the operation. In the program code the macro READ is called twice, first to move a byte into D0 (line 21) and then a byte into D1 (line 22). At the position of the call the macro code is generated, replacing the parameters \1, \2, etc., with the parameter arguments specified in the macro call (following lines 21 and 22 are unnumbered lines which are the expansion of the macro code). The second example of macro use shows the definition a file control area (i.e. where the filename, size, etc. is defined - many operating systems provide libraries of macros for this and similar functions).

In Program 27.1a (below) the pseudo-operator OPT MEX (display macro expansions) is specified (line 3) and following the macro calls in the program listing is the expanded macro code. Program 27.1b (next page) is identical to Program 27.1a except that the OPT MEX pseudo-operator is not specified and the internal code of the macros is not shown (this is the default). In general the pseudo-operator OPT MEX would be used when debugging macros (so that the generated code can be seen and any errors determined).

```
-----------------------------------------------------------------------------------
1                                  * Program 27.1 - Example Macros
2                                  *
3                                            OPT        MEX                     expand macro calls
4                                  * a macro that reads a byte from I/O register $50000
5                                  READ      MACRO
6                                            CLR.L      \1           clear register
7                                            MOVE.B     $50000,\1    get data byte
8                                            ENDM
9                                  *
10                                 * a macro which defines a file control area
11                                 FILE:     MACRO
12                                           DC.W       '\1'        device name (four characters)
13                                           DC.L       '\2'        file name (eight characters)
14                                           DC.W       '\3'        file name extension (4 chars)
15                                           DC.W       \4          0 if new file, 1 if it exists
16                                           DC.W       \5          file length in blocks
17                                           DC.W       \6          block size
18                                           ENDM
19                                 *
20           00001000                        ORG        $1000
21   00001000                      START:    READ       D0           read a byte into D0
     00001000 4280                           CLR.L      D0           clear register
     00001002 103900050000                   MOVE.B     $50000,D0    get data byte
22   00001008                                READ       D1           read a byte into D1
     00001008 4281                           CLR.L      D1           clear register
     0000100A 123900050000                   MOVE.B     $50000,D1    get data byte
23   00001010 4E4E                           TRAP       #14
24   00001012                      FILE1:    FILE       SYS0,NAMES,DATA,0,100,256
     00001012 53595330                       DC.W       'SYS0'      device name (four characters)
     00001016 4E414D455300                   DC.L       'NAMES'     file name (eight characters)
     0000101E 44415441                       DC.W       'DATA'      file name extension (4 chars)
     00001022 0000                           DC.W       0           0 if new file, 1 if it exits
     00001024 0064                           DC.W       100         file length in blocks
     00001026 0100                           DC.W       256         block size
25   00001028                      FILE2:    FILE       SYS1,ADDRESS,DATA,1,500,512
     00001028 53595331                       DC.W       'SYS1'      device name (four characters)
     0000102C 414444524553                   DC.L       'ADDRESS'   file name (eight characters)
     00001034 44415441                       DC.W       'DATA'      file name extension (4 chars)
     00001038 0001                           DC.W       1           0 if new file, 1 if it exits
     0000103A 01F4                           DC.W       500         file length in blocks
     0000103C 0200                           DC.W       512         block size
26                                           END

SYMBOL NAME      SECT    VALUE              SYMBOL NAME      SECT    VALUE
FILE        MACR   *                        READ        MACR   *
FILE1                    00001012           START                    00001000
FILE2                    00001028
-----------------------------------------------------------------------------------
```

Program 27.1a Macro examples with expansion of macro code shown

```
--------------------------------------------------------------------------------
 1                              * Program 27.1 - Example Macros
 2                              *
 3                              * a macro that reads a byte from I/O register $5000
 4                              READ     MACRO
 5                                       CLR.L     \1          clear register
 6                                       MOVE.B    $50000,\1   get data byte
 7                                       ENDM
 8                              *
 9                              * a macro which defines a file control area
10                              FILE:    MACRO
11                                       DC.W      '\1'        device name (four characters)
12                                       DC.L      '\2'        file name (eight characters)
13                                       DC.W      '\3'        file name extension (4 chars)
14                                       DC.W      \4          0 if new file, 1 if it exists
15                                       DC.W      \5          file length in blocks
16                                       DC.W      \6          block size
17                                       ENDM
18                              *
19              00001000                 ORG       $1000
20    00001000                  START:   READ      D0                   read a byte
21    00001008                           READ      D1                   read a byte
22    00001010 4E4E                      TRAP      #14
23    00001012                  FILE1:   FILE      SYS0,NAMES,DATA,0,100,256
24    00001028                  FILE2:   FILE      SYS1,ADDRESS,DATA,1,500,512
25                                       END
--------------------------------------------------------------------------------
```

Program 27.1b Macro examples without macro expansion

```
--------------------------------------------------------------------------------
 1                              * PROGRAM 27.2a
 2                              * example of condition assembly structures
 3                              *
 4              00000001        PIA:     EQU       1
 5                                       IFNE      PIA       if PIA assemble next section
 6 0 00000000 4E71                       NOP
 7 0 00000002 4E71                       NOP
 8                                       ENDC
 9                                       IFEQ      PIA       if PIT assemble next section
10                                       NOP
11                                       NOP
12                                       ENDC
13                                       END
SYMBOL NAME      SECT    VALUE         SYMBOL NAME      SECT    VALUE
PIA                     00000001
--------------------------------------------------------------------------------
```

Program 27.2a Conditional assembly with PIA = 1

```
--------------------------------------------------------------------------------
 1                              * PROGRAM 27.2b
 2                              * example of condition assembly structures
 3                              *
 4              00000000        PIA:     EQU       0
 5                                       IFNE      PIA       if PIA assemble next section
 6                                       NOP
 7                                       NOP
 8                                       ENDC
 9                                       IFEQ      PIA       if PIT assemble next section
10 0 00000000 4E71                       NOP
11 0 00000002 4E71                       NOP
12                                       ENDC
13                                       END
SYMBOL NAME      SECT    VALUE         SYMBOL NAME      SECT    VALUE
PIA                     00000000
--------------------------------------------------------------------------------
```

Program 27.2b Conditional assembly with PIA = 0

27.2 Conditional Assembly

Conditional assembly allows the programmer to write a comprehensive program that can cover many conditions. For example, a program may have to be used with target system equipped with an MC6821 PIA or an MC68230 PIT. At assembly time the user specifies, by using the EQU pseudo-operator, which device is to be used and conditional assembly directives determine which section of code is assembled. The form of a conditional clause depends upon the assembler being used (a powerful disk based assembler may have several), but a common form is:

```
        IFxx        expression
        .....                           body of structure
        ENDC
```

where xx is a condition tested and ENDC terminates the conditional clause. Valid values of xx include (Motorola 1983b):

EQ expression 0
NE expression not 0
LT expression less than 0
LE expression less than or equal to 0
GT expression greater than 0
GE expression greater than or equal to 0

Programs 27.2a and 27.2b (previous page) show a simple example of conditional assembly. Depending upon the value of PIA (if PIA = 1 a PIA is fitted otherwise a PIT is fitted) one or other of the conditional sections will be assembled (the addresses and generated machine code in the listings show which sections are generated in each case). Note that the only difference in the source code is the value specified for PIA. In practice a complex program may have to take account of many conditions when it is assembled. Do not confuse conditional assembly, which is used to control that action of the assembler in assembling sections of code, and conditional program structures which control the action of the program at run time.

28

Programming time intervals

Timing events or the programming of time intervals is an important requirement in many computer applications, e.g.:

1 to delay program execution for a few seconds while a message is displayed;
2 to sample data from an experiment 10000 times per second for analysis.

It is possible to time events and delays using:

(a) a sequence of NOP instructions;
or (b) a program delay loop;
or (c) a programmable timer chip, e.g. MC6840 PTM or MC68230 PIT.

Each instruction performed by the processor takes a certain time depending upon:

1 The complexity of the instruction, e.g. a divide takes longer than an add.
2 How the operands are obtained (i.e. addressing modes used).
3 The speed of the processor as set by the clock.
4 The speed of primary memory, e.g. if WAIT states are used.

A number of MC68000 microprocessors are available which use different clock speeds. A clock is an electronic circuit which generates a pulse at regular intervals (e.g. every 0.125μSec for an 8MHz clock) and is use to time/synchronise events within the computer system. Although primary memory speed tends to match the processor speed, the memory circuits can use WAIT states to make the processor wait until it is ready to transfer the data (the hardware manuals should indicate if WAIT states are used). The number of clock cycles each instruction takes can be determined from tables in Appendix B.

28.1 The NOP instruction and delay loops

The NOP instruction does nothing except increment the program counter. The fetch/execute takes 4 clock cycles (no WAIT states) and a sequence of NOP instructions can be used to give a small delay, e.g. using an MC68000 with an 8 MHz clock (clock period 0.125μSec) twenty NOP instructions will delay program execution by 10μSec.

If a delay longer than a few microseconds is required a simple way to achieve this is by using a program loop. Program 28.1 (next page) displays an * every second and a newline every minute using subroutine DELAY (a copy of which is in the subroutine library in Appendix D), which contains two loops:

(a) an inner loop giving a delay of one millisecond (line 49);
and (b) an outer loop that sets the number of milliseconds to be delayed (lines 48 to 52).

```
-------------------------------------------------------------------------------
 1                         * Program 28.1: Timing using delay loops
 2                         * Display * every second & newline every minute
 3                         *
 4                         * WRITE('display * every second & newline every minute')
 5                         * LOOP FOR ever
 6                         *     REPEAT
 7                         *     UNTIL one second elapsed
 8                         *     WRITE('*')
 9                         *     IF one minute elapsed
10                         *         WRITE(newline)
11                         *     END IF
12                         * END LOOP
13                         *
14                                   XREF      WRLINE,WRTEXT,WRCHAR
15 001000 4EB9 00028038 START:       JSR       WRTEXT
16 001006 64 69 73 70 6C             DC.W      ' * every second & newline every minute$'
17 001034 203C 2020202A              MOVE.L    #'   *',D0     * character into D0
18 00103A 723B         LOOP1:        MOVEQ     #59,D1         initialize SECONDS counter
19 00103C 4EB9 0002802C              JSR       WRLINE         display a newline
20                         * LOOP for ever
21 001042 6100      0010 LOOP2:      BSR       DELAY          delay for one second
22 001046 03E8                       DC.W      1000           delay count in milliseconds
23 001048 4EB9 00028028              JSR       WRCHAR         display an *
24 00104E 51C9      FFF2             DBRA      D1,LOOP2       every 60 characters
25 001052 60E6                       BRA.S     LOOP1
26                         *
27                         * SUBROUTINE DELAY: delay for a time period in milliseconds
28                         *   The delay count follows JSR in memory (word value) so:-
29                         *         JSR       DELAY
30                         *         DC.W      10000          delay_count is one second
31                         *
32                         * SUBROUTINE DELAY
33                         * IF delay_count > 0
34                         *     LOOP FOR delay_count in milliseconds
35                         *         LOOP for one millisecond
36                         *         END LOOP
37                         *     END LOOP
38                         * RETURN
39                         *
40                         * 1mSec delay count for an 8 MHz 68000 (no WAIT sates)
41        0000031C         DELAYM:   EQU.W     796            1 mSec count for 8MHz 68000
42 001054 48E7      C080 DELAY:      MOVEM.L   D0-D1/A0,-(A7)
43 001058 206F      000C             MOVE.L    12(A7),A0      get address of delay_count
44 00105C 3018                       MOVE.W    (A0)+,D0       get delay_count into D0
45 00105E 2F48      000C             MOVE.L    A0,12(A7)      restore return address
46 001062 5340                       SUBQ.W    #1,D0          delay_count -1 for DBRA
47 001064 6D10                       BLT.S     DELAY3         if delay_count <= 0 exit
48 001066 323C      031C DELAY1:     MOVE.W    #DELAYM,D1     one millisecond delay count
49 00106A 51C9      FFFE DELAY2:     DBRA      D1,DELAY2      INNER DELAY LOOP !!
50 00106E 4E71                       NOP                      NOP delay to make DELAYM
51 001070 4E71                       NOP                      calculation more accurate
52 001072 51C8      FFF2             DBRA      D0,DELAY1      OUTER DELAY LOOP !!
53 001076 4CDF      0103 DELAY3:     MOVEM.L   (A7)+,D0-D1/A0
54 00107A 4E75                       RTS
55                                   END
-------------------------------------------------------------------------------
```

Program 28.1 Timing events using a polling loop

In Program 28.1 the number of milliseconds to delay is passed to the subroutine by a word parameter following the JSR in memory. For example, to delay for one second:

```
        JSR     DELAY
        DC.W    1000            delay_count is one second
```

	instruction		clock cycles
DELAY1:	MOVE.W	#DELAYM,D1	8
DELAY2:	DBRA	D1,DELAY2	10 (14 on exit from loop)
	NOP		4
	NOP		4
	DBRA	D0,DELAY1	10 (14 on exit from loop)

Table 28.1 Instruction timing for the delay loop in Program 28.1

The Program 28.1 version of the subroutine DELAY assumes an MC68000 with an 8MHz clock using memory without WAIT states. The number of clock cycles taken for each instruction in the main loop are shown in Table 28.1 (values from Appendix B).

Assuming a clock speed of 8MHz a delay of 100 milliseconds (i.e. D0 = 99 on entry to the algorithm of Table 28.1) will require a delay of 800000 clock cycles.

Thus: $(8 + DELAYM*10 + 14 + 4 + 4 + 10) * 100 + 4 = 800000$

Hence: $DELAYM = (((800000 - 4)/100) - 10 - 4 - 4 - 14 - 8)/10$

This calculation gives a value for the inner loop count of DELAYM equal to 795.996 or 796 as an integer (the NOP instructions in the outer loop are used to increase the accuracy of the calculation). The delays obtained using the routine DELAY are only approximate.

1 No account is taken of other instructions within subroutine DELAY (e.g. saving and restoring registers) and instructions outside the subroutine.
2 The clock speed may vary with temperature, power supply levels, etc.
3 The memory may use WAIT states.

Although this technique is acceptable where approximate delays are required, e.g. to display a message on the screen for 10 seconds, it is very difficult to time events accurately if anything except trivial processing is to be carried out. Any modification to the program is likely to change total instruction timings and a recalculation of the delay loop parameters will be required. For accurate timing of events a timer chip which is external to the processor and therefore independent of program instruction execution is required.

28.2 The MC6840 PTM (Programmable Timer Module)

The MC6840 PTM (Motorola 1983c) contains three **independent** 16-bit counters under program control. The counters decrement on each clock pulse with a flag being set in the PTM Status Register when a counter value becomes zero. The flag(s) may be polled or interrupts enabled. The clock used can be selected by the program to be either:

(a) the MC6800 microprocessor equivalent clock generated by the MC68000 processor (this is output on the E or Enable line) with a period equal to ten MC68000 clock periods (for use by the slower MC6800 family chips),

or (b) a clock supplied by external equipment (e.g. to synchronise with an experiment).

28.2.1 MC6840 PTM signal lines

The MC6840 has a number of I/O lines which can be used by external equipment:

External RESET. A low signal on this input line completely resets the PTM (stops counter/timers, clears interrupt enables, clears any interrupts pending, etc.). This is usually connected to the MC68000 RESET line so that the MC6840 is reset with all the other I/O devices. Remember when programming the MC6840 to either:

(a) use the RESET switch to stop the program;

or (b) reset the PTM using program instructions before stopping the program.

Otherwise the PTM may continue interrupting the operating system or other programs.

Clock inputs. A clock is an electronic circuit which generates a pulse at regular intervals, e.g. every microsecond. Such a clock can be used to synchronise or trigger events within the computer or with external equipment. Each MC6840 counter/timer can use either the MC6800 equivalent clock generated by the MC68000 processor (called the **Enable** clock) or a clock supplied by external equipment, e.g. from an experiment (each counter/timer can be programmed independently).

Gate inputs. There are three gate inputs (one for each counter/timer) which can be used by external equipment to start the timers. Using the 'wave measurement' modes the MC6840 can be used to make measurements on external signals applied to the gate inputs (for further details see Motorola 1983c).

Timer Outputs. Associated with each counter/timer is an output line which can be used to control external equipment. The MC6840 can generate output waveforms; either square wave (using sixteen bit counting) or pulse width modulated (8-bit counting mode) in either continuous or single shot mode (Motorola 1983c).

Each counter/timer can be programmed independently of the others using different clocks, gate inputs and with interrupts and outputs enabled or disabled. The following sections will introduce the programming of the MC6840 PTM using polled I/O techniques.

28.2.2 The MC6840 PTM registers

The MC6840 has a number of registers which, on the Force board (Force 1984), are mapped to the alternate memory mapped addresses shown in Table 28.2.

Force board address	MC6840 PTM Register
$4CF41 (byte, write)	Control Register 1 (selected by bit 0 of)
" (byte, write)	Control Register 3 (Control Register 2)
$4CF43 (byte, write)	Control Register 2
" (byte, read)	Status Register
$4CF45 (and $4CF47) (word, write)	Timer 1 latch (initial value of counter)
" " (word, read)	Timer 1 counter (current value of counter)
$4CF49 (and $4CF4B) (word, write)	Timer 2 latch
" " (word, read)	Timer 2 counter
$4CF4D (and $4CF4F) (word, write)	Timer 3 latch
" " (word, read)	Timer 3 counter

Table 28.2 MC6840 PTM memory mapped register address (Force board)

The alternate bytes used to access the 16-bit values of the latches and counters (Table 28.2) are actually separate 8-bit registers (for full details see Motorola 1983c). When programming the MC6840 from an MC68000 a MOVEP.W instruction (see Chapter 18.6 and Program 28.2) is used to load 16-bit values directly into the counters.

Control Registers

There are three **write only** control registers used to control the actions of the three counter/timers. Bit 0 of each register has a special function with the remainder (bits 1 to 7) being identical in the three cases.

7	6	5	4	3	2	1	0
output control	INT enable	operating mode			counter mode	clock source	see below

bit **7** **Output control** - enables and disables the timer output pin
- 0 output disabled
- 1 output enabled

bit **6** **INT enable** enables and disables interrupts from the timer
- 0 interrupts disabled
- 1 interrupts enabled (the Force board used Autovectored level 4 interrupts)

bits **5 4 3** **operating mode** see note 1 below
- 0 0 0 continuous operating mode
- ? ? ? for other modes see Motorola 1983c

bit **2** **counter mode** see note 2 below
- 0 16-bit counting mode
- 1 8-bit counting mode

bit **1** **clock source** selects between the Enable and external clock sources
- 0 use External clock (e.g. from external equipment)
- 1 use Enable clock (i.e. the 6800 equivalent clock generated by the 68000)

bit **0** **Control Register 1**
- 0 start **all** counter/timers to operate in the modes specified by the other Control Register bits;
- 1 internal RESET: stops counter/timers, resets the outputs, clears any interrupt flags set in the Status Register and then loads the three latches into the associated counters (internal reset does **not** affect the contents of the latches or the Control Registers; unlike external RESET).

bit **0** **Control Register 2** controls access to CR1 & CR3 (which share an address).
- 0 Access CR3, i.e. at address $4CF41 on FORCE board
- 1 Access CR1, i.e. at address $4CF41 on FORCE board

bit **0** **Control Register 3** enables the use of a divide by 8 prescaler **for timer 3** only.
- 0 Counter 3 uses clock specified by CR3 bit 1
- 1 Counter 3 uses clock specified by CR3 bit 1 divided by 8, e.g. if the Enable clock is 800KHz (i.e. 8MHz 68000 clock) this allows timer 3 to be clocked at 100KHz.

Notes:

1 There are a number of counting modes available (Motorola 1983c) of which only *continuous mode* will be considered in this chapter:
  (a) the counter counts down and when 0 is reached the corresponding timer flag bit is set in the Status Register and, if enabled, an interrupt is generated;
  then (b) the value in the latch is automatically loaded into the counter and counting starts again.

2 Each counter/timer contains a sixteen bit counter which is used either in:
  (a) in 16-bit counting mode: as a positive 16-bit integer (value 0 to 65535);
  or (b) in 8-bit counting mode: as two eight bit counters used to produce pulse width modulated outputs.

3 It is worth clearing all the Control Registers bits and executing an internal reset when the program starts and terminates (to ensure that no odd interrupt occurs).

Status Register

7	6	5	4	3	2	1	0
INT Flag	*	*	*	*	timer 3 flag	timer 2 flag	timer 1 flag

In the Status Register counter/timers 1, 2 and 3 are assigned bits 0, 1 and 2 respectively. When a counter reaches 0 the corresponding bit is set in the Status Register. If the interrupt enable (bit 6) is set in the corresponding Control Register, bit 7 of the Status Register will also be set and an interrupt request sent to the processor. The appropriate Interrupt Mask bits in the 68000 Status Register must be set to the correct priority for this to have any effect. On the Force boards the MC6840 PTM is at priority 4 using the autovectored interrupt response (vector address $70). All the interrupt flags are cleared by the following:

1 an external RESET signal applied to RESET input;
2 an internal RESET, i.e. setting bit 0 of CR1 to 1;
3 the sequence (a) read Status Register, and (b) read a Timer Counter.

The latter will only clear those flag bits which were set when the Status Register was read. This prevents missing interrupts which might occur after the Status Register is read, but prior to reading a Timer Counter. An individual interrupt flag can be cleared by:

  (a) a write to the corresponding Timer Latch;
then (b) a counter initialisation sequence for the corresponding timer (see Motorola 1983c for further details of using the Gate inputs).

28.2.3 Programming the MC6840 PTM

In outline the MC6840 PTM is programmed as follows:

1 The program RESETs the PTM to clear any existing interrupts, etc.
2 The program sets up the Control Register for each timer specifying:
   (a) which clock is to be used (Enable or external);
   (b) the counting mode: 16-bit or 8-bit;
   (c) the operating mode: continuous, single-shot, etc.;
   (d) if interrupts are enabled;
   (e) if the output line is enabled;
3 The program loads the initial count value into the latch.
4 The program gives the command to reset the PTM and load the contents of the latches into the corresponding counters (Control Register 1 bit 0 = 1) and then the command to start the counter/timers (Control Register 1 bit 0 = 0).
5 On each 'tick' (pulse) of the appropriate clock a counter is decremented.
7 The event (when a counter reaches 0) can be serviced:
   (a) the flag bit can be polled by a program waiting for the event;
   or (b) if interrupts are enabled an interrupt request is sent to the processor and an interrupt service routine can service the event.

Program 28.2 (next page) uses an MC6840 PTM to display an * every second and a newline every minute by timing intervals using polled I/O techniques.

line
25-26 display start up message
27-31 clear all the Control Registers and do an internal RESET
36-37 sets up counter/timer 1 latch to 40000 (i.e. using a 800KHz clock on the Enable line this will set the interrupt flag every 0.05 second)
42 sets up timer/counter 1 Control Register to use Enable clock, 16-bit counting mode, continuous mode, with interrupts and outputs disabled; bit 0 = 1 to reset the device and load all counters from the corresponding latches
44 similar to line 42 but clears bit 0 of Control Register 1 to start the counter/timers.
45 initialise counter for 1 minute (59 when using DBRA)
46 do a newline
47 initialise counter for 1 second (i.e. as the PTM interrupts every 0.05 second to time one second requires a count of 19 when using the DBRA instruction)

There is then a program loop:

48 tests bit 0 of the Status Register to see if counter/timer 1 interrupt flag is set;
49 if no interrupt flag is set branch back to line 48
50 read counter 1 to clear the interrupt flag (i.e. a read of the Status Register followed by a read of a Timer Counter clears the flag)
51 loop control for 1 second count
52-53 display an * to indicate that a second has elapsed
54 loop control for 1 minute
55 loop back to display a newline and wait for the next second

```
--------------------------------------------------------------------------------
 1                          * Program 28.2 - Test MC6840 PTM using polled I/O
 2                          * Display * every second & newline every minute
 3                          *
 4                          * WRITE('display * every second & newline every minute')
 5                          * Initialise PTM to count 40000 using Enable clock
 6                          * LOOP FOR ever
 7                          *     REPEAT
 8                          *     UNTIL one second elapsed
 9                          *     WRITE('*')
10                          *     IF one minute elapsed
11                          *         WRITE(newline)
12                          *     END IF
13                          * END LOOP
14                          *
15                          * MC6840 PTM addresses (offsets) for Force board
16         0004CF41         PTM:      EQU.L    $4CF41    base address for force board
17         00000000         PTMCR1:   EQU      $0        (byte, write) control reg 1
18         00000000         PTMCR3:   EQU      $0        (byte, write) control reg 3
19         00000002         PTMCR2:   EQU      $2        (byte, write) control reg 2
20         00000002         PTMSTAT:  EQU      $2        (byte, read)  status register
21         00000004         PTMT1:    EQU      $4        (word, write) timer 1 latch
22                          *                            (word, read)  timer 1 counter
23                          *
24                                    XREF     WRCHAR,WRLINE,WRTEXT
25 001000 4EB9 00028038 START:        JSR      WRTEXT
26 001006 64 69 73 70 6C              DC.W     ' * every second & newline every minute$'
27 001034 207C 0004CF41               MOVEA.L  #PTM,A0          base address of PTM
28 00103A 117C 0000 0002              MOVE.B   #0,PTMCR2(A0)    clear CR2 (& access CR3)
29 001040 117C 0000 0000              MOVE.B   #0,PTMCR3(A0)    clear CR3
30 001046 117C 0001 0002              MOVE.B   #1,PTMCR2(A0)    set CR2 to access CR1
31 00104C 117C 0001 0000              MOVE.B   #1,PTMCR1(A0)    reset PTM and clear CR1
32                          *
33                          * Set counter/timer 1 to set interrupt flag every 0.05 sec
34                          * With an 8MHz 68000 clock the Enable clock is 800KHz
35                          * a count of 40000 will set flag in 0.05 of a second
36 001052 303C      9C40              MOVE.W   #40000,D0        set count for 0.05 second
37 001056 0188      0004              MOVEP.W  D0,PTMT1(A0)     load timer 1 latch
38                          *
39                          * Timer 1 control register: Enable clock, 16-bit counting,
40                          *    continuous mode, interrupts & output disabled
41                          *    bit 0 = 1 to RESET and load all counters from latches
42 00105A 117C 0003 0000              MOVE.B   #3,PTMCR1(A0)  load CR1
43                          * Start timers (as above but bit 0 = 0 to start timers)
44 001060 117C 0002 0000              MOVE.B   #2,PTMCR1(A0)  load CR1
45 001066 743B              LOOP1:    MOVEQ    #59,D2         set up count for 1 minute
46 001068 4EB9 0002802C               JSR      WRLINE         do a newline
47 00106E 7213              LOOP2:    MOVEQ    #19,D1         set up count for 1 second
48 001070 0828 0000 0002 LOOP3:       BTST.B   #0,PTMSTAT(A0) wait for timer 1 flag
49 001076 67F8                        BEQ.S    LOOP3
50 001078 0108      0004              MOVEP.W  PTMT1(A0),D0   read counter to clear flag
51 00107C 51C9      FFF2              DBRA     D1,LOOP3       loop for 1 second
52 001080 103C      002A              MOVE.B   #'*',D0        then display a *
53 001084 4EB9 00028028               JSR      WRCHAR
54 00108A 51CA      FFE2              DBRA     D2,LOOP2       loop for 1 minute
55 00108E 60D6                        BRA.S    LOOP1
56                                    END
--------------------------------------------------------------------------------
```

Program 28.2 Timing events using an MC6840 PTM, polled I/O

28.3 The MC68230 PIT (Parallel Interface/Timer)

The MC68230 PIT (Motorola 1983a) contains a 24-bit counter which can be clocked using:

(a) the MC68000 clock (on the CLK input line) with a divide by 32 (5-bit) prescaler,
or (b) a clock supplied by external equipment (on the PC2/TIN line) which may optionally be used with a divide by 32 (5-bit) prescaler .

The counter or prescaler decrements on each clock pulse and when the counter becomes zero a flag is set in the Timer Status Register (TSR) (which can be polled) and if interrupts are enabled the program being executed is interrupted.

Once set up, the MC68230 is independent of the program in the processor, thus allowing real-time acquisition and processing of data (e.g. for reading data from an analogue to digital converter). This assumes that there is **sufficient** time between the interrupts to process the data.

28.3.1 MC68230 PIT timer signal lines

Three of the Port C alternate function pins (see Chapter 21) can be used by the timer:

PC2/TIN	PC2 or as the external clock input for the timer (Timer IN)
PC3/TOUT	PC3 or the timer interrupt request output or timer output (Timer OUT)
PC7/TIACK	PC7 or the interrupt acknowledge to a timer interrupt request

The Timer OUT signal can be used either as the interrupt request signal into the MC68000 processor or to control external equipment.

28.3.2 The MC68230 PIT registers

offset	MC68230 timer register	program name
20	Timer Control Register	TCR
22	Time interrupt Vector Register	TIVR
24	null register (used by MOVEP)	TCPR
26	Counter Preload Register (high)	TCPRH
28	Counter Preload Register (medium)	TCPRM
2A	Counter Preload Register (low)	TCPRL
2C	null register (used by MOVEP)	TCREG
2E	Count Register (high)	TCRH
30	Count Register (medium)	TCRM
32	Count Register (low)	TCRL
34	Timer Status Register	TSR

Table 28.3 Memory mapped register offsets for the MC68230 PIT timer
Base addresses are $130001 for the Bytronic board and $A00001 for the Kaycomp

Timer Interrupt Vector Register (TIVR)

This register contains the 8-bit vector supplied in response to timer interrupt acknowledge (TIACK) when vectored interrupts are being used. On RESET the register is initialised to $0F, which, if an interrupt occurs will cause the uninitialised interrupt vector.

Timer Control Register (TCR) controls the operation of the timer

7	6	5	4	3	2	1	0
TOUT/TIACK control			ZD control	*	Clock control		Timer enable

bits **7 6 5** **TOUT/TIACK control** sets the PC3/TOUT & PC7/TIACK pin functions
- 0 0 x the PC3/TOUT & PC7/TIACK pins are used for port C I/O **(see note 1)**
- 0 1 x the PC3/TOUT pin is used for the TOUT signal; in the halt state this pin is high and in the run state it is toggled when the counter decrements to zero (i.e. a square wave is output)
  the PC7/TIACK pin is used for normal port C bit 7 I/O (PC7)
- 1 0 1 the PC3/TOUT pin is used as the timer interrupt request (TOUT)
  the PC7/TIACK pin is used as the timer interrupt acknowledge (TIACK)
  this supports vectored interrupts (see Chapter 23.7)
- 1 1 1 the PC3/TOUT pin is used as the timer interrupt request (TOUT)
  the PC7/TIACK pin is used for normal port C bit 7 I/O (PC7)
  this supports autovectored interrupts (see Chapter 23.7)
- ? ? ? for other pin functions see Motorola 1983a

bit **4** **ZD control** determines what happens after zero detect (i.e. counter reaches 0)
- 0 after zero detect the counter is reloaded from the counter preload register
- 1 after zero detect the counter continues counting

bit **3** is unused

bit **2 1** **clock control** selects the clock source (see note 2)
- 0 0 the PC2/TIN pin is used for normal port C I/O (PC2) and the 68000 clock (CLK) and the divide by 32 prescaler are used - see note 3
  The timer enable bit determines whether the timer is in the run or halt state.
- 0 1 the PC2/TIN pin is used for the timer input (TIN) and the 68000 clock (CLK) and the divide by 32 prescaler are used - see note 3
  The timer is in the run state when the timer enable bit is one and the TIN pin is high; otherwise the timer is in the halt state.
- 1 0 the PC2/TIN pin is used for the timer input (TIN) and the divide by 32 prescaler is used.
  The 5-bit prescaler is decremented on the rising edge of the TIN pin. The 24-bit counter is decremented or rolled over or reloaded from the prescaler register when the prescaler rolls over from $00 to $1F.
  The timer enable bit determines whether the timer is in the run or halt state.
- 1 1 the PC2/TIN pin is used for the timer input (TIN) and the prescaler is not used.
  The 24-bit counter is decremented or rolled over or reloaded from the prescaler register on the rising edge of the TIN pin.
  The timer enable bit determines whether the timer is in the run or halt state.

bit **0** **Timer enable**
- 0 disabled - the timer is in the halt state
- 1 enabled - the timer is in the run state

Notes:

1 **TOUT/TIACK control (bits 7, 6 & 5)** specify whether the PC3/TOUT and PC7/TIN pins are used for port C bits 3 and 7 or for interrupt control. Autovectored interrupts will use TOUT and vectored interrupts will use TOUT and TIN (see Chapter 23.7).
2 When using the 68000 clock (on the CLK pin) the prescaler is always used (i.e. CLK is divided by 32). When using TIN is the prescaler can be used or disabled.
3 The 5-bit prescaler is decremented on the falling edge of the clock input (CLK). The 24-bit counter is decremented or rolled over or reloaded from the prescaler register when the prescaler rolls over from $00 to $1F.

Timer Counter Preload Registers (TCPRH, TCPRM, TCPRL)

These registers hold the 24-bit counter value (bits 0-7 in TCPRL, 8-15 in TCPRM and 16-23 in TCPRH). The null register labelled TCPR in Table 28.3 is used with a MOVEP.L instruction:

```
        MOVEP.L    D0,TCPR(A0)           load preload register
```

where A0 would contain the base address of the PIT and D0 the 24-bit value to be loaded into the counter preload register (the bits 24-31 of D0 which are written to the null register are ignored). On a read the null register returns zeros.

Timer Count Registers (TCRH, TCRM, TCRL)

The Count Registers return the current value of the 24-bit counter (can be read at any time). The null register TCREG in Table 28.3 is used with a MOVEP.L instruction:

```
        MOVEP.L    TCREG(A0),D0         read the counter
```

would return the current value of the counter (bits 24-31 of D0 read from the null register are returned as zeros). Data written to the Count Registers are ignored.

Timer Status Register (TSR)

7	6	5	4	3	2	1	0
*	*	*	*	*	*	*	ZDS

The Zero Detect Status (ZDS) flag is set when the 24-bit counter decrements from $000001 to $000000. The ZDS is cleared by the *direct method of resetting* (i.e. a 1 is written to ZDS) or when the timer is halted (i.e. on RESET or timer disabled).

28.3.3 Programming the MC68230 PIT

In outline the MC68230 PIT timer is programmed as follows:

1 The Timer Control Register is loaded (setting TOUT/TIACK control, ZD control & Clock control) with timer enable (bit 0) reset to 0. Thus the timer is halted.
2 The Counter Preload Register is loaded with the count value.
3 Timer Control Register bit 0 is set (to 1) to start the timer (bits 1-7 stay the same).

The timer is running and on each pulse of the clock the counter is decremented. The program can then detect when the ZDS flag is set:

(a) the ZDS flag bit in the Timer Status Register can be polled by the program;
or (b) if interrupts are enabled an interrupt request is sent to the processor and an interrupt service routine can service the event.

Program 28.3 (next page) uses an MC68230 PIT to display an * every second and a newline every minute by timing intervals using polled I/O techniques.

line
23-25 display start up message and set up base address of PIT in A0
31 set up the Timer Control Register:
(a) TOUT/TIACK control: PC3/TOUT & PC7/TIACK pin carry PC3 & PC7
(b) the counter is reloaded from preload register after zero detect (bit 4 is 0)
(c) use 68000 clock on CLK pin (bits 2 and 1 are 0)
(d) disable timer (bit 0 is 0), i.e. timer is in halt state
34-35 set up the Counter Preload Register to count one second, i.e. with an 8MHz clock using the divide by 32 prescaler the counter decrements 250000 times per second
37 sets bit 0 of the Timer Control Register to 1 to start the timer
38-39 initialise counter for 1 minute (59 when using DBRA) and do a newline

There is then a program loop:

41 tests bit 0 of the Timer Status Register to see if the ZDS flag is set
42 if ZDS is not set branch back to line 41
44 clear the ZDS flag in the Timer Status Register by writing a 1 to it (i.e. the *direct method of resetting*)
45-46 display an * to indicate a second has elapsed
47 loop control for 1 minute
48 loop back to line 38 to reset minute counter and then wait for the next second

Problem for Chapter 28

1 Write a program to check the instruction timings shown in Appendix B , e.g.:

```
MOVEQ      #10,D0        is 4 clock cycles
MOVE.L     #10,D0        is 12 clock cycles
MOVE.L     (A0)+,(A1)+   is 20 clock cycles
```

Explain discrepancies (if any). Remember when using the 68000 clock for timing:

(a) the MC6840 PTM Enable clock is a tenth the 68000 clock (i.e. 800KHz for an 8MHz 68000 clock)

(b) the MC68230 PIT is clocked at one thirtysecond of the 68000 clock (i.e. 250KHz for an 8MHz 68000 clock)

Extend the program to check the delay loop timing of Table 28.1. If available replace the MC68000 with an MC68010 and check the timing of loops such as:

```
LOOP:     CLR.W     (A0)+
          DBRA      D0,LOOP
```

2 Depending on the timer available modify Program 28.2 (for the MC6840 PTM) or 28.3 (for the MC68230 PIT) to use interrupts.

3 Using a suitable motor circuit (see Appendix F.7) implement the PWM (pulse width modulation) motor speed control using timer interrupts described in Appendix F.7.4. Extend the program to measure the motor speed in RPM.

```
 1                         * Program 28.3 - MC68230 PIT timer polled I/O
 2                         * Display * every second & newline every minute
 3                         *
 4                         * WRITE('display * every second & newline every minute')
 5                         * set up timer control register
 6                         * set timer counter to count 1 second (count=250000)
 7                         * start timer (by setting bit in control register to 1)
 8                         * LOOP FOR ever
 9                         *     REPEAT
10                         *     UNTIL one second elapsed
11                         *     WRITE('*')
12                         *     IF one minute elapsed
13                         *         WRITE(newline)
14                         *     END IF
15                         * END LOOP
16                         *
17        00A00001         PIT:     EQU     $A00001     PIT base address for Kaycomp
18        00000020         TCR:     EQU     $20         timer control register
19        00000024         TCPR:    EQU     $24         counter preload MOVEP address
20        00000034         TSR:     EQU     $34         timer status register
21                         *
22                                  XREF    WRCHAR,WRTEXT,WRLINE,DELAY
23 400400 4EB9 00008038    START:   JSR     WRTEXT
24 400406 64 69 73 70 6C            DC.W    '* every second & newline every minute$'
25 400434 207C 00A00001             MOVEA.L #PIT,A0         base address of PIT
26                         * set up Timer Control Register
27                         * (a) TOUT/TIACK control to 000
28                         * (b) counter reloaded from preload register after ZDS detect
29                         * (c) use 68000 clock (on CLK pin)
30                         * (d) bit 0 is 0 to disable timer
31 40043A 117C 0000 0020            MOVE.B  #$0,TCR(A0)     Timer Control Register
32                         * Counter Preload Register for 1 second (8MHz clock & divide
33                         *  by 32 prescaler the counter decrements 250000 times/sec)
34 400440 203C 0003D090             MOVE.L  #250000,D0      counter preload register
35 400446 01C8      0024            MOVEP.L D0,TCPR(A0)         to time 1 sec
36                         * set time control register bit 0 to 1 to start timer
37 40044A 08E8 0000 0020            BSET.B  #0,TCR(A0)      timer enable
38 400450 723B             LOOP1:   MOVEQ   #59,D1          counter for 1 minute
39 400452 4EB9 0000802C             JSR     WRLINE          do a newline
40                         * loop for 1 second
41 400458 0828 0000 0034   LOOP2:   BTST    #0,TSR(A0)      wait for ZDS flag to set
42 40045E 67F8                      BEQ.S   LOOP2
43                         * one second elapsed, clear ZDS (direct method of resetting)
44 400460 117C 0001 0034            MOVE.B  #1,TSR(A0)      clear the ZDS flag
45 400466 103C      002A            MOVE.B  #'*',D0         display an * every second
46 40046A 4EB9 00008028             JSR     WRCHAR
47 400470 51C9      FFE6            DBRA    D1,LOOP2        decrement 1 minute count
48 400474 6000      FFDA            BRA     LOOP1
49                                  END
```

Program 28.3 Timing events using an MC68230 PIT, polled I/O

29

Interfacing to high level languages

Major requirements of modern software systems include reliability, flexibility, reusability and maintainability (see Chapter 17 for a discussion of these points). Although it is possible to implement assembly language modules which satisfy these requirements it is very difficult due to the long winded machine dependent nature of the code. Today the majority of a complex software system would be implemented in a high-level language with assembly language used for critical modules:

1 where speed and efficiency are very important, e.g. in real time control systems;
2 where implementation would be very difficult in a high level language, e.g. a memory test program (see the Problem for Chapter 17).

When implementing systems where the majority of the code is in a high-level language the assembly language modules have to adhere to the rules imposed by the high-level language which calls it, e.g. register usage, memory allocation, subroutine calling conventions, etc. (which vary from compiler to compiler). This chapter will present as an example the conventions of the Whitesmiths C cross compiler for MC68000 target computers (Whitesmiths 1986, 1987a, 1987b).

29.1 Interfacing Whitesmiths C to 68000 assembly language

The rules defining the C interface to 68000 assembly language are complex (Whitesmiths 1986) and include (only integer and character data are considered):

1 Module executable code is generated into the **.text** segment (see Program 8.1b).
2 Literal and global data is generated into the **.data** segment (see Program 8.1b).
3 Short integers are stored as 16-bit words, integers and long integers as long words.
4 Function calls are performed so (in C all modules are implemented as functions):
   (a) in a function call arguments are moved onto the stack **right to left**, i.e. the last argument is moved onto the stack first;
   (b) when used as function arguments character and short data is sign extended to a long word (integer and long integer data is already long word);
   (c) the function is called via `jsr _func` (an _ prefixes the function name specified in the program);
   (d) the function may use registers D0, D1, D2, D6, D7, A0, A1 and A2 without problems, if any others are used they must be **preserved** (by using MOVEM);
   (e) the function result (if any) is returned in D7.

5 C functions (including the main program) maintain a data area (called a stack frame by Whitesmiths) on the stack using `LINK A6,#n` where n is the number of bytes reserved and A6 holds the base address of the data area (see Chapter 14.5 for a full description of LINK):

   (a) if used, non-volatile registers (D3, D4, D5, A3, A4 and A5) are then stacked;

   (b) the top of the stack may be used as scratch storage; if so D0 will also be pushed onto the stack or the stack pointer adjusted.

   Within the function A6 points to the base of the data area with local variables positioned at -2(A6), -4(A6), etc. and function arguments at 8(A6), 12(A6), etc.

6 To return, non-volatile registers are restored, the data area removed with `UNLK A6` and an RTS executed. The function arguments will be removed from the stack on return to the calling program.

Program 29.1 (next page) shows a C program (main program and a function). Fig. 29.2 is the corresponding compiler listing which shows the C source code and the corresponding assembly language code generated. The program demonstrates:

1 The scope of variables (see Chapter 16.2), i.e.:

   (a) g_int and g_char are named global variables which could be accessed by any function within the current program code file (e.g. g_char is accessed in both the main program and the function sub_1);

   (b) the use of the LINK and UNLK instructions to allocate and deallocate local data areas on the stack (both within the main program and the function; see Chapter 14.5).

2 The parameters are passed to the function `sub_1` on the stack (see Chapter 16.5). Note that the first and second parameters are passed by **value** (the value of g_char, l_char, 10 and 20 are pushed onto the stack) and the third is passed by reference (a **pointer** to the variable is pushed onto the stack. i.e. the addresses of g_int and l_int).

Fig. 29.1 shows a run of the program on the Bytronic 68000 board. The function is called with three arguments, i.e. character (byte value), multiplier (integer word) and result (integer word):

```
sub_1(character,multiplier,&result)
```

where the value returned in result = multiplier * (character + 10) + g_char

i.e. on the first call 10 * ('A' + 10) + 'A' = 815
  on the second call 20 * ('B' + 10) + 'A' = 1585

```
--------------------------------------------------------------
/* Program 29.1 - simple C program, main with a function */

#include <stdio.h>

short int g_int=10;                    /* define global variables */
char g_char='A';

void main()                                       /* main program */
{
    void sub_1(char, short int, short int *);  /* prototype */

    short int l_int;                   /* define local variables */
    char l_char;

    l_int = g_int;                             /* local = global */
    l_char = 'B';                            /* local = constant */
    sub_1(g_char, 10, &g_int);               /* call subroutine */
    printf("g_int %8d \n", g_int);              /* print result */
    sub_1(l_char, 20, &l_int);               /* call subroutine */
    printf("l_int %8d \n", l_int);              /* print result */
}

/* function sub_1 with three parameters */
void sub_1
      (char arg_char, short int arg_mul, short int *arg_int)
{
    short int s_int;                    /* define local variable */

    s_int = arg_char + 10;                       /* set up local */
    *arg_int = s_int * arg_mul + g_char;   /* return result */
}
--------------------------------------------------------------
```

Program 29.1 C program to demonstrate function calling

```
--------------------------------------------------------------
M68> go/run program, from start address (<CR> for 001000) ? $
Whitesmiths C for Bytronic MC68000 microcomputer
g_int      815
l_int     1585
Program terminated by RTS instruction!
M68>
--------------------------------------------------------------
```

Fig. 29.1 Run of the C program on the Bytronic 68000 board

Program 29.1 was executed on the Bytronic single board computer where the memory addresses were (from the linker map):

_main = $1368 Whitesmiths C prefixes names with an _ character
_sub_1 = $13E2
_g_int = $376C
_g_char = $376E

To place a breakpoint on line 30 (see Fig. 29.2 for the line numbers) the address would be:

address of _sub_1 - offset of _sub_1 + offset of first instruction of line 30
= $13E2 - $92 +$A4 = $13F4

When the breakpoint was inserted the stack appeared so on the first call of sub_1 (values on stack are long words unless otherwise stated):

		address	contents		comments on contents
A7	@line 27 →	000FDA	00	4B	s_int='A'+10 (word)
A6	@line 27 →	000FDC	00	00	A6 on entry to sub_1
A6+2	@line 27 →	000FDE	0F	F4	"
A6+4	@line 27 →	000FE0	00	00	return address
A6+6	@line 27 →	000FE2	13	94	"
A6+8	@line 27 →	000FE4	00	00	arg_char=g_char='A'
A6+10	@line 27 →	000FE6	00	41	"
A6+12	@line 27 →	000FE8	00	00	arg_mul=10
A6+14	@line 27 →	000FEA	00	0A	"
A7	@line 8 →	000FEC	00	00	arg_int=address of g_int*
A6-6	@line 8 →	000FEE	37	6C	"
A6-4	@line 8 →	000FF0	00	42	l_char='B' (Byte)
A6-2	@line 8 →	000FF2	00	0A	l_int=10 (word)
A6	@line 8 →	000FF4	00	00	A6 on entry to main()
A6+2	@line 8 →	000FF6	00	00	"

* the top of the stack is (sometimes) used as a long word scratch stored for the first parameter of a function call, i.e. the `link a6,#-8` in the main program reserves a long word scratch store in addition to storage for l_int and l_char.

Fig. 29.3 shows a run of the program with the breakpoint inserted and a display of the information on the stack similar to the above diagram (with explanatory notes).

```
-------------------------------------------------------------------------------------
 1  /* Program 29.1 - simple C program, main with a function */
 2
 3  #include <stdio.h>
 4
 5  short int g_int=10;               /* define global variables */
                                      .data                       open data segment
                                      .even
  00000                     _g_int:
  00000  000a                         .word   10                  define g_int value 10
 6  char g_char='A';
  00002                     _g_char:
  00002  41                           .byte   65                  define g_char value 'A'
                                      .text                       open code segment
  00000                     L5:                                   define text string
  00000  67 5f 69 6e 74 20            .byte   103,95,105,110,116,32    "g_int %8d \n"
  00006  25 38 64 20 0a 00            .byte   37,56,100,32,10,0
  0000c                     L51:                                  define text string
  0000c  6c 5f 69 6e 74 20            .byte   108,95,105,110,116,32    "l_int %8d \n"
  00012  25 38 64 20 0a 00            .byte   37,56,100,32,10,0
 7
 8  void main()                                  /* main program */
 9  {
                                      .even                       force even word boundary
  00018                     _main:                                entry of main program
  00018  4e56 fff8                    link    a6,#-8              set up local storage
10      void sub_1(char, short int, short int *);  /* prototype */
11
12      short int l_int;              /* define local variables */
13      char l_char;
14
15      l_int = g_int;                          /* local = global */
  0001c  3d79 00000000 D              move.w  _g_int,-2(a6)       copy global to local
  00022  fffe
16      l_char = 'B';                         /* local = constant */
  00024  1d7c 0042 fffd               move.b  #66,-3(a6)          copy constant to local
17      sub_1(g_char, 10, &g_int);            /* call subroutine */
  0002a  2ebc 00000000 D              move.l  #_g_int,(sp)        push address of g_int
  00030  4878 000a                    pea     10                  push 10 (constant)
  00034  7e00                         moveq.l #0,d7
  00036  1e39 00000002 D              move.b  _g_char,d7          get g_char as long word
  0003c  2f07                         move.l  d7,-(sp)            push g_char
  0003e  4eb9 00000092 T              jsr     _sub_1              call function sub_1
  00044  508f                         addq.l  #8,sp               remove parameters
18      printf("g_int %8d \n", g_int);           /* print result */
  00046  3e39 00000000 D              move.w  _g_int,d7           get g_int
  0004c  48c7                         ext.l   d7                  make long word
  0004e  2e87                         move.l  d7,(sp)             push it
  00050  2f3c 00000000 T              move.l  #L5,-(sp)push text address
  00056  4eb9 00000000 U              jsr     _printf             call printf
  0005c  588f                         addq.l  #4,sp               remove parameters
-------------------------------------------------------------------------------------
```

Fig. 29.2a C main program with generated assembly code (text in bold added later)

```
19      sub_1(l_char, 20, &l_int);              /* call subroutine */
  0005e  486e fffe              pea     -2(a6)          push address of l_int
  00062  4878 0014              pea     20              push 20 (constant)
  00066  7e00                   moveq.l #0,d7
  00068  1e2e fffd              move.b  -3(a6),d7       get l_char as long word
  0006c  2f07                   move.l  d7,-(sp)        push l_char
  0006e  4eb9 00000092 T        jsr     _sub_1          call function sub_1
  00074  4fef 000c              lea     12(sp),sp       remove parameters
20      printf("l_int %8d \n", l_int);             /* print result */
  00078  3e2e fffe              move.w  -2(a6),d7       get l_int
  0007c  48c7                   ext.l   d7              make long word
  0007e  2e87                   move.l  d7,(sp)         push it
  00080  2f3c 0000000c T        move.l  #L51,-(sp)      push text address
  00086  4eb9 00000000 U        jsr     _printf         call printf
  0008c  588f                   addq.l  #4,sp           remove parameters
  0008e  4e5e                   unlk    a6              deallocate local storage
  00090  4e75                   rts                     return to monitor
                        *fnsize=61
21  }
22
23  /* function sub_1 with three parameters */
24  void sub_1
25      (char arg_char, short int arg_mul, short int *arg_int)
26  {
                                .even
  00092                 _sub_1:
  00092  4e56 fffe              link    a6,#-2          set up local data area
27      short int s_int;                /* define local variable */
28
29      s_int = arg_char + 10;                     /* set up local */
  00096  7e0a                   moveq.l #10,d7          load 10 (constant)
  00098  7c00                   moveq.l #0,d6
  0009a  1c2e 000b              move.b  11(a6),d6       get arg_char
  0009e  de86                   add.l   d6,d7           arg_char + 10
  000a0  3d47 fffe              move.w  d7,-2(a6)       move into s_int
30      *arg_int = s_int * arg_mul + g_char;    /* return result */
  000a4  226e 0010              move.l  16(a6),a1       get address of arg_int
  000a8  3e2e fffe              move.w  -2(a6),d7       get s_int
  000ac  cfee 000e              muls    14(a6),d7       s_int * arg_mul
  000b0  7c00                   moveq.l #0,d6
  000b2  1c39 00000002 D        move.b  _g_char,d6      get g_char
  000b8  de86                   add.l   d6,d7           (s_int * arg_mul)+g_char
  000ba  3287                   move.w  d7,(a1)         move into arg_int
  000bc  4e5e                   unlk    a6              deallocate data area
  000be  4e75                   rts                     return to main program
                        *fnsize=84
                                .globl  _main           define global names
                                .globl  _sub_1
                                .globl  _g_char
                                .globl  _g_int
  00003  00                     .globl  _printf
31  }
```

Fig. 29.2b C function with generated assembly code (text in bold added later)

```
--------------------------------------------------------------------------------
M68> Breakpoint set
Enter address to place new breakpoint ? $13F4
Breakpoint set OK

M68> go/run program, from start address (<CR> for 00001000) ? $
Whitesmiths C for FORCE MC68000 microcomputer

Program breakpoint (exception vector 47)
Program Counter (PC) = 000013F4  Status Register (SR) = 2700  CCR = -----
D0  = 00000000 D1  = 00000000 D2  = 00000000 D3  = 00000000
D4  = 00000000 D5  = 00000000 D6  = 00000041 D7  = 0000004B
A0  = 00001043 A1  = 00000000 A2  = 00000000 A3  = 00000000
A4  = 00000000 A5  = 00000000 A6  = 00000FDC A7  = 00000FDA


MC68000 monitor V1.1, please enter command (<ESC> to abort, ? for help)
M68> Memory modify word values starting at address $FDA
Enter word values (<CR> next value, < previous value, <ESC> to exit)
00000FDA 004B ? $               s_int='A'+10 (word)
00000FDC 0000 ? $               A6 on entry to sub_1
00000FDE 0FF4 ? $                       "
00000FE0 0000 ? $               return address
00000FE2 1394 ? $                       "
00000FE4 0000 ? $               arg_char=g_char='A'
00000FE6 0041 ? $                       "
00000FE8 0000 ? $               arg_mul=10
00000FEA 000A ? $                       "
00000FEC 0000 ? $               arg_int = address  of g_int
00000FEE 376C ? $                       "
00000FF0 0042 ? $               l_char='B' (Byte)
00000FF2 000A ? $               l_int=10 (word)
00000FF4 0000 ? $               A6 on entry to main()
00000FF6 0000 ? $                       "
M68>
--------------------------------------------------------------------------------
```

Fig. 29.3 Run of program 29.1 with breakpoint showing stack within function _sub_1 (explanatory text in **bold** added later)

30

Review of the MC68000 Family

The MC68000 (Stritter & Gunter 1979, Motorola 1989) appeared in 1979 and followed the successful Motorola MC6800 family of 8-bit microprocessors. It had 16-bit data and address buses (allowing up to 16Mbyte of memory), an extensive set of addressing modes and a large instruction set which could operate on 8-bit, 16-bit and 32-bit data.

The MC68008 appeared in 1982 as a low cost alternative to the MC68000 (Motorola 1984b, 1989). From a programmer's point of view it is identical to the MC68000 with the exception that:

1 The address bus is reduced to 20 bits (maximum 1Mbyte of memory).
2 The data bus is 8 bits wide (thus reducing performance).

The MC68010 ((MacGregor et al 1983, Motorola 1989) appeared in 1982. It is upward compatible with the MC68000 and supports:

1 Faster execution of tightly looped software via *loop mode* (Motorola 1989).
2 Virtual memory when used with the MC68451 Memory Management Unit.

The MC68020 (MacGregor et al 1984) appeared in 1984. It is upward compatible with the MC68000 and MC68010 and has enhancements including:

1 All instructions extended to 32-bit operations.
2 A larger register set, more instructions, more data types and addressing modes.
3 An on-chip instruction cache to speed up processing.
4 32-bit data and address buses (allowing addressing of 4Gbytes of memory) with dynamic bus sizing.
5 Co-processor interface (e.g. for an MC68881 floating point co-processor)

The MC68030 appeared in 1987. Enhancements include:

1 256 byte on-chip data and instruction caches.
2 An on-chip memory management unit.

The MC68040 (Edenfield et al 1990, Motorola 1991) was announced in 1990. Enhancements include:

1 Large on-chip data and instruction caches using advanced pipelining techniques.
2 An on-chip floating point co-processor.

Motorola provide an extensive range of support chips, e.g. a floating point co-processor (**MC68881**), Memory Management Units (**MC6845, MC68551, etc.**), DMA controllers (**MC68440, MC68450, etc.**) and peripheral devices such as the **MC68681** DUART and **MC68230** PIT.

Appendix A
Representation of Data Within the Computer

When humans use numeric data they usually represent the numbers using the decimal system, i.e. base 10. Working in decimal numbers requires the ability to differentiate between ten different states, the digits 0 through 9. For the human brain this is straightforward, and may even be extended to take account of alphabetic information (i.e. the characters a to z and A to Z). Computer systems are built from large numbers of similar electronic circuits. Although it is possible to build electronic circuits which can store and manipulate ten states, it is easier and cheaper to build electronic switches that may be in one of two states, either ON or OFF. Such circuits can therefore be used to represent binary data (base 2) with, for example, binary 1 and 0 being represented by the ON and OFF states respectively.

Computer systems internally represent **all** information, data and instructions, in binary form, with conversion between binary and human readable forms for input and output. When working in machine code or assembly language it is sometimes necessary to use binary or some similar number system. Binary numbers tend to be very long and hence it is easy to make mistakes when dealing with such data. In such situations the hexadecimal (base 16) number system is commonly used (it is very easy to convert between binary and hexadecimal).

A.1 Decimal and binary integer numbers

	Decimal Digit	Binary Bit
number base	10	2
possible states	0,1,2,3,4,5,6,7,8,9	0 or 1

The above table shows that a decimal digit can represent one of ten states, 0 through 9, and a binary bit (a single binary digit is called a bit) can represent two states, 0 or 1. It is possible, however, to represent more states by joining a sequence of digits or bits together, and in such a case it is assumed that the least significant digit or bit is the rightmost and the most significant is the leftmost. The bits or digits are generally numbered starting with the least significant from 0.

A.1.1 An eight digit decimal number

digit	**7**	**6**	**5**	**4**	**3**	**2**	**1**	**0**
digit value	10^7	10^6	10^5	10^4	10^3	10^2	10^1	10^0

In the decimal system the least significant (rightmost) digit represents units (10^0), the next tens (10^1), the next hundreds (10^2), etc., therefore the above eight digit number can represent values in the range 0 (all digits 0) to 99999999 (all digits 9).

A.1.2 An eight-bit binary number

bit	7	6	5	4	3	2	1	0
bit value	2^7	2^6	2^5	2^4	2^3	2^2	2^1	2^0

In the binary system the least significant (rightmost) bit represents units (2^0), the next twos (2^1), the next fours (2^2), etc., therefore the above 8-bit binary number can represent values in the range 0 (all bits 0) to 11111111 binary (all bits are 1). It is possible to convert between number bases and 11111111 binary is equivalent to 255 decimal. Larger values can be represented by more bits, for example a 16-bit binary number can represent 0 to 65535 decimal, and a 32-bit number 0 to 4294967295.

Within a computer system a memory word is built up from a number of bits. Typical word sizes are eight bits (usually called a **byte**), 16 bits, 32 bits or 64 bits. In practice the majority of modern computer systems use a memory based on bytes of storage, with sequences of bytes being used to store 16-, 32- or 64-bit numeric data.

A.2 Binary Addition

The following truth tables show all the possible combinations of the addition of:

(a) two bits A and B

and (b) two bits A and B plus a carry in from a previous addition.

In both cases the addition results in a SUM and a carry out.

A	+	B	SUM	carry
0		0	0	0
0		1	1	0
1		0	1	0
1		1	0	1

A	+	B	+ carry in	SUM	carry out
0		0	0	0	0
0		0	1	1	0
0		1	0	1	0
0		1	1	0	1
1		0	0	1	0
1		0	1	0	1
1		1	0	0	1
1		1	1	1	1

The following are examples of decimal and binary addition:

```
decimal  binary        decimal binary          decimal binary
   5       101            10      1010            27     11011
  +2     + 10           + 9     +1001            +15    + 1111
   7       111            19     10011            42    101010
```

The rightmost bits are added using the left hand table above. This results in a SUM and a carry bit which is carried out to be added into the addition of the next two bits (using the right hand table above). This addition then results in a sum and a carry out, etc.

The majority of modern computer systems store numeric values in sequences of bytes, i.e. 8-bit words of storage. A single byte is limited to representing a number in the range 0 to 255 decimal. If the addition of two bytes results in a carry out of bit 7, the result is greater than 255, and an error has occurred. When carrying out integer arithmetic on a computer system care must be taken to ensure that the results will fit the word size being used (generally 16 or 32 bits are used for integer number calculations).

A.3 Signed Binary Numbers

Mathematical and scientific calculations require the storage of negative as well as positive integer numbers. To represent a positive or negative number using the binary system one bit, usually the leftmost bit, is reserved for the sign. A negative number can then be represented in a number of forms, e.g. to represent -10 decimal as an eight bit signed binary number:

(a) sign-true magnitude	(b) ones-complement	(c) twos complement
10001010	11110101	11110110

Sign-True Magnitude Form. The leftmost bit holds the sign of the number, 0 for positive and 1 for negative, and the other seven bits represent the magnitude. In the example (a) 0001010 is the magnitude equivalent to 10 decimal, and the leftmost bit is 1 to indicate that the value is negative. This system is not commonly used in computer systems because it requires separate addition and subtraction circuits.

Ones Complement Form. To obtain the negative of a number each bit of the positive binary value is complemented, i.e. 0s are replaced with 1s and 1s with 0s. In example (b) +10 decimal, 00001010 binary, is complemented to form -10 decimal, i.e. 11110101 binary. This form is used in some computer systems, e.g. CDC 7600 series, but it has the problem that 0 can take two forms +0 (00000000) or -0 (11111111).

Twos Complement Form. To obtain the negative value of a number the ones complement is obtained, and then 1 added, i.e. in (c) above the value of +10 decimal, 00001010, is ones complemented to obtain 11110101, and then 1 added to obtain 11110110 (-10 decimal).

The advantage of complemented numbers is that separate addition and subtraction circuits are not required. To subtract a number, its complement is formed (a very easy operation), and the result added (using the normal adder circuits) to the other number. The majority of modern computer systems use twos complement form to represent signed binary numbers. In practice signed numbers are used for normal arithmetic calculations, and unsigned numbers for addresses, e.g. in assembly language programs. The range that can be represented by signed and unsigned 8-bit, 16-bit and 32-bit binary numbers is shown in Chapter 1 Table 1.1.

A.4 Overflow

Overflow occurs if the number of bits is too small to store the result of an arithmetic operation. For example, when using 8-bit signed numbers the binary addition 01101110 + 00101101 (decimal: 110 + 45) would result in the value 10011011 binary. It can be seen that the addition of the two positive numbers has resulted in the incorrect negative value -101 decimal. After the computer hardware has carried out an arithmetic operation it sets **condition code** bits that indicate if:

(a) the result was 0;
(b) the result was negative;
(c) a carry resulted from the operation;
or (d) an overflow occurred during the operation.

The condition code bits can be used in program control structures and for checking for error conditions. Many high-level language run-time systems automatically check for overflow errors, and special instructions can be used by assembly language programs to test the condition code bits.

A.5 Hexadecimal Numbers

When working in assembly languages it is often necessary to specify memory addresses and bit patterns. To do this using binary numbers would be cumbersome and error prone, i.e. to represent a 16-bit binary number sixteen 0s and 1s would have to be entered. In practice, the hexadecimal (base 16) number system is commonly used:

(a) it is a very concise way to represent numbers (each hexadecimal digit represents four binary bits);

and (b) it is easy to convert between binary and hexadecimal.

decimal	hexadecimal	binary
0	0	0000
1	1	0001
2	2	0010
3	3	0011
4	4	0100
5	5	0101
6	6	0110
7	7	0111
8	8	1000
9	9	1001
10	A	1010
11	B	1011
12	C	1100
13	D	1101
14	E	1110
15	F	1111

decimal	hexadecimal	binary
16	10	00010000
17	11	00010001
18	12	00010010
19	13	00010011
20	14	00010100
21	15	00010101
22	16	00010110
23	17	00010111
24	18	00011000
25	19	00011001
26	1A	00011010
27	1B	00011011
28	1C	00011100
29	1D	00011101
30	1E	00011110
31	1F	00011111

Table A.1 Decimal, hexadecimal and binary numbers

A.6 Conversion between the binary and hexadecimal number systems

To convert a binary number to hexadecimal:

(a) working from the least significant (rightmost) bit split the binary number up into groups of four bits;

then (b) using Table A.1 convert each group of four bits into the equivalent hexadecimal digit.

For example:

```
0001001110011110 to 0001 0011 1001 1110 = 139E hexadecimal
```

To convert from hexadecimal to binary, replace each hexadecimal digit with the equivalent four bit binary value.

A.7 Conversion of Decimal Numbers to Binary

To convert a positive decimal integer the following algorithm starts by generating the least significant binary bit, then the next, etc.:

```
LOOP
    next binary bit = remainder of DECIMAL_VALUE/2
    DECIMAL_VALUE = DECIMAL_VALUE/2 (ignoring remainder)
UNTIL DECIMAL_VALUE=0
```

e.g. convert decimal 38 to binary

```
                                                     result
38/2 = 19 remainder 0 gives binary 0                      0
19/2 = 9  remainder 1 gives binary 1                     10
9/2  = 4  remainder 1 gives binary 1                    110
4/2  = 2  remainder 0 gives binary 0                   0110
2/2  = 1  remainder 0 gives binary 0                  00110
1/2  = 0  remainder 1 gives binary 1                 100110
```

To obtain the binary equivalent of a negative decimal number, convert the absolute value to binary then take the twos complement.

A.8 Conversion of binary numbers to decimal

The following algorithm converts a binary number into decimal:

```
DECIMAL_VALUE=0
LOOP starting with the most significant binary bit
    BIT_VALUE = value of current binary bit
    DECIMAL_VALUE = DECIMAL_VALUE*2 + BIT_VALUE
UNTIL current bit is the least significant
```

For example, convert 100110 binary (remember the least significant bit is bit 0):

```
bit processed      5      4       3       2       1       0
DECIMAL_VALUE   (((1*2 + 0)*2 + 0)*2 + 1)*2 + 1)*2 + 0 = 38
```

A.9 Conversion of decimal numbers to hexadecimal

The following algorithm generates the least significant (rightmost) hexadecimal digit, then the next digit, etc.:

```
LOOP
    REMAINDER = remainder of DECIMAL_VALUE/16
    next hexadecimal digit =
            hexadecimal equivalent of REMAINDER
    DECIMAL_VALUE = DECIMAL_VALUE/16 (ignoring remainder)
UNTIL DECIMAL_VALUE=0
```

e.g. convert 1567 to hexadecimal

```
                                                        result
1567/16 = 97 remainder 15 gives hexadecimal F              F
97/16   = 6  remainder 1  gives hexadecimal 1             1F
6/16    = 0  remainder 6  gives hexadecimal 6            61F
```

A.10 Conversion of hexadecimal numbers to decimal

```
DECIMAL_VALUE=0
LOOP starting with the most significant hexadecimal digit
    DIGIT_VALUE = decimal value of current hexadecimal digit
    DECIMAL_VALUE = DECIMAL_VALUE*16 + DIGIT_VALUE
UNTIL current hexadecimal digit is the least significant
```

For example, convert 61F hexadecimal to decimal:

```
hex digit processed    2        1         0
DECIMAL_VALUE        ((6*16) + 1)*16 + 15 = 1567
```

A.11 Fixed Point Real Numbers

So far only integer numbers have been considered. Such numbers are useful when calculations on whole number values are required, e.g. for loop control in programs. In practice, however, it is necessary to be able to represent fractional components of numbers as well. These are called real numbers and one means by which these may be represented is in fixed point form. The following shows a 16-bit binary value in which the whole number part (with sign) is stored in eight bits and the fractional component in eight bits:

15	14	13	12	11	10	9	8	7	6	5	4	3	2	1	0
sign	2^6	2^5	2^4	2^3	2^2	2^1	2^0	2^{-1}	2^{-2}	2^{-3}	2^{-4}	2^{-5}	2^{-6}	2^{-7}	2^{-8}

← whole number part → ← factional number part →

For example decimal 10.75 would be 1010.11 binary. The major limitations of this system of real number representation are that:

1 The maximum absolute size of numbers is limited by the number of bits assigned to hold the whole number part (as with a normal binary integer number).

2 When dealing with small fractional components accuracy is lost, and very small values cannot be represented at all, i.e. the smallest value that can be represented by the above fixed point number is 0.00390625.

In practice these restrictions on fixed point numbers do not make it worth while providing the extra software or hardware within the computer system to process them.

A.12 Floating Point Real Numbers

In many scientific and engineering applications very small or very large numbers have to be represented, e.g. from the sizes of atomic particles to intergalactic distances. In the floating point number system the real value is represented by a signed fractional component called the mantissa and a signed exponent. For example, decimal floating point numbers (using base 10) can be represented:

$$\text{mantissa} * 10^{\text{exponent}} \quad \text{where } 0.1 >= \text{mantissa} < 1.0$$

To maintain accuracy the absolute value of the mantissa is maintained within the range shown (this process is called normalisation), e.g. e.g 6520000.0 would be $0.652*10^7$ and -0.00000000652 would be $-0.652*10^{-8}$. In practice many printers cannot print superscripts so the above examples would be printed as follows: 6520000.0 as 0.652E7 and -0.00000000652 as -0.652E-8 where the E indicates an exponent of 10.

Within computer systems the fractional component is held as a binary fraction and the exponent is a power of 2 (or possibly 16). A typical system may store each floating point number in 32 bits with 24 bits to hold the signed mantissa and 8 bits for the signed exponent. In this case the accuracy of the mantissa is 23 binary bits (which is equivalent to 6 or 7 decimal figures of accuracy), and the range of the exponent would be -128 to +127. Greater accuracy can be obtained by using 64-bit storage in which 53 bits may be used to store the signed mantissa (giving 15 to 17 decimal figures of accuracy) and 11 bits for the exponent.

Floating point calculations can be carried out using floating point co-processor chips (Huntsman & Cawthron 1983), or emulated in software that uses the integer arithmetic operations of a computer. The advantage of floating point hardware is that it can be several orders of magnitude faster than software emulation, but it requires more complex and expensive hardware.

A.13 ASCII Character Code

Table A.2 lists the ASCII character codes (American Standard Code for Information Interchange), with the columns being the decimal value, the hexadecimal value, then the corresponding character. ASCII is the most widely used character code for data transmission between computers, terminals and printers. As with all information within the computer system, characters are represented by binary patterns. In the ASCII code each character is represented by a seven bit code that is stored one character per byte (with bit 7 set to 0 or used as a parity check).

The characters below 32 decimal (20 hexadecimal) are non-printing control characters. These are used to control the action of printers, display screens, communications systems, etc. Important control characters are:

NUL	null: no action (used as a fill or delay character)
BEL	bell: rings the keyboard bell or buzzer
BS	backspace: move back one character width
HT	horizontal tabulate: move horizontally to next tabulate position
LF	line feed: move page vertically one character height
FF	form feed: new page on printer, clear display screen
CR	carriage return: move to start of current line
ESC	escape: used in many systems as a program control character
SP	space: move horizontal by one character width

For example to move a printer or a display screen to a new line position the characters CR (carriage return) then LF (line feed) will be output. In addition some of the printable characters will depend upon the printer font being used.

It is worth noting that the ASCII codes for the numeric characters 0 to 9, and alphabetic characters A to Z and a to z, are arranged in ascending numerical order. This property can be used for:

1 Testing if a character is within a range, i.e. in the range A to Z.
2 The conversion of numeric decimal data, entered at a keyboard, into internal binary form.

Do not confuse the code for a numeric character with the equivalent numeric binary value, i.e the code for the **character 1** is 31 hexadecimal (49 decimal). When a number composed of several digits is read from a keyboard the character codes are read, turned into the equivalent binary numeric value and then added to any previous total. The following algorithm reads a decimal number from a keyboard (until a non-digit is entered):

```
NUMBER=0
READ(character)
LOOP WHILE character is in the range '0' to '9'
    DIGIT_VALUE = character - '0'
    NUMBER = NUMBER*10 + DIGIT_VALUE
    READ(character)
END LOOP
```

In the majority of programming languages a character code value is specified by enclosing it in quote marks. In the above algorithm characters are read from the keyboard until a non-digit character is hit. If the character is a digit, say 7 was hit, the ASCII code for 0 is subtracted from it to get the equivalent numeric value DIGIT_VALUE, i.e. in this case 30 hexadecimal (the code for '0'), will be subtracted from 37 hexadecimal (the code for '7'), to give DIGIT_VALUE=7. The NUMBER entered so far is then multiplied by ten and the current DIGIT_VALUE added.

0	00	NUL
1	01	SOH
2	02	STX
3	03	ETX
4	04	EOT
5	05	ENQ
6	06	ACK
7	07	BEL
8	08	BS
9	09	HT
10	0A	LF
11	0B	VT
12	0C	FF
13	0D	CR
14	0E	SO
15	0F	SI
16	10	DLE
17	11	DC1
18	12	DC2
19	13	DC3
20	14	DC4
21	15	NAK
22	16	SYN
23	17	ETB
24	18	CAN
25	19	EM
26	1A	SUB
27	1B	ESC
28	1C	FS
29	1D	GS
30	1E	RS
31	1F	US

32	20	SP
33	21	!
34	22	"
35	23	#
36	24	$
37	25	%
38	26	&
39	27	'
40	28	(
41	29	)
42	2A	*
43	2B	+
44	2C	,
45	2D	-
46	2E	.
47	2F	/
48	30	0
49	31	1
50	32	2
51	33	3
52	34	4
53	35	5
54	36	6
55	37	7
56	38	8
57	39	9
58	3A	:
59	3B	;
60	3C	<
61	3D	=
62	3E	>
63	3F	?

64	40	@
65	41	A
66	42	B
67	43	C
68	44	D
69	45	E
70	46	F
71	47	G
72	48	H
73	49	I
74	4A	J
75	4B	K
76	4C	L
77	4D	M
78	4E	N
79	4F	O
80	50	P
81	51	Q
82	52	R
83	53	S
84	54	T
85	55	U
86	56	V
87	57	W
88	58	X
89	59	Y
90	5A	Z
91	5B	[
92	5C	\
93	5D	]
94	5E	^
95	5F	_

96	60	`
97	61	a
98	62	b
99	63	c
100	64	d
101	65	e
102	66	f
103	67	g
104	68	h
105	69	i
106	6A	j
107	6B	k
108	6C	l
109	6D	m
110	6E	n
111	6F	o
112	70	p
113	71	q
114	72	r
115	73	s
116	74	t
117	75	u
118	76	v
119	77	w
120	78	x
121	79	y
122	7A	z
123	7B	{
124	7C	|
125	7D	}
126	7E	~
127	7F	DEL

Table A.2 The ASCII Character Codes: columns are decimal and hexadecimal numeric character code value followed by the character

When character information is transmitted over a noisy communications channel a parity bit can replace bit 7 (which is not used in the ASCII code) or be added to make the total character length of 9-bits (for more details of parity checking see the Problem for Chapter 12).

Appendix B

MC68000 Information

B.1 Addressing Modes

Table B.1 presents details of the MC68000 addressing modes in the following format:

Addressing Mode. This is the name of the mode.

Mode and Register Fields. Within the generated machine code a number of bits define the addressing mode or modes that are used (for the source and destination operands). This information is specified as follows:

1 If the addressing mode can only be a register (e.g. EXG exchange instruction), a three-bit **Register** field specifies the operand.
2 If a number of addressing modes can be used (e.g. source or destination of an ADD), the operand is specified by a three-bit **Mode** field and a three-bit **Register** field. In cases when the the Mode value is 111 the three Register bits supply further addressing mode information, not a register number.

Effective Address. This entry shows how the effective address (EA) is obtained.

Addressing Categories. This entry lists the address categories that the addressing mode can be used for:

D **Data References**: the effective address refers to a data operand (this in fact includes everything except address registers).

M **Memory References**: the effective address refers to an operand in memory (i.e. not in a register).

C **Control References**: are those that can specify the destination of a jump (JMP) or jump to subroutine (JSR) instruction.

A **Alterable References**: an operand is alterable if it can be written to.

Time. Each memory read/write cycle takes four clock cycles (assuming that the memory is not using WAIT states). Thus the more complex addressing modes that have to access memory a number of times can use several read/write cycles to obtain the operand. This column lists, for each addressing mode, the number of clock periods required to:

(a) fetch any extension words that are part of the machine code (e.g. address register indirect with displacement, requires a fetch of the displacement (d_{16}) from the word following the op-code);
(b) calculate the effective address (e.g. address register indirect with displacement, requires Effective Address = (An) + displacement);
and (c) reading or writing the memory operand (i.e. accessing the Effective Address calculated above).

The number of clock periods for byte (**B**) and word (**W**), and long word (**L**) operands are shown. These values have to be added to the basic instruction execution times to obtain the full execution time of the complete instruction.

Assembler Syntax: is then shown for each addressing mode.

Page: the page in this book where this mode is described in detail.

Table B.1 Addressing Modes

EA = Effective Address,
An = Address Register (n = 0 to 7)
Dn = Data Register (n = 0 to 7)
Xn = Address or Data Register used as an Index Register
SR = Status Register
CCR = Condition Code Register
nnn = Absolute Address

d_8 = 8-bit displacement or offset
d_{16} = 16-bit displacement or offset
N = 1 for byte, 2 for word & 4 for long word
( ) = contents of, e.g. (A1) contents of A1
PC = Program Counter
SSP = System Stack Pointer
USP = User Stack Pointer
xxx = numeric value

				Categories				Time		Assembler	Page
Addressing Mode	Mode	Register	Effective Address	D	M	C	A	B,W	L	Syntax	
Data Register Direct	000	reg. no.	EA = Dn	*			*	0	0	Dn	40
Address Register Direct	001	reg. no.	EA=An				*	0	0	An	100
Address Register Indirect	010	reg. no.	EA=(An)	*	*	*	*	4	8	(An)	100
Address Register Indirect with Postincrement	011	reg. no.	EA=(An), An=An+N	*	*		*	4	8	(An)+	101
Address Register Indirect with Predecrement	100	reg. no.	An=An-N, EA=(An)	*	*		*	6	10	-(An)	101
Address Register Indirect with Displacement	101	reg. no.	EA=(An)+d_{16}	*	*	*	*	8	12	d_{16}(An)	103
Address Register Indirect with Index & Displacement	110	reg. no.	EA=(An)+(Xn)+d_8	*	*	*	*	10	14	d_8(An,Xn)	104
Absolute Short	111	000	EA=(next word)	*	*	*	*	8	12	nnn.W	60
Absolute Long	111	001	EA=(next two words)	*	*	*	*	12	16	nnn.L	60
Program Counter Relative with Displacement	111	010	EA=(PC)+d_{16}	*	*	*		8	12	d_{16}(PC)	117
PC Relative with Index and Displacement	111	011	EA=(PC)+(Xn)+d_8	*	*	*		10	14	d_8(PC,Xn)	117
Immediate	111	100	Data=Next Word	*	*			4	8	#xxx	41

For example, (using Table B.1) *address register indirect addressing mode* using A2:

1 The machine code Mode and Register bits are 101 and 010 respectively.
2 The EA (Effective Address) is (A2) (A2 contains the operand address).
3 It can be used for all addressing categories: data, memory, control and alterable.
4 To calculate the effective address and fetch the operand takes 4 clock cycles for byte and word (one memory access), and 8 for long word (two memory accesses).
5 The assembler syntax is (An) where n is the register number (0 to 7).

For example, *Program Counter (PC) relative with displacement addressing mode*:

1 The machine code Mode and Register bits are 111 and 010 respectively.
2 The EA (Effective Address) is (PC) + d_{16} (PC contents plus a 16-bit displacement value).
3 It can be used for addressing categories: data, memory and control, i.e. **not** alterable.
4 To calculate effective address and fetch the operand takes 8 clock cycles for a byte or word and 12 for a long word. In the case of long word this is made up from 4 cycles to fetch the displacement and 8 cycles to fetch the long word operand.
5 The assembler syntax is LABEL(PC) where LABEL will be a 16-bit displacement.

B.2 MC68000 Instructions

The following section presents a number of tables that contain the following information:

Mnemonic: the operation mnemonic used by the assembler.

Size: the operand size(s) that can be used with the operation.

Address Mode: this column and the following columns show the valid source (**s** =) and destination (**d** =) operand addressing modes that can be used with the operation. If an addressing mode column is blank the mode may not be used. Under each addressing mode column the following information is given:

(a) # shows the length in bytes of the machine code instruction;

and (b) % shows the number of clock cycles to execute the instruction (includes instruction fetch, extension word fetches (if any), effective address calculation, and operand fetches (if any)).

Operation-code bit pattern: lists the bit patterns of the machine code instruction; a complete instruction consists of the operation-code word followed by any operand words:

```
bit          15 14 13 12 11 10  9  8  7  6  5  4  3  2  1  0
            +-----------------------------------------------+
            |              Operation-Code Word              |
            |    (specifies operation and addressing modes  |
            +-----------------------------------------------+
            |               Immediate Operand               |
            |          (if any,  one or two words)          |
            +-----------------------------------------------+
            |        Source Effective Address Extension     |
            |          (if any,  one or two words)          |
            +-----------------------------------------------+
            |     Destination Effective Address Extension   |
            |          (if any,  one or two words)          |
            +-----------------------------------------------+
```

Depending upon the complexity of the instruction its length may be from one word (2 bytes, e.g. CLR of a register), to 5 words (10 bytes, e.g. MOVE using long absolute addressing for both source and destination operands).

Within the operation-code word the addressing modes are specified by a six-bit field, consisting of three mode bits and three register bits (as listed in Table B.1). For example, an operation word with a single effective address (source or destination) would appear:

bit	15	14	13	12	11	10	9	8	7	6	5	4	3	2	1	0
	x	x	x	x	x	x	x	x	x	x	mode			register		

An operation word with source and destination effective addresses would appear, e.g. `MOVE.W (A0)+,-(A1)`:

bit	15	14	13	12	11	10	9	8	7	6	5	4	3	2	1	0
					Destination						Source					
	x	x	x	x	register			mode			mode			register		

CCR: shows how the Condition Code bits (X N Z V C) are affected by the operations.

For example, ADD.L when the source is in a data register:

1 s (source) = Dn
2 d = Dn, (An), (An)+, -(An), d_{16}(An), d_8(An,Xn), Abs.W or Abs.L (e.g.in the case of d = d_{16}(An), the instruction length is 4 bytes and execution time is 24 clock cycles).
3 The operation-code bit pattern is 1101 DDD1 10EE EEEE where DDD is the register number and EEEEEE is the destination effective address, e.g. ADD.L D3,LABEL(A1) becomes 1101 0111 1010 1001, followed by a word containing the displacement.
4 All the CCR bits are affected by the operation.

Notes:

*	Word operand only
#	Number of Bytes in Instruction
%	Execution Times in Clock Cycles
s	Source (S10 base 10 operand)
d	Destination (D10 base 10 operand)
<	value is maximum number
d8	8-bit Displacement
d16	16-bit Displacement
Imm	Immediate Data
Imm3	3-bit Immediate Data
Imm8	8-bit Immediate Data

A	Address Register Number
C	Test Condition
D	Data Register Number
E	Destination Effective Address
e	Source Effective Address
f	Direction: 0 = right, 1 = left
M	Destination EA Mode
P	Displacement
Q	Quick Immediate Data
R	Destination Register
r	Source Register
S	Size: 00 = byte 01 = Word 01 = Long Word
V	Vector Number
XX	Move size 01 = Byte 11 = Word 01 = long

CCR	Condition Code Notation
*	Set according to result of operation
-	Not affected by operation
0	Always cleared to 0
1	Always set to 1
U	Undefined after operation

Table B.2 MC68000 Operation Codes

Mnemonic	Size	Address Mode	Dn		An		(An)		(An)+		-(An)		d16(An)		d8(An,Xn)		AbsW		AbsL		d16(PC)		d8(PC,Xn)		Immed		op-code bit pattern 1111 11 / 5432 1098 7654 3210	CCR XNZVC	page
			#	%	#	%	#	%	#	%	#	%	#	%	#	%	#	%	#	%	#	%	#	%	#	%			
ABCD	B	s=Dn d=	2	6																							1100 RRR1 0000 0rrr	*U*U*	56
		s=-(An) d=									2	18															1100 RRR1 0000 1rrr		
ADD	BW	s=Dn d=	2	4			2	12	2	12	2	14	4	16	4	18	4	16	6	20							1101 DDD1 SSEE EEEE	*****	50
		d=Dn s=	2	4	2	*4	2	8	2	8	2	10	4	12	4	14	4	12	6	16	4	12	4	14	4	8	1101 DDD0 SSee eeee		
	L	s=Dn d=	2	8			2	20	2	20	2	22	4	24	4	26	4	24	6	28							1101 DDD1 10EE EEEE		
		d=Dn s=	2	8	2	8	2	14	2	14	2	16	4	18	4	20	4	18	6	22	4	18	4	20	6	16	1101 DDD0 10ee eeee		
ADDA	W	d=An s=	2	8	2	8	2	12	2	12	2	14	4	16	4	18	4	16	6	20	4	16	4	18	4	12	1101 AAA0 11ee eeee	-----	100
	L	d=An s=	2	8	2	8	2	14	2	14	2	16	4	18	4	20	4	18	6	22	4	18	4	20	6	16	1101 AAA1 11ee eeee		
ADDI	BW	s=Imm d=	4	8			4	16	4	16	4	18	6	20	6	22	6	20	8	24							0000 0110 SSEE EEEE	*****	50
	L	s=Imm d=	6	16			6	28	6	28	6	30	8	32	8	34	8	32	1	36									
ADDQ	BW	s=Imm3 d=	2	4	2	*8	2	12	2	12	2	14	4	16	4	18	4	16	6	20							0101 QQQ0 SSEE EEEE	*****	51
	L	s=Imm3 d=	2	8	2	8	2	20	2	20	2	22	4	24	4	26	4	24	6	28									
ADDX	BW	s=Dn d=	2	4																							1101 RRR1 SS00 0rrr	*****	56
		s=-(An) d=									2	18															1101 RRR1 SS00 1rrr		
	L	s=Dn d=	2	8																							1101 RRR1 1000 0rrr		
		s=-(An) d=									2	30															1101 RRR1 1000 1rrr		
AND	BW	s=Dn d=	2	4			2	12	2	12	2	14	4	16	4	18	4	16	6	20							1100 DDD1 SSEE EEEE	-**00	91
		d=Dn s=	2	4			2	8	2	8	2	10	4	12	4	14	4	12	6	16	4	12	4	14	4	8	1100 DDD0 SSee eeee		
	L	s=Dn d=	2	8			2	20	2	20	2	22	4	24	4	26	4	24	6	28							1100 DDD1 10EE EEEE		
		d=Dn s=	2	8			2	14	2	14	2	16	4	18	4	20	4	18	6	22	4	18	4	20	6	16	1100 DDD0 10ee eeee		
ANDI	BW	s=Imm d=	4	8			4	16	4	16	4	18	6	20	6	22	6	20	8	24							0000 0010 SSEE EEEE	-**00	92
	L	s=Imm d=	6	16			6	28	6	28	6	30	8	32	8	34	8	32	10	36									
ANDI CCR	B	d=CCR s=																							4	20	0000 0010 0011 1100	*****	92
ANDI SR	W	d=SR s=																							4	20	0000 0010 0111 1100	*****	199
ASL,ASR		see Table B.3																											93
Bcc		see Table B.5																											80
BCHG/CLR		see Table B.4																											95
BRA	B	d8=																							2	10	0110 0000 PPPP PPPP	-----	77
	W	d16=																							4	10			
BSET		see Table B.4																											95
BSR	B	d8=																							2	18	0110 0001 PPPP PPPP	-----	109
	W	d16=																							4	18			
BTST		see Table B.4																											95
CLR	BW	d=	2	4			2	12	2	12	2	14	4	16	4	18	4	16	6	20							0100 0010 SSEE EEEE	-0100	40
	L	d=	2	6			2	20	2	20	2	22	4	24	4	26	4	24	6	28									
CMP	BW	d=Dn s=	2	4	2	*4	2	8	2	8	2	10	4	12	4	14	4	12	6	16	4	12	4	14	4	8	1011 DDD0 SSee eeee	-****	82
	L	d=Dn s=	2	6	2	6	2	14	2	14	2	16	4	18	4	20	4	18	6	22	4	18	4	20	6	14			
CMPA	W	d=An s=	2	6	2	6	2	10	2	10	2	12	4	14	4	16	4	12	6	18	4	14	4	16	4	10	1011 AAA0 11ee eeee	-****	100
	L	d=An s=	2	6	2	6	2	14	2	14	2	16	4	18	4	20	4	18	6	22	4	18	4	20	6	14	1011 AAA1 11ee eeee		
CMPI	BW	s=Imm d=	4	8			4	12	4	12	4	14	6	16	6	18	6	16	8	20							0000 1100 SSEE EEEE	-****	82
	L	s=Imm d=	6	14			6	20	6	20	6	22	8	24	8	26	8	24	10	28									
CMPM	BW	s=(An)+ d=							2	12																	1011 RRR1 SS00 1rrr	-****	82
	L	s=(An)+ d=							2	20																			
DBcc		see Table B.6																											88
DIVS/U		see Table B.7																											52
EOR	BW	s=Dn d=	2	4			2	12	2	12	2	14	4	16	4	18	4	16	6	20							1011 rrr1 SSEE EEEE	-**00	91
	L	s=Dn d=	2	8			2	16	2	16	2	18	4	20	4	22	4	20	6	24									
EORI CCR	B	d=CCR s=																							4	20	0000 1010 0011 1100	*****	92

Table B.2 MC68000 Operation Codes

Mnemonic	Size	Address Mode	Dn #	Dn %	An #	An %	(An) #	(An) %	(An)+ #	(An)+ %	-(An) #	-(An) %	d16(An) #	d16(An) %	d8(An,Xn) #	d8(An,Xn) %	AbsW #	AbsW %	AbsL #	AbsL %	d16(PC) #	d16(PC) %	d8(PC,Xn) #	d8(PC,Xn) %	Immed #	Immed %	op-code bit pattern 1111 5432	11 1098	7654	3210	CCR XNZVC	page
EORI SR	W	d=SR s=																							4	20	0000	1010	0111	1100	*****	199
EORI	BW	s=Imm d=	4	8			4	16	4	16	4	18	6	20	6	22	6	20	8	24							0000	1010	SSEE	EEEE	-**00	92
	L	s=Imm d=	6	16			6	28	6	28	6	30	8	32	8	34	8	32	10	36												
EXG	L	s=Dn d=	2	6																							1100	DDD1	0100	0DDD	-----	51
		s=An d=			2	6																					1100	AAA1	0100	1AAA		
		s=Dn d=			2	6																					1100	DDD1	1000	1AAA		
EXT	W	d=	2	4																							0100	1000	1000	0DDD	-**00	52
	L	d=	2	4																							0100	1000	1100	0DDD		
ILLEGAL			2	34																							0100	1010	1111	1100	-----	201
JMP		d=					2	8					4	10	4	14	4	10	6	12	4	10	4	14			0100	1110	11EE	EEEE	-----	78
JSR		d=					2	16					4	18	4	22	4	18	6	20	4	18	4	22			0100	1110	10EE	EEEE	-----	109
LEA	L	d=An s=					2	4					4	8	4	12	4	8	6	12	4	8	4	12			0100	AAA1	11ee	eeee	-----	116
LINK		d16=Imm s=			4	16																					0100	1110	0101	0AAA	-----	111
LSL,LSR		see Table B.3																														94
MOVE	BW	d=Dn s=	2	4	2	4	2	8	2	8	2	10	4	12	4	14	4	12	6	16	4	12	4	14	4	8	00XX	RRRM	MMee	eeee	-**00	41
		d=(An) s=	2	8	2	8	2	12	2	12	2	14	4	16	4	18	4	16	6	20	4	16	4	18	4	12						
		d=(An)+ s=	2	8	2	8	2	12	2	12	2	14	4	16	4	18	4	16	6	20	4	16	4	18	4	12						
		d=-(An) s=	2	8	2	8	2	12	2	12	2	14	4	16	4	18	4	16	6	20	4	16	4	18	4	12						
		d=d16(An) s=	4	12	4	12	4	16	4	16	4	18	6	20	6	22	6	20	8	24	6	20	6	22	6	16						
		d=d8(An,Xn) s=	4	14	4	14	4	18	4	18	4	20	6	22	6	24	6	22	8	26	6	22	6	24	6	18						
		d=Abs.W s=	4	12	4	12	4	16	4	16	4	18	6	20	6	22	6	20	8	24	6	20	6	22	6	16						
		d=Abs.L s=	6	16	6	16	6	20	6	20	6	22	8	24	8	26	8	24	10	28	8	24	8	26	8	20						
MOVE	L	d=Dn s=	2	4	2	4	2	12	2	12	2	14	4	16	4	18	4	16	6	20	4	16	4	18	6	12	0010	RRRM	MMee	eeee	-**00	41
		d=(An) s=	2	12	2	12	2	20	2	20	2	22	4	24	4	26	4	24	6	28	4	24	4	26	6	20						
		d=(An)+ s=	2	12	2	12	2	20	2	20	2	22	4	24	4	26	4	24	6	28	4	24	4	26	6	20						
		d=-(An) s=	2	12	2	12	2	20	2	20	2	22	4	24	4	26	4	24	6	28	4	24	4	26	6	20						
		d=d16(An) s=	4	16	4	16	4	24	4	24	4	26	6	28	6	30	6	28	8	32	6	28	6	30	8	24						
		d=d8(An,Xn) s=	4	18	4	18	4	26	4	26	4	28	6	30	6	32	6	30	8	34	6	30	6	32	8	26						
		d=Abs.W s=	4	16	4	16	4	24	4	24	4	26	6	28	6	30	6	28	8	32	6	28	6	30	8	24						
		d=Abs.L s=	6	20	6	20	6	28	6	28	6	30	8	32	8	34	8	32	10	36	8	32	8	34	10	28						
MOVE CCR	W	d=CCR s=	2	12			2	16	2	16	2	18	4	20	4	22	4	20	6	24	4	20	4	22	4	16	0100	0100	11ee	eeee	*****	79
MOVE SR	W	d=SR s=	2	12			2	16	2	16	2	18	4	20	4	22	4	20	6	24	4	20	4	22	4	16	0100	0110	11ee	eeee	*****	199
		s=SR d=	2	6			2	12	2	12	2	14	4	16	4	18	4	16	6	20							0100	0000	11EE	EEEE	-----	137
MOVE USP	L	s=USP d=			2	4																					0100	1110	0110	1AAA	-----	200
		d=USP s=			2	4																					0100	1110	0110	0AAA		
MOVEA	W	d=An s=	2	4	2	4	2	8	2	8	2	10	4	12	4	14	4	12	6	16	4	12	4	14	4	8	0011	AAA0	01ee	eeee	-----	100
	L	d=An s=	2	4	2	4	2	12	2	12	2	14	4	16	4	18	4	16	6	20	4	16	4	18	6	12	0010	AAA0	01ee	eeee	-----	
MOVEM		see Table B.8																														110
MOVEP	W	s=Dn d=											4	16													0000	DDD1	1000	1AAA	-----	152
		s=d16(An) d=	4	16																							0000	DDD1	0000	1AAA		
	L	s=Dn d=											4	24													0000	DDD1	1100	1AAA		
		s=d16(An) d=	4	24																							0000	DDD1	0100	1AAA		
MOVEQ	L	s=Imm8 d=	2	4																							0111	DDD0	QQQQ	QQQQ	-**00	42
MULS/U		see Table B.7																														52
NBCD	B	d=	2	6			2	12	2	12	2	14	4	16	4	18	4	16	6	20							0100	1000	00EE	EEEE	*U*U*	56
NEG	BW	d=	2	4			2	12	2	12	2	14	4	16	4	18	4	16	6	20							0100	0100	SSEE	EEEE	*****	51
	L	d=	2	6			2	20	2	20	2	22	4	24	4	26	4	24	6	28												

Table B.2 MC68000 Operation Codes

Mnemonic	Size	Address Mode	Dn #	Dn %	An #	An %	(An) #	(An) %	(An)+ #	(An)+ %	-(An) #	-(An) %	d16(An) #	d16(An) %	d8(An,Xn) #	d8(An,Xn) %	AbsW #	AbsW %	AbsL #	AbsL %	d16(PC) #	d16(PC) %	d8(PC,Xn) #	d8(PC,Xn) %	Immed #	Immed %	op-code bit pattern 1111 11 5432 1098 7654 3210	CCR XNZVC	page
NEGX	BW	d=	2	4			2	12	2	12	2	14	4	16	4	18	4	16	6	20							0100 0000 SSEE EEEE	*****	56
	L	d=	2	6			2	20	2	20	2	22	4	24	4	26	4	24	6	28									
NOP			2	4																							0100 1110 0111 0001	-----	246
NOT	BW	d=	2	4			2	12	2	12	2	14	4	16	4	18	4	16	6	20							0100 0110 SSEE EEEE	-**00	91
	L	d=	2	6			2	20	2	20	2	22	4	24	4	26	4	24	6	28									
OR	BW	s=Dn d=					2	12	2	12	2	14	4	16	4	18	4	16	6	20							1000 DDD1 SSEE EEEE	-**00	91
		d=Dn s=	2	4			2	8	2	8	2	10	4	12	4	14	4	12	6	16	4	12	4	14	4	8	1000 DDD0 SSee eeee		
	L	s=Dn d=					2	20	2	20	2	22	4	24	4	26	4	24	6	28							1000 DDD1 10EE EEEE		
		d=Dn s=	2	8			2	14	2	14	2	16	4	18	4	20	4	18	6	22	4	18	4	20	6	16	1000 DDD0 10ee eeee		
ORI	BW	s=Imm d=	4	8			4	16	4	16	4	18	6	20	6	22	6	20	8	24							0000 0000 SSEE EEEE	-**00	92
	L	s=Imm d=	6	16			6	28	6	28	6	30	8	32	8	34	8	32	10	36									
ORI CCR	B	d=CCR s=																							4	20	0000 0000 0011 1100	*****	92
ORI SR	W	d=SR s=																							4	20	0000 0000 0111 1100	*****	199
PEA	L	s=					2	12					4	16	4	20	4	16	6	20	4	16	4	20			0100 1000 01ee eeee	-----	116
RESET			2	132																							0100 1110 0111 0000	-----	151
ROL,ROR		see Table B.3																											94
RTE			2	20																							0100 1110 0111 0011	*****	211
RTR			2	20																							0100 1110 0111 0111	*****	111
RTS			2	16																							0100 1110 0111 0101	-----	109
SBCD	B	s=Dn d=	2	6																							1000 RRR1 0000 0rrr	*U*U*	56
	L	s=-(An) d=									2	18															1000 RRR1 0000 1rrr		
Scc	B	cc=True d=	2	6			2	12	2	12	2	14	4	16	4	18	4	16	6	20							0101 CCCC 11EE EEEE	-----	84
		cc=False d=	2	4			2	12	2	12	2	14	4	16	4	18	4	16	6	20									
STOP																									4	4	0100 1110 0111 0010	*****	212
SUB	BW	s=Dn d=	2	4			2	12	2	12	2	14	4	16	4	18	4	16	2	20							1001 DDD1 SSEE EEEE	*****	50
		d=Dn s=	2	4	2	*4	2	8	2	8	2	10	4	12	4	14	4	12	6	16	4	12	4	14	4	8	1001 DDD0 SSee eeee		
	L	s=Dn d=	2	8			2	20	2	20	2	22	4	24	4	26	4	24	6	28							1001 DDD1 10EE EEEE		
		d=Dn s=	2	8	2	8	2	14	2	14	2	16	4	18	4	20	4	18	6	22	4	18	4	20	6	16	1001 DDD0 10ee eeee		
SUBA	W	d=An s=	2	8	2	8	2	12	2	12	2	14	4	16	4	18	4	16	6	20	4	16	4	18	6	12	1001 AAA0 11ee eeee	-----	100
	L	d=An s=	2	8	2	8	2	14	2	14	2	16	4	18	4	20	4	18	6	22	4	18	4	20	6	16	1001 AAA1 11ee eeee		
SUBI	BW	s=Imm d=	4	8			4	16	4	16	4	18	6	20	6	22	6	20	8	24							0000 0100 SSEE EEEE	*****	50
	L	s=Imm d=	6	16			6	28	6	28	6	30	8	32	8	34	8	32	10	36									
SUBQ	BW	s=Imm3 d=	2	4	2	*8	2	12	2	12	2	14	4	16	4	18	4	16	6	20							0101 QQQ1 SSEE EEEE	*****	51
	L	s=Imm3 d=	2	8	2	8	2	20	2	20	2	22	4	24	4	26	4	24	6	28									
SUBX	BW	s=Dn d=	2	4																							1001 RRR1 SS00 0rrr	*****	56
		s=-(An) d=									2	18															1001 RRR1 SS00 1rrr		
	L	s=Dn d=	2	8																							1001 RRR1 1000 0rrr		
		s=-(An) d=									2	30															1001 RRR1 1000 1rrr		
SWAP	W	d=	2	4																							0100 1000 0100 0DDD	-**00	52
TAS	B	d=	2	4			2	14	2	14	2	16	2	18	4	20	4	18	6	22							0100 1010 11EE EEEE	-**00	96
TRAP			2	38																							0100 1110 0100 VVVV	-----	201
TRAPV			2	34	trap taken																						0100 1110 0111 0110	-----	201
			2	4	trap not taken																								
TST	BW	d=	2	4			2	8	2	8	2	10	4	12	4	14	4	12	6	16							0100 1010 SSEE EEEE	-**00	81
	L	d=	2	4			2	12	2	12	2	14	4	16	4	18	4	16	6	20									
UNLK					2	12																					0100 1110 0101 1AAA	-----	113

Table B.3 Shift and rotate instructions (page 92)

Mnemonic	Size	Address Mode	Dn #	Dn %	(An) #	(An) %	(An)+ #	(An)+ %	-(An) #	-(An) %	d16(An) #	d16(An) %	d8(An,Xn) #	d8(An,Xn) %	AbsW #	AbsW %	AbsL #	AbsL %	op-code bit pattern 1111 11 5432 1098 7654 3210	CCR XNZVC	page
ASL,ASR	BW	count=Dn d=	2	6+2n															1110 rrrf SS10 0DDD	*****	81
		count=#1-8 d=	2	6+2n															1110 QQQf SS00 0DDD		
	L	count=Dn d=	2	8+2n															1110 rrrf 1010 0DDD		
		count=#1-8 d=	2	8+2n															1110 QQQf 1000 0DDD		
memory	W	count=1 d=			2	12*	2	12*	2	14*	4	16*	4	18*	4	16*	6	20*	1110 000f 11EE EEEE		81
LSL,LSR	BW	count=Dn d=	2	6+2n															1110 rrrf SS10 1DDD	***0*	82
		count=#1-8 d=	2	6+2n															1110 QQQf SS00 1DDD		
	L	count=Dn d=	2	8+2n															1110 rrrf 1010 1DDD		
		count=#1-8 d=	2	8+2n															1110 QQQf 1000 1DDD		
memory	W	count=1 d=			2	12*	2	12*	2	14*	4	16*	4	18*	4	16*	6	20*	1110 001f 11EE EEEE		
ROL,ROR	BW	count=Dn d=	2	6+2n															1110 rrrf SS11 1DDD	-**00	82
		count=#1-8 d=	2	6+2n															1110 QQQf SS01 1DDD		
	L	count=Dn d=	2	8+2n															1110 rrrf 1011 1DDD		
		count=#1-8 d=	2	8+2n															1110 QQQf 1001 1DDD		
memory	W	count=1 d=			2	12*	2	12*	2	14*	4	16*	4	18*	4	16*	6	20*	1110 011f 11EE EEEE		
ROXL,	BW	count=Dn d=	2	6+2n															1110 rrrf SS11 0DDD	***0*	83
ROXR		count=#1-8 d=	2	6+2n															1110 QQQf SS01 0DDD		
	L	count=Dn d=	2	8+2n															1110 rrrf 1011 0DDD		
		count=#1-8 d=	2	8+2n															1110 QQQf 1001 0DDD		
memory	W	count=1 d=			2	12*	2	12*	2	14*	4	16*	4	18*	4	16*	6	20*	1110 010f 11EE EEEE		

Table B.4 Bit change, clear, set and test instructions (page 95)

Mnemonic	Size	Address Mode	Dn #	Dn %	(An) #	(An) %	(An)+ #	(An)+ %	-(An) #	-(An) %	d16(An) #	d16(An) %	d8(An,Xn) #	d8(An,Xn) %	AbsW #	AbsW %	AbsL #	AbsL %	d16(PC) #	d16(PC) %	d8(PC,Xn) #	d8(PC,Xn) %	op-code bit pattern 1111 11 5432 1098 7654 3210	CCR XNZVC	page
BCHG	B	bit#=Dn d=			2	12	2	12	2	14	4	16	4	18	4	16	6	20					0000 rrr1 01EE EEEE	--*--	83
		bit#=Imm d=			4	16	4	16	4	18	6	20	6	22	6	20	8	24					0000 1000 01EE EEEE		
	L	bit#=Dn d=	2	<8																			0000 rrr1 01EE EEEE		
		bit#=Imm d=	4	<12																			0000 1000 01EE EEEE		
BCLR	B	bit#=Dn d=			2	12	2	12	2	14	4	16	4	18	4	16	6	20					0000 rrr1 10EE EEEE	--*--	83
		bit#=Imm d=			4	16	4	16	4	18	6	20	6	22	6	20	8	24					0000 1000 10EE EEEE		
	L	bit#=Dn d=	2	<10																			0000 rrr1 10EE EEEE		
		bit#=Imm d=	4	<14																			0000 1000 10EE EEEE		
BSET	B	bit#=Dn d=			2	12	2	12	2	14	4	16	4	18	4	16	6	20					0000 rrr1 11EE EEEE	--*--	83
		bit#=Imm d=			4	16	4	16	4	18	6	20	6	22	6	20	8	24					0000 1000 11EE EEEE		
	L	bit#=Dn d=	2	<8																			0000 rrr1 11EE EEEE		
		bit#=Imm d=	4	<12																			0000 1000 11EE EEEE		
BTST	B	bit#=Dn d=			2	8	2	8	2	10	4	12	4	14	4	12	6	16	4	12	4	14	0000 rrr1 00EE EEEE	--*--	83
		bit#=Imm d=			4	12	4	12	4	14	6	16	6	18	6	16	8	20	6	16	6	18	0000 1000 00EE EEEE		
	L	bit#=Dn d=	2	<6																			0000 rrr1 00EE EEEE		
		bit#=Imm d=	4	<10																			0000 1000 00EE EEEE		

Table B.5 MC68000 Condition Branch Operation (see Table 10.3 for condition code encoding CCCC) (page 80)

Mnemonic	Size	Address Mode	#	%	Branch	op-code bit pattern	XNZVC
Bcc	B	d8(PC)	2 2	10 8	YES NO	0110 CCCC PPPP PPPP	-----
	W	d16(PC)	4 4	10 12	YES NO	0110 CCCC 0000 0000	-----

Table B.6 MC68000 DBcc Operation (see Table 10.3 for condition code encoding CCCC) (page 88)

Mnemonic	Size	Address Mode	#	%	cc	counter	Branch	op-code bit pattern	XNZVC
DBcc	W	d16(PC) counter=Dn	4	10 12 14	false true false	<> -1 N/A = -1	YES NO NO	0101 CCCC 1100 1DDD	-----

Table B.7 MC68000 Divide and Multiply Operation Codes (page 52)

Mnemonic		Address	Dn		(An)		(An)+		-(An)		d16(An)		d8(An,Xn)		Abs.W		Abs.L		d16(PC)		d8(PC,Xn)		Immed		op-code bit pattern 1111 11	CCR
	Size	Mode	#	%	#	%	#	%	#	%	#	%	#	%	#	%	#	%	#	%	#	%	#	%	5432 1098 7654 3210	XNZVC
DIVS	W	d=Dn s=	2	<158	2	<162	2	<162	2	<164	4	<166	4	<168	4	<166	6	<170	4	<166	4	<168	4	<162	1000 DDD1 11ee eeee	-***0
DIVU	W	d=Dn s=	2	<140	2	<144	2	<144	2	<146	4	<148	4	<150	4	<148	6	<152	4	<148	4	<150	4	<144	1000 DDD0 11ee eeee	-***0
MULS	W	d=Dn s=	2	<70	2	<74	2	<74	2	<76	4	<78	4	<80	4	<78	6	<82	4	<78	4	<80	4	<74	1100 DDD1 11ee eeee	-**00
MULU	W	d=Dn s=	2	<70	2	<74	2	<74	2	<76	4	<78	4	<80	4	<78	6	<82	4	<78	4	<80	4	<74	1100 DDD0 11ee eeee	-**00

Table B.8 MC68000 Move Multiple Operation Codes (page 110)

Mnemonic		Address	(An)		(An)+		-(An)		d16(An)		d8(An,Xn)		Abs.W		Abs.L		d16(PC)		d8(PC,Xn)		op-code bit pattern 1111 11	CCR
	Size	Mode	#	%		%	#	%	#	%	#	%	#	%	#	%	#	%	#	%	5432 1098 7654 3210	XNZVC
MOVEM	BW	s=Xn d=	4	8+4n			4	8+4n	6	12+4n	6	14+4n	6	12+4n	8	16+4n					0100 1000 10EE EEEE a7 a0 d7 d0	-----
		d=Xn s=	4	12+4n	4	12+4n			6	16+4n	6	18+4n	6	16+4n	8	20+4n	6	16+4n	6	18+4n	0100 1100 10ee eeee a7 a0 d7 d0	
	L	s=Xn d=	4	8+8n			4	8+8n	6	12+8n	6	14+8n	6	12+8n	8	16+8n					0100 1000 11EE EEEE a7 a0 d7 d0	-----
		d=Xn s=	4	12+8n	4	12+8n			6	16+8n	6	18+8n	6	16+8n	8	20+8n	6	16+8n	6	18+8n	0100 1100 11ee eeee a7 a0 d7 d0	

Appendix C

Using the system monitor and assembler

There are many microcomputer systems based on the Motorola MC68000 microprocessor chip. Problems may arise when implementing the sample programs presented in this book due to differences in system hardware configuration and the assembler and operating system.

Students on formal courses will be given guidance on these problems, otherwise, after reading the following notes, reference should be made to the manuals (hardware and software) of the target 68000 based microcomputer.

C.1 Hardware configurations

In terms of the hardware configuration problems arise in the following areas (see Chapter 2.4):

1 Memory mapping of RAM, ROM and I/O device registers (see Chapter 4.4 and 18.4) which effects:
   (a) the position of the user program areas within RAM;
   (b) if the exception vector table is in RAM or ROM (see Charter 23.4);
   (c) which areas of the map (if any) are privileged (see Chapter 23.1).

2 I/O programming:
   (a) position of I/O registers in memory map;
   (b) vectored or autovectored interrupts (see Chapter 23.7);
   (c) priority of interrupts (see Chapter 23.7).

C.2 Different operating systems and monitors

The operating system or monitor of the microcomputer affects the whole program development environment. For example, a system running a sophisticated operating system, e.g. UNIX, will place severe restrictions on low-level programming (see Chapter 2.4):

1 assembly language programming may not be possible at all;
2 assembly language modules may have to be called from a high-level language;
3 the operating system will position the executable code in memory;
4 user programs will run in *user mode* (see Chapter 23.1) and be severely restricted, i.e. not able to access I/O registers, etc.

A computer system designed for low-level development will be less restrictive, e.g. single board systems place no restrictions on user programs (see Chapter 2.4). Even then the following points need to be considered:

1 What RAM areas can be used by user programs. More sophisticated environments will organise the positioning of program and data, requiring user programs to generate relocatable code (see Chapter 3.4 and 15.1). Otherwise an absolute address may have to be specified, e.g.:

```
        ORG        $1000          !! program start address
```

The assembler used for the majority of the sample programs in this book did not need this; the program was automatically placed at the start of the user RAM area.

2 The sample programs in this book assumed that the stack pointer (address register A7) had been set up before the program starts. If the stack has not been set up a memory area should be assigned and A7 initialised with a statement such as:

```
        MOVEA.L    #$1000,A7      set the top of the stack
```

3 The sample programs in this book are terminated with the instruction:

```
        TRAP       #14            !! STOP PROGRAM
```

using the TRAP #14 instruction to stop the program and return control to the monitor. The microcomputer manual should state the preferred instruction that is used to stop programs and in such cases the TRAP #14 should be replaced, e.g. it could be the return from subroutine (RTS) instruction:

```
        RTS                       !! STOP PROGRAM
```

Another way to stop a program is to insert a breakpoint at the final instruction.

4 The run-time support provided for programs (see Chapter 3.5) varies widely from system to system. The sample programs in this book made use of the subroutine library of Appendix D (see Chapter 9.3). The majority of monitors provide basic character I/O facilities and the remainder can be implemented from this (see subroutines RDCHAR and WRCHAR in Appendix D).

5 The programming of I/O devices requires access to input/output device registers, the exception vector table, etc. On multi-user microcomputers and microcomputers which run user programs in User Mode (see Chapter 23.1 for details of processor modes) this will not be allowed. In such cases it may be possible to switch user programs into Supervisor Mode by means of an operating system call (refer to microcomputer manuals). In general it is expected that I/O techniques will be taught on single user machines running in Supervisor Mode all the time.

C.3 Different assemblers

Very few problems should be encountered when implementing the sample programs in this book using a native Motorola 68000 assembler (allowing for variations in hardware and monitor; see C.1 and C.2 above).

When using a cross assembler (see Chapter 3.3) modifications will have to be made when implementing programs (for example, AS68K requires the source code to be in lower case characters). In the main the instruction operation-code mnemonics and addressing modes will be the same with variations occurring the data specifications and pseudo-operators. The following outlines data specification (see Chapter 5.1) and pseudo-operators (see Chapter 5.2) for the XA8 and AS68K cross assemblers.

C.3.1 Real Time Systems XA8 cross assembler

The XA8 cross assembler (Real Time Systems 1986) operates with a wide range of target processors. Programs 5.1c, 8.1a, 9.2, 13.2 and 16.1 were generated using the version for the MC68000. For example, specification of numeric and character data (see Chapter 5.1):

Decimal	default
Binary	by postfixing the value with B, e.g. 10101B
Hexadecimal	by postfixing the value with H, e.g. 1AFH
Character	characters enclosed in ' marks, e.g. 'A'
Text strings	characters in " marks, e.g. "Hello user"

Examples of Pseudo-operators are (comments start with a ;):

```
.PROCESSOR M68000    ;specify processor
.END                 ;end of program code
.PSECT   _text       ;start program area in segment _text
.PSECT   _data       ;start data area in segment _data
.BYTE    10          ;define a byte of storage
.WORD    456         ;define a word of storage
.DOUBLE  5678        ;define long word of storage
.TEXT    "hello"     ;define a text string
.EVEN                ;align to even word boundary
```

C.3.2 Whitesmiths AS68K cross assembler

The AS68K cross assembler (see Programs 5.1d and 8.1b) is generally used as the assembler pass of the Whitesmiths C cross compiler (Whitesmiths 1987a, 1987b). For example, specification of numeric and character data (see Chapter 5.1):

Decimal	default
Binary	not allowed
Hexadecimal	by prefixing with 0x or 0X (as in C), w.g. 0x1af
Character	a character prefixed with a ' mark, i.e. 'a
Text strings	characters in " marks, e.g. "Hello user"

Examples of pseudo-operators are (program code must be lower case letters and comments start with an *):

```
.end                 *end of program code
.text                *start of program segment
.data                *start of data segment
.byte    10          *define a byte of storage
.word    456         *define a word of storage
.long    5678        *define long word of storage
```

C.4 Using the 68000 Assembler

The exact way the 68000 assembler is invoked depends upon the particular program development environment, for example:

Traditional environment in which the editor and assembler are separate programs (possibly from different software houses), i.e.:

(a) when editing is finished the program is saved to disk and the editor terminated;

(b) the assembler is invoked and assembles the program, then:

  (i) if no errors are found the object code file can be linked and executed.

else (ii) if errors are found return to step (a) to correct the errors.

The main problem with this environment is that the assembly errors are either reported on the screen or within a listing or error file and when the editor is invoked (to correct the errors) the error reports are lost. The user is faced with the choice of trying to remember the lines in error and the corresponding error messages, or, getting a hardcopy printout of the error report.

Integrated environment in which the assembler can be invoked from within the editor, e.g. by hitting function key 1 (Bramer 1989b). If assembly errors are found the editor displays (for each error in turn):

(a) the page of program text containing the error;
(b) the error message (or messages) for a particular line of code;
and (c) the screen cursor is positioned on the line or at the start of the line.

The user can see the error message(s) and the line (and page) where the assembler found the error. The program would be edited to remove the cause of the error and then, by hitting function key 1, the next error message and corresponding page of text displayed. The process would be repeated until all error messages had been displayed and the program is assembled again. In this way a sequence of errors can be stepped through and dealt with interactively, and, when the assembly is without errors, the resultant object code file would be linked and executed.

C.4.1 Assembly errors

If an error occurs during assembly a message will be displayed (on the editor screen or the listing or error file):

Errors: a message will indicate the type of error which must be corrected before the program can be run.
Warnings: a warning indicates a possible error, e.g. that an alternative and more efficient op-code could be used.

C.4.2 **Integrated environment** with an editor invoking the assembler.

Fig. 6.2 in Chapter 6 shows an assembly error message generated by the M68 ROM based editor/assembler where function key 1 is used to invoke the assembler:

(a) the first line of the error message (below the menu) indicates the error:
illegal operation code
and (b) the second line shows the position on the source line where the error was found.

The program source follows with the line in error at the top. If the error was corrected and the program assembled again (function key 1) the next error found would be due to the invalid size extension .F in the following line.

Fig. C.1 (next page) shows an assembly error message from the editor ED68K (Bramer 1989b) which has invoked (by function key 1) the Whitesmiths AS68K assembler (Whitesmiths 1986). Following the menu there are two error messages:

3: unknown instruction
3: missing newline

The program source follows with the line in error (line 3 of the program) at the top. If function key 1 was keyed again the next message would be displayed.

C.4.3 **Traditional environment** with separate editor and assembler.

Listing C.1 shows an assembler listing, generated by a Motorola assembler (Motorola 1983b), of a version of Program 6.1 (see Chapter 6) with deliberate errors. Following the listings is a discussion of the errors and their meaning. Listing C.2 and Listing C.3 show similar assembler listings generated by the Whitesmiths AS68K and Real Time Systems XA8 cross assemblers respectively.

C.5 Assembly language errors - discussion

If an error is obvious to the assembler it can give a clear message, but in situations where the error is unclear a number of messages may be displayed, i.e. at line 10 of Listing C.1. An error in one line can cause errors to be generated for other lines, some of which may be correct (i.e. in listing C.1 lines 3 and 8 display an error although only line 8 is incorrect). In cases where the cause of the error is not clear, correct the obvious errors and assemble again. The resultant assembler listing can then be examined and the remaining errors corrected.

The error messages in the examples are relatively clear giving a good indication of where in the statement the error is (i.e. in the label, op-code or operand). By referring to the error message and examining the statement it should be possible to use the system editor to correct the statement and then assemble again. In general, correction will be iterative with correction of obvious errors followed by the more obscure (as shown in Listing C.1 one error may generate several messages or spurious errors which will all disappear when a single error is corrected). After a few weeks one learns the idiosyncrasies of the assembler and error correction becomes 'relatively' straightforward.

For example, sometimes an assembler will generate an error message for a statement which appears to have no syntax errors, and, where the error message does not appear to be caused by an error on another line. In such a case try deleting the complete line and typing it in again. A common problem is that invalid control characters hit by mistake when typing in the program are placed in the program text, i.e. characters with the ASCII character code less than 32 decimal (the exception is Carriage Return which is used as the line terminator, see Table A.2 of Appendix A). These are not visible on the screen when the text is displayed, but the assembler finds the faulty character and rejects the line with an error message. A good editor will not allow the entry of control characters (other than the line terminator Carriage Return) except in special circumstances (if allowed such control characters are often highlighted in some way).

```
F1:AS68K  F2:FileIO  F3:DELline  F4:Find     F5:Copy   F6:Paste F8:Exit/Save
   Help      DELall     UnDel       Replace     Cut               Exit/Abort

3:       unknown instruction
3:       missing newline

           cla.l      d1            *clear d1
           clr.f      d4            *clear d4
           clr.l      d5            *clear d5
start:     clr.l      d6            *clear d6
           move.w     #56f,d0       *move word  to d0
           move.w     d0,d1         *move word from d0 to d1
           move.l     #0x12ag,d2    *move long word  to d2
           move.l     d2,d3         *move long word d2 to d3
           move.w     #0x12345,d4   *move wordb to d4
           move.b     #0x12,d5      *move byte  to d5
           move.b     #0x123,d5     *move byte  to d5
           move.w     #'a,d6        *move word  to d6
           move.w     #chara,d6     *move chara to d6
           move.l     #'a,d7        *move long word to d7
           trap       #14           *!! stop program !!
           .end                     *end of program

 Cursor is on line 3 where the error occurred
```

Fig. C.1 Assembly error message from ED68K invoking Whitesmiths AS68K assembler

```
------------------------------------------------------------------------------
 1                              * Program 6.1A - various assembler errors
 2              00001000             ORG      $1000
 3   00001000 4280         START:    CLR.L    D0              clear D0
****** ERROR   - redefined symbol
 4              4AFB4E714E71         CLA.L    D1              clear D1
****** ERROR   - undefined operation (op-code)
 5   00001008 4AFB                   CLR.F    D2              clear D2
****** ERROR   - size code/extension is invalid
 6   0000100A 4284                   CLR.L    D4              clear D4
 7   0000100C 4285                   CLR.L    D5              clear D5
 8   0000100E 4286         START:    CLR.L    D6              clear D6
****** ERROR   - redefined symbol
 9   00001010 4AFB4E71               MOVE.W   #56F,D0         move word 56 into D0
****** ERROR   - illegal character (in context)
10   00001014 4AFB4E714E71           MOVE.W   #$0FFG,D1       move word FFF hex
****** ERROR   - undefined symbol
****** ERROR   - syntax error
****** ERROR   - too many operands for this instruction
****** ERROR   - illegal address mode for this operand
11   0000101A 4AFB4E71               MOVE.W   #'ABC',D2       move A & B
****** ERROR   - value too large
12   0000101E 383C4100               MOVE.W   #'A',D4         move the ASCII code A
13   00001022 4AFB4E71               MOVE.B   #'AB',D5        move the ASCII code A
****** ERROR   - value too large
14   00001026 4AFB4E71               MOVE.W   #CHARA,D6       move CHARA into D6
****** ERROR   - undefined symbol
15   0000102A 4AFB4E714E71           MOVE.W   D1,D8           move word D1 to D3
****** ERROR   - undefined symbol
16   00001030 4E4E                   TRAP     #14
17                                   END

SYMBOL TABLE LISTING
SYMBOL NAME      SECT    VALUE        SYMBOL NAME      SECT    VALUE
START       MULT
------------------------------------------------------------------------------
```

Listing C.1 An assembler listing with errors (Motorola assembler - Motorola 1983b)

line **error message and explanation**

3,8: **redefined symbol**: the label START is being used on both lines. One of the labels will have to be changed which will correct both errors

4: **undefined operation (op-code)**: the op-code CLA is invalid

5: **size code/extension is invalid**: the operand length extension .F is illegal

9: **illegal character (in context)**: the value 56F is specified (i.e. an illegal decimal number)

10: A whole list of errors are generated by the value $0FFG being specified (i.e. G is not allowed in a hexadecimal number)

11,13: **value too large**: the character strings 'ABC' and 'AB' are too large to fit into word and byte operands respectively

14: **undefined symbol**: the symbol CHARA (a label) is not defined anywhere in the program

15: **undefined symbol**: D8 is not a valid destination data register or a defined label or name

```
--------------------------------------------------------------------------------
~~1WSL 3.0 as68k
 1                              * program 6.1a - with various errors
 2 00000  4280                  start:    clr.l     d0            *clear d0
 3                                        cla.l     d1            *clear d1
          *** unknown instruction ***
          *** missing newline ***
 4 00002  42b9 00000000 U                 clr.f     d4            *clear d4
          *** missing newline ***
 5 00008  4285                            clr.l     d5            *clear d5
 6 0000a  4286                  start:    clr.l     d6            *clear d6
          *** redefinition of start ***
 7 0000c  303c 000c T                     move.w    #56f,d0       *move word  to d0
          *** bad #f or #b ***
 8 00010  3200                            move.w    d0,d1         *move word from d0 to d1
 9 00012  203c 0000012a                   move.l    #0x12ag,d2    *move long word  to d2
          *** bad operand(s) ***
          *** missing newline ***
10 00018  2602                            move.l    d2,d3         *move long word d2 to d3
11 0001a  383c 2345                       move.w    #0x12345,d4   *move wordb to d4
12 0001e  1a3c 0012                       move.b    #0x12,d5      *move byte  to d5
13 00022  1a3c 0023                       move.b    #0x123,d5     *move byte  to d5
14 00026  3c3c 0061                       move.w    #'a,d6        *move word  to d6
15 0002a  3c3c 0000 U                     move.w    #chara,d6     *move chara to d6
16 0002e  2e3c 00000061                   move.l    #'a,d7        *move long word to d7
17 00034  4e4e                            trap      #14           *!! stop program !!
18                                        .end                    *end of program

          number of assembler errors = 7

               code segment size = 54        data segment size = 0

3:            unknown instruction
3:            missing newline
4:            missing newline
6:            redefinition of start
7:            bad #f or #b
9:            bad operand(s)
9:            missing newline
--------------------------------------------------------------------------------
```

Listing C.2 An assembler listing with errors (Whitesmiths AS68K assembler)

line	**error message and explanation**
3:	**unknown instruction & missing newline:** the op-code cla is invalid
4:	**missing newline:** the operand length extension .f is illegal
6:	**redefinition of start:** the label START is used on lines 2 and 6
7:	**bad #f or #b:** the value 56F is specified (i.e. an illegal decimal number)
9:	**bad operand(s) and missing newline:** g is not a valid hexadecimal character

The AS68K assembler is generally used as the assembler pass of the Whitesmiths C cross compiler. The compiler would therefore reduce assembly time errors to a minimum. Therefore care has to be taken when using the assembler by itself, e.g.:

1 in line 11 the value 0x12345 has been truncated to 0x2345 to fit a word;
2 in line 13 the value 0x123 has been truncated to 0x23 to fit a byte;
3 in line 15 the symbol CHARA has been assumed to be external and would generate an error at link time if it was not defined in another module.

```
-------------------------------------------------------------------------------
  1                          ; Program 6.1a - with various errors
  2                                  .PROCESSOR M68000
  3   000000  4280      START:       CLR.L      D0          ;clear D0
  4                                  CLA.L      D1          ;clear D1
  5                                  CLR.F      D4          ;clear D4
  6   000002  4285                   CLR.L      D5          ;clear D5
  7   000004  4286      START:       CLR.L      D6          ;clear D6
  8   000006  303C                   MOVE.W     #56F,D0     ;move word 56 decimal to D0
              0000
  9   00000A  3200                   MOVE.W     D0,D1       ;move word from D0 to D1
 10   00000C  243C                   MOVE.L     #12AGH,D2   ;move long word 12AB hex to D2
              00000000
 11   000012  2602                   MOVE.L     D2,D3       ;move long word D2 to D3
 12   000014  1A3C                   MOVE.B     #1000,D5    ;move byte  to D5
              03E8
 13   000018  3C3C                   MOVE.W     #CHARA,D6   ;move CHARA to D6
              0000
 14   00001C  2E3C                   MOVE.L     #'A',D7     ;move long word 'A' to D7
              00000041
 15   000022  33C1                   MOVE.W     D1,D8       ;???
              00000000
 16   000028  4E4E                   TRAP       #14         ;!! STOP PROGRAM !!
 17                                  .END                   ;end of program
p6_1a.s: 10 errors detected; no obj o/p; no nm o/p; ls = p6_1a.ls;
-------------------------------------------------------------------------------

-------------------------------------------------------------------------------
x68000 (1):
     p6_1a.s 4:           unknown op-code CLA.L
     p6_1a.s 5:           unknown op-code CLR.F
     p6_1a.s 7:           attempt to redefine START
     p6_1a.s 5:           CLR.F not defined
     p6_1a.s 10:          12AGH not defined
     p6_1a.s 4:           CLA.L not defined
     p6_1a.s 15:          D8 not defined
     p6_1a.s 8:           56F not defined
     p6_1a.s 13:          CHARA not defined
     p6_1a.s 7:           phase error START = 00000000H pass1 00000004H pass2
-------------------------------------------------------------------------------
```

Listing C.3 An assembler listing and error file (Real Time Systems XA8 cross assembler)

line	error message and explanation
4	**unknown op-code CLA.L**: CLA is not a valid op-code
5	**unknown op-code CLR.F**: .F is not a valid operand size
7	**attempt to redefine START**: START is defined in lines 3 and 7
5	**CLR.F not defined**: .F is not a valid operand size
10	**12AGH not defined**: G is not allowed in a hexadecimal number
4	**CLA.L not defined**: CLA is not a valid op-code
15	**D8 not defined**: D8 is not a data register or a defined label
8	**56F not defined**: F is not allowed in a decimal number
13	**CHARA not defined**: the label CHARA is not defined within the program
7	**phase error**: due to the errors in pass 1 a phase error occurs in pass 2

Appendix D

Assembly Language Subroutine Library

This appendix describes how to implement the MC68000 assembly language subroutine library (Chapter 9 described the use of the subroutine library by user programs).

D.1 Test Program 1

The first thing is to get the basic single character input and output subroutines (RDCHECK, RDCHAR and WRCHAR) working. Enter the code of Test Program 1 (edit as required for particular assembler). The code of the subroutines RDCHECK, RDCHAR and WRCHAR will require modifying to suit the particular monitor or operating system being used. The subroutines call the monitor/operating system to:

1 RDCHECK check if the keyboard has been hit, clear the Z flag if so;
2 RDCHAR read a single character from the keyboard into D0;
3 WRCHAR write a single character from D0 to the display screen.

The subroutines in Test Program 1 include code to access the M68 monitor (statements in which the comments start with the character !). This code should be removed and code to suit the target microcomputer system substituted, e.g. calls to the monitor of the target computer or code to access the I/O device directly (as in Chapter 25 or 26).

Once Test Program 1 has been entered and modified it should be executed, tested (see program description for its function) and modified until it is working correctly. The program writes a newline on the display by writing the characters CR (carriage return) and LF (line feed). If the target display uses other characters for a new line the code should be modified as required (when implementing the rest of the library the code for WRLINE will require a similar modification).

D.2 Enter the Library

When Test Program 1 is working correctly the library code can be entered. Substitute, from Test Program 1, the working code for RDCHECK, RDCHAR and WRCHAR, and make any modifications to WRLINE that were found necessary. Additional modifications may be required to the library to suit the assembler/microcomputer being used. Assemble the library and fix any errors reported.

D.3 Test Program 2

Enter Test Program 2 carrying out modifications as required. Link this to the library (see Chapter 9 for a discussion of the techniques used) and test the complete program (see the program description). The value of DELAYM in the subroutine DELAY may require modification to suit the clock speed of the target microcomputer.

Test Program 1

```
-----------------------------------------------------------------
* Test Program 1 - for 68000 assembly library routines
*
* Test character I/O routines RDCHECK, RDCHAR and WRCHAR.
*
* This program should display and interact with the user:
* 1. display the characters ABCD
* 2. do a newline on the display
* 3. read characters and echo them until an x is hit
* 4. test RDCHECK and read/write characters until an X is hit
*
* WRITE('ABCD',newline)
* LOOP
*    READ(character)
*    WRITE(character)
* UNTIL character = 'x'
* LOOP
*    DELAY for short period
*    IF keyboard hit
*        READ(character)
*    WRITE(character)
* UNTIL character = 'X'
* STOP
*
START:     CLR.L      D0
           MOVE.B     #'A',D0
           JSR        WRCHAR           display test character A
           MOVE.B     #'B',D0
           JSR        WRCHAR           display test character B
           MOVE.B     #'C',D0
           JSR        WRCHAR           display test character C
           MOVE.B     #'D',D0
           JSR        WRCHAR           display test character D
           MOVE.B     #$0D,D0
           JSR        WRCHAR           display carriage return
           MOVE.B     #$0A,D0
           JSR        WRCHAR           display line feed
* LOOP UNTIL an 'X; is hit
LOOP:      JSR        RDCHAR           read character from keyboard
           JSR        WRCHAR
           CMPI.B     #'x',D0          IF 'x' hit exit from loop
           BNE.S      LOOP
* now test RDCHECK
LOOP1:     MOVE.L     #250000,D1           delay for short time
DELAY:     SUBQ       #1,D1
           BNE.S      DELAY
           JSR        RDCHECK          keyboard hit ?
           BEQ.S      WRITE            if NOT go
           BSR        RDCHAR           read the character
WRITE:     BSR        WRCHAR           and echo it
           CMPI.B     #'X',D0          IF 'X' hit exit from loop
           BNE.S      LOOP1
           TRAP       #14              !! STOP PROGRAM
*
*----------------------------------------------------------------
*
* ***** machine dependent code for 68000 library ******
* ***** replace with new code from test program 1
*
```

```
* code for single board computer systems with M68 monitor
*   code with comment starting with ! is M68 dependent
*
*
* SUBROUTINE WRCHAR: display character in D0 onto screen
* save all registers
* rmove bits 7 to 31 to leave ASCII characters
* CALL M68 monitor to transmit character to host computer
* restore registers
* RETURN
*
WRCHAR:  MOVEM.L   D0-D7/A0-A6,-(A7)  save registers
         ANDI.L    #$7F,D0            bits 0-6 of ASCII code
         TRAP      #11                ! M68 system call
         DC.W      2                  ! transmit character
         MOVEM.L   (A7)+,D0-D7/A0-A6  restore registers
         RTS
*
* SUBROUTINE RDCHAR: read character from the keyboard into D0
* save registers except for D0 (returns the character read)
* CALL M68 monitor to read character from keyboard
* remove any odd bits to form a 7 bit ASCII character
* restore all registers
* RETURN
*
RDCHAR:  MOVEM.L   D1-D7/A0-A6,-(A7)  save registers
         TRAP      #11                ! M68 system call
         DC.W      $0E                ! read a character
         ANDI.L    #$07F,D0           get bits 0 to 6
         MOVEM.L   (A7)+,D1-D7/A0-A6  restore registers
         RTS
;
; rdcheck - clear z bit if a character received from terminal
RDCHECK:  TRAP     #11                ! M68 system call
          DC.W     $1B                ! check keyboard
          RTS
*
          END
-------------------------------------------------------------
```

Test Program 2

```
-------------------------------------------------------------
* Test Program 2 - for 68000 assembly library routines
*
* Before running this program the character input and output
*  routines RDCHAR and WRCHAR should be tested and working.
*
* WRITE(newline,'test for WRTEXT odd characters')
* WRITE(newline,'test for WRTEXT even characters')
* LOOP
*     WRITE(newline,'enter decimal number (0 to STOP) ? ')
*     READ(NUMBER)
*     WRITE(newline,'value in decimal ',NUMBER)
*     WRITE(' and in hex byte, word and long word ')
*     CALL WRHEXB, WRHEXW and WRHEXL to display in hexadecimal
* UNTIL NUMBER=0
* display register values
* WRITE(newline,'dispay an * every second ',newline)
* LOOP FOR ever
*     WRITE('*')
*     DELAY for one second
* END LOOP
```

```
*
* !! Reference external library subroutines using XREF. Modify
* !! the statement if required for the computer being used.
          XREF     WRCHAR,WRLINE,WRSPAC,WRTEXT,WRTXTA
          XREF     WRHEXD,WRHEXB,WRHEXW,WRHEXL,WRDECW
          XREF     RDCHAR,RDDECW,WRREGS,DELAY,WRTXTS
*
START:    MOVEA.L  #MESS1,A0      use WRTXTA to
          JSR      WRTXTA         display first message
          PEA.L    MESS2          use WRTXTS to
          JSR      WRTXTS         display second message
          JSR      WRLINE         newline
          JSR      WRTEXT         display even text
          DC.W     'test for WRTEXT even characters$'
          JSR      WRLINE
          JSR      WRTEXT          display odd text
          DC.W     'test for WRTEXT odd characters$'
LOOP:     JSR      WRLINE
          JSR      WRTEXT
          DC.W     'enter decimal number (0 to STOP) ? $'
          JSR      RDDECW          read a decimal number
          JSR      WRLINE
          JSR      WRTEXT
          DC.W     'value in decimal $'
          JSR      WRDECW          display number in decimal
          JSR      WRLINE
          JSR      WRTEXT          and then in hexadecimal
          DC.W     'and in hex byte, word and long word $'
          JSR      WRHEXB
          JSR      WRSPAC
          JSR      WRHEXW
          JSR      WRSPAC
          JSR      WRHEXL
          TST.W    D0              IF D0 = 0 end of program
          BNE.L    LOOP
* use WRREGS to display register contents
          JSR      WRLINE
          JSR      WRREGS          display all registers
* display * on screen every second
          JSR      WRLINE
          JSR      WRTEXT
          DC.W     'display an * every second$'
          JSR      WRLINE
          MOVE.L   #'   *',D0     * character into D0
*
* LOOP for ever
LOOPD:    JSR      WRCHAR         display an *
          JSR      DELAY          delay for one second
          DC.W     1000           delay count in milliseconds
          BRA.S    LOOPD
          TRAP     #14             !! STOP PROGRAM
*
* test data for WRTXTA and WRTXTS
MESS1:    DC.W     'Test program 2'
          DC.B     $0A,$0D,'$'
MESS2:    DC.W     '  for assembly language routines'
          DC.B     $0A,$0D,'$'
          END
------------------------------------------------------------
```

Assembly Language Library for MC68000 Microprocessor

```
---------------------------------------------------------------------
* ASSEBMBLY LIBRARY FOR 68000 BASED MICROCOMPUTER SYSTEMS.
*
* !! Define entry points of subroutines using XDEF.
         XDEF      WRCHAR,WRLINE,WRSPAC,WRTEXT,WRTXTA
         XDEF      WRHEXD,WRHEXB,WRHEXW,WRHEXL,WRDECW
         XDEF      RDCHAR,RDDECW,WRREGS,DELAY,RDCHECK
*
*--------------------------------------------------------------------
*
* ***** machine dependent code for 68000 library ******
*
* ***** include working code from test program 1 ******
*
*--------------------------------------------------------------------
* ****** start of machine independent routines ******
*
* SUBROUTINE WRLINE: display a new line on the screen
* save register D0
* WRITE(carriage return character)
* WRITE(line feed character)
* delay (required for some systems to allow VDU to scroll)
* restore registers
* RETURN
*
WRLINE:  MOVE.L    D0,-(A7)      save D0
         MOVEQ     #$0D,D0       display carriage return
         BSR.S     WRCHAR
         MOVEQ     #$0A,D0       display line feed
         BSR.S     WRCHAR
         MOVE.W    #10000,D0     set delay approx 10mSec
WRLIN1:  DBRA      D0,WRLIN1     DELAY !!
         MOVE.L    (A7)+,D0       restore D0
         RTS
*
*--------------------------------------------------------------------
* SUBROUTINE WRTEXT display text: text follows JSR in memory:
*        JSR       WRTEXT
*        DC.W      'this is a text message $'
*        .....
* text MUST be terminated by $ or program will crash on RTS
*
* SUBROUTINE WRTEXT
* save registers
* get address of start of text off stack into A0
* CALL WRTXTA to display the text
* restore RETURN address from A0 onto stack
* restore registers
* RETURN
*
* on entry:-
*     top of stack contains address of start of text
*
* internal registers:-  A0 holds address of text
*
WRTEXT:  MOVE.L    A0,-(A7)       save registers
         MOVEA.L   4(A7),A0       move address of text into A0
         BSR.S     WRTXTA         display the text using WRTXTA
* NOTE: WRTXTA sets A0 to word after text end (RETURN address)
WRTXT1:  MOVE.L    A0,4(A7)       reset return address
         MOVE.L    (A7)+,A0       restore register*
         RTS                      RETURN to calling program
*
```

```
*-------------------------------------------------------------
* SUBROUTINE WRTXTA display text: address of text in A0:
*         MOVEA.L   #TEXT,A0       move address into A0
*         JSR       WRTXTA
*         ......
*
* * text of message (must be terminated by $)
* TEXT:   DC.W       'this is a text message $'
*
* SUBROUTINE WRTXTA
* save register D0
* LOOP UNTIL (character='$')
*     display next character using WRCHAR
* END LOOP
* IF A0 is odd
*     make return address even A0=A0+1
* END IF
* restore register D0
* RETURN
*
* on entry:-
*     A0 holds the start address of the text (terminated by $)
* on exit:-
*     A0 contains address of word following end of text
*
* internal registers:-
*     D0 holds each character to be output
*
WRTXTA:  MOVEM.L   D0,-(A7)        save registers
*
* LOOP UNTIL a $ is found
WRTXT2:  MOVE.B    (A0)+,D0        get the next character
         CMPI.B    #'$',D0         if it is a $ end of loop
         BEQ.S     WRTXT3
         BSR.L     WRCHAR          display the character
         BRA.S     WRTXT2
*
* IF A0 is odd, i.e. D1 is even, A0=A0+1 to make even
WRTXT3:  MOVE.L    A0,D0           get last text address
         BTST      #0,D0           IF bit 0 <> 0 A0 is odd
         BEQ.S     WRTXT4
         ADDQ      #1,A0           so add 1 to make it even
WRTXT4:  MOVEM.L   (A7)+,D0        restore registers
         RTS                       RETURN to calling program
*
*-------------------------------------------------------------
* SUBROUTINE WRTXTS display text: address of text is on stack
*         PEA.L     TEXT         move address into on stack
*         JSR       WRTXTS
*         ......
*
* TEXT:   DC.W       'this is a text message $'
*
* message must be terminated by $
*
* SUBROUTINE WRTXTS
* save register A0
* get text string address off stack in A0
* call WRTXTA to display the text
* restore register A0
* remove parameter off stack
* RETURN
*
```

```
* on entry:-
*       top of stack holds the start address of the text
*
* internal registers:-
*       A0 holds text address for WRTXTA
*
WRTXTS:   MOVEM.L   A0,-(A7)        save registers
          MOVE.L    8(A7),A0        get text address
          BSR.S     WRTXTA          and display text
          MOVEM.L   (A7)+,A0        restore registers
          MOVE.L    (A7),4(A7)      return address
          ADDQ.L    #4,A7           reset stack
          RTS                       RETURN to calling program
*
*-------------------------------------------------------------
* SUBROUTINE WRSPAC: display a space on the screen
*
CHARSP:   EQU.B     ' '             define the ASCII for space
WRSPAC:   MOVE.L    D0,-(A7)        save D0
          MOVEQ     #CHARSP,D0      display a space
          BSR.S     WRCHAR
          MOVE.L    (A7)+,D0        restore D0
          RTS
*
*-------------------------------------------------------------
* SUBROUTINE WRHEXD: display a single hex digit in D0
*
* SUBROUTINE WRHEXD
* save D0
* DIGIT = D0 AND $0F         (to remove unwanted bits)
* IF DIGIT < 10
*       character = DIGIT + ASCII '0'
* ELSE
*       character = DIGIT - 10 + ASCII 'A'
* END IF
* WRITE(character)
* RETURN
*
WRHEXD:   MOVE.L    D0,-(A7)           save D0
          ANDI.B    #$0F,D0            get a single hex value
          CMPI.B    #$0A,D0            0 to 9 or A to F ?
          BLT.S     WRHEX1             if D0>10
          ADDI.B    #'A'-'0'-$0A,D0    set up offset for A to F
WRHEX1:   ADDI.B    #'0',D0            convert to ASCII character
          BSR.L     WRCHAR             display it
          MOVE.L    (A7)+,D0           restore D0
          RTS
*
*-------------------------------------------------------------
* SUBROUTINE WRHEXB: display byte value in D0 in hexadecimal
*
* SUBROUTINE WRHEXB
* CALL WRHEXD to display bits 4 to 7
* CALL WRHEXD to display bits 0 to 3
* RETURN
*
* on entry D0 holds value to display in Hexadecimal
* internally D1 holds the loop count
*
WRHEXB:   ROL.B     #4,D0           get bits 4 to 7
          BSR.S     WRHEXD          and display them
          ROL.B     #4,D0           get bits 0 to 3
          BSR.S     WRHEXD          and display them
          RTS
```

```
*-------------------------------------------------------------
* SUBROUTINE WRHEXW: display word value in D0 in hexadecimal
*
* SUBROUTINE WRHEXW
* CALL WRHEXB to display bits 8 to 15
* CALL WRHEXB to display bits 0 to 7
* RETURN
*
* on entry D0 holds value to display in Hexadecimal
* internally D1 holds the loop count
*
WRHEXW:   ROL.W      #8,D0           display bits 8 to 15
          BSR.S      WRHEXB
          ROL.W      #8,D0           display bits 0 to 7
          BSR.S      WRHEXB
          RTS
*
*-------------------------------------------------------------
* SUBROUTINE WRHEXL: display 32 bit value in D0 in hexadecimal
*
* SUBROUTINE WRHEXL
* CALL WRHEXW to display bits 16 to 31
* CALL WRHEXW to display bits 0 to 15
* RETURN
*
WRHEXL:   SWAP       D0
          BSR.S      WRHEXW          display top 16 bits
          SWAP       D0
          BSR.S      WRHEXW          display bottom 16 bits
          RTS
*
*-------------------------------------------------------------
* SUBROUTINE WRDECW: display word D0 as signed decimal NUMBER
*
* SUBROUTINE WRDECW
* save registers
* IF NUMBER=0
*     WRITE('0')
* ELSE
*     IF NUMBER<0
*         WRITE('-')
*         NUMBER=-NUMBER
*     END IF
*     PRINT_0 = FALSE  (is used to supress leading zeros)
*     DIVISOR=10000
*     LOOP
*        QUOTIENT=NUMBER/DIVISOR
*        IF QUOTIENT<>0 OR (QUOTIENT=0 AND PRINT_0)
*            PRINT_0=TRUE
*            character=QUOTIENT + ASCII '0'
*            WRITE(character)
*        END IF
*        NUMBER= REMAINDER OF NUMBER/DIVISOR
*        DIVISOR=DIVISOR/10
*     UNTIL DIVISOR=0
* END IF
* restore registers
* RETURN
*
* on entry D0 holds word NUMBER to display
*
```

```
* internal registers:-
*       D0 holds the ASCII characters to be displayed
*       D1 (word) DIVISOR
*       D2 (word) NUMBER to be displayed
*       D3 (logical) PRINT_0 (1 = TRUE)
*
CHARMI:   EQU.B      '-'              define ASCII characters
CHAR0:    EQU.B      '0'
WRDECW:   MOVEM.L    D0-D3,-(A7)      save registers
          CLR.L      D2
          MOVE.W     D0,D2            move NUMBER into D2
*
* IF NUMBER = 0 display '0' and RETURN
          BNE.S      WRDEC1           IF NUMBER=0
          MOVEQ      #CHAR0,D0        display a 0
          BSR.L      WRCHAR
          BRA.S      WRDEC6           and RETURN to calling program
*
* IF NUMBER < 0 display a '-' sign
WRDEC1:   BPL.S      WRDEC2           IF NUMBER < 0
          MOVEQ      #CHARMI,D0       display a - size
          BSR.L      WRCHAR
          NEG.W      D2               and make NUMBER positive
WRDEC2:   MOVE.L     #10000,D1        DIVISOR = 10000
          CLR.L      D3               PRINT_0 = FALSE
*
* LOOP
WRDEC3:   DIVU.W     D1,D2            NUMBER=NUMBER/DIVISOR
          MOVE.W     D2,D0            get QUOTIENT
          BNE.S      WRDEC4           if QUOTIENT = 0
          TST        D3               and PRINT_0 is FALSE
          BEQ.S      WRDEC5           dont print leading zeros
WRDEC4:   MOVEQ      #1,D3            PRINT_0 = TRUE
          ADDI.B     #'0',D0          character=QUOTIENT+ASCII '0'
          BSR.L      WRCHAR           display character
*  printer numeric character - get remainder of division
WRDEC5:   CLR.W      D2               clear quotient- bits 0 to 15
          SWAP       D2               NUMBER=remainder bits 0 to 15
          DIVU.W     #10,D1           DIVISOR=DIVISOR/10
          BNE.S      WRDEC3           LOOP UNTIL DIVISOR = 0
* UNTIL DIVISOR = 0
WRDEC6:   MOVEM.L    (A7)+,D0-D3      restor registers
          RTS                         RETURN to calling program
```

```
*
*-------------------------------------------------------------
* SUBROUTINE RDDECW: read a signed decimal word value into D0
*  terminate on a non-digit character entered.
*   Note that sign is extended to bits 16 to 31 of D0.
*
* SUBROUTINE RDDECW
* save registers
* NEGATIVE=FALSE       (used to indicate if number is negative)
* NUMBER=0
* read character
* IF character='-'
*    NEGATIVE=TRUE
*    READ(character)
* END IF
* LOOP WHILE character in range '0' to '9'
*    WRITE(character)
*    DIGIT=character - ASCII '0'
*    NUMBER=NUMBER*10+DIGIT
*    READ(character)
* END LOOP
* IF NEGATIVE
*    NUMBER:=-NUMBER
* END IF
* restore registers
* RETURN
*
* internal registers:-
*    D0 holds characters and DIGIT
*    D1 (word) holds NUMBER
*    D2 (logical) holds NEGATIVE  (1 for TRUE)
*
* on exit D0 holds the number (word)
*
RDDECW:  MOVEM.L   D1-D2,-(A7)    save registers
         CLR.L     D1             clear NUMBER to 0
         CLR.L     D2             NEGATIVE = FALSE
         BSR.L     RDCHAR         read a character
* IF character = '-' set NEGATIVE = TRUE
         CMPI.B    #'-',D0        IF character = '-'
         BNE.S     RDDEC1
         BSR.L     WRCHAR         echo - character
         MOVEQ     #1,D2          NEGATIVE = TRUE
         BSR.L     RDCHAR         read a character
* LOOP WHILE character is in range '0' to '9'
RDDEC1:  CMPI.B    #'0',D0        IF character < '0'
         BLT.S     RDDEC3            EXIT from LOOP
         CMPI.B    #'9',D0        IF character > '9'
         BGT.S     RDDEC3            EXIT from LOOP
         BSR.L     WRCHAR         echo character
         SUBI.B    #'0',D0        DIGIT= character - ASCII '0'
         MULU.W    #10,D1         NUMBER=NUMBER*10
         ADD.W     D0,D1          NUMBER=NUMBER+DIGIT
         BSR.L     RDCHAR         read character
         BRA.S     RDDEC1         LOOP again
* END LOOP
RDDEC3:  MOVE.L    D1,D0          return result in D0
* IF NEGATIVE set NUMBER=-NUMBER
         TST       D2             IF NEGATIVE
         BEQ.S     RDDEC4
         NEG.L     D0             NUMBER was a negative value
RDDEC4:  MOVEM.L   (A7)+,D1-D2    restore regiaters
         RTS                      RETURN to calling program
*
```

```
*-----------------------------------------------------------------
*
* subroutine to display registers contents D0 to D7, A0 to A7
*
* SUBROUTINE WRREGS
* save D0 on stack
* WRITE(neline,'Register contents at PC =')
* display RETURN address to give indication of where JSR is
* set up text 'D0' to display data registers
* display contents of registers D0 to D7 using WRREG1
* set up text 'A0' to display address registers
* display contents of registers A0 to A7 using WRREG1
* restore D0
* RETURN
*
WRREGS:  MOVEM.L   D0-D1,-(A7)
         MOVE.L    8(A7),D0          get RETURN address
         BSR.L     WRLINE
         BSR.L     WRTEXT
         DC.W      'Register contents at PC =$'
         BSR.L     WRHEXL            display it in hex
         MOVE.L    (A7),D0           get D0 back off stack
         BSR.L     WRLINE
         MOVE.L    #'D0 =',D1        text for display to D1
         BSR.S     WRREG1            display register D0
         MOVE.L    4(A7),D0
         BSR.S     WRREG1            display register D1
         MOVE.L    D2,D0
         BSR.S     WRREG1            display register D2
         MOVE.L    D3,D0
         BSR.S     WRREG1            display register D3
         BSR.L     WRLINE
         MOVE.L    D4,D0
         BSR.S     WRREG1            display register D4
         MOVE.L    D5,D0
         BSR.S     WRREG1            display register D5
         MOVE.L    D6,D0
         BSR.S     WRREG1            display register D6
         MOVE.L    D7,D0
         BSR.S     WRREG1            display register D7
         BSR.L     WRLINE
         MOVE.L    #'A0 =',D1        text for display to D1
         MOVE.L    A0,D0
         BSR.S     WRREG1            display register A0
         MOVE.L    A1,D0
         BSR.S     WRREG1            display register A1
         MOVE.L    A2,D0
         BSR.S     WRREG1            display register A2
         MOVE.L    A3,D0
         BSR.S     WRREG1            display register A3
         BSR.L     WRLINE
         MOVE.L    A4,D0
         BSR.S     WRREG1            display register A4
         MOVE.L    A5,D0
         BSR.S     WRREG1            display register A5
         MOVE.L    A6,D0
         BSR.S     WRREG1            display register A6
         MOVE.L    A7,D0
         BSR.S     WRREG1            display register A7
         BSR.L     WRLINE
         MOVEM.L   (A7)+,D0-D1
         RTS
```

```
*
* display register value held in D0
* SUBROUTINE WRREG1
* WRITE('D? = ') or WRITE('A? = ')
* WRITE(contents of D0 in hexadecimal (long word))
* add 1 to digit character for next register display
* RETURN
*
WRREG1:   EXG       D0,D1            exchange data and text
          ROL.L     #8,D0            diaplay D or A to screen
          BSR.L     WRCHAR
          ROL.L     #8,D0            display register digit
          BSR.L     WRCHAR
          ROL.L     #8,D0            display space
          BSR.L     WRCHAR
          ROL.L     #8,D0            display =
          BSR.L     WRCHAR
          ADDI.L    #$10000,D0       increment digit value
          EXG       D0,D1            exchange text and data
          BSR.L     WRHEXL           display register value
          BSR.L     WRSPAC
          RTS
*
*---------------------------------------------------------------
*
* SUBROUTINE DELAY: delay for a time period in milliseconds
*   The delay count follows JSR in memory (word value) so:-
*         JSR       DELAY
*         DC.W      10000            delay_count is one second
*
* SUBROUTINE DELAY
* save registers
* IF delay_count > 0
*     LOOP FOR delay_count in milliseconds
*         LOOP for one millisecond
*         END LOOP
*     END LOOP
* restore registers
* RETURN
*
* The following is the delay count for an 8 MHz 68000 without
*  wait states.  This may require modification.
DELAYM:   EQU.W     796              1 mSec delay count
DELAY:    MOVEM.L   D0-D1/A0,-(A7)
          MOVE.L    12(A7),A0       get address of delay_count
          MOVE.W    (A0)+,D0        get delay_count into D0
          MOVE.L    A0,12(A7)       restore return address
          SUBQ.W    #1,D0           delay_count -1 for DBRA
          BLT.S     DELAY3          if delay_count <= 0 exit
DELAY1:   MOVE.W    #DELAYM,D1      one millisecond delay count
DELAY2:   DBRA      D1,DELAY2       INNER DELAY LOOP !!
          NOP                       NOP delay to make DELAYM
          NOP                       calculation more accurate
          DBRA      D0,DELAY1       OUTER DELAY LOOP !!
DELAY3:   MOVEM.L   (A7)+,D0-D1/A0
          RTS
          END

---------------------------------------------------------------
```

Appendix E

Record data structures

Chapter 13 described the use of arrays, which are data structures in which all the elements are of the same type (e.g. arrays of numbers or strings of characters). A record is a data structure in which the elements can be of different types.

Consider a stock control program used in a factory which uses a stock record that contains information on items in stock and their cost:

Record element	Contents
1	item name (8 characters)
2	item cost (integer word)
3	number of items in stock (integer word)

Program E.1a is a Pascal program which uses a record of this structure and calls a subroutine to display the total value of a particular item in stock. Program E.1b is an assembly language equivalent. On entry to the subroutine SVALUE, line 32 in Program E.1b, A0 contains the address of the start of a record, and using the correct offset (defined by the values ITNAME, ITCOST and ITSTOCK in lines 28, 29 and 30), the element values can be accessed using *address register indirect with displacement addressing mode* (lines 36, 43 and 44). This program shows how a complex data structure could be built up, i.e. the main program could have an array of these item records.

In Program E.1b address register A0 is used to pass the start address of the record to the subroutine. This type of structure is sometimes known as a parameter block (i.e. an area of memory which contains a number of parameters to be passed to a subroutine), the start address of which can be passed to the subroutine in a register or on the stack.

Using *address register indirect with index and displacement addressing mode*, arrays of records can be processed. Program E.2a is a Pascal program which uses an array of records and Program E.2b is an assembly language equivalent. On entry to the subroutine SVALUE, line 34 in Program E.2b, A0 contains the address of the start of the array of item records, and D1 contains the number of the record to be processed (it is assumed that the first record is number 0). The record number is multiplied by 12 (line 35) to form the index into the array (i.e. each record is 12 bytes long). The elements of the record can then be accessed by giving the correct displacement (lines 39, 46 and 47).

The sample Pascal and assembly language equivalent programs show the advantages and disadvantages of high and low-level programming. In general the advantages of using Pascal are:

1 Pascal programs are relatively easy to design and code (in many cases, as in the examples, there may be no need to use Structured English).
2 Error detection at compile and run time is generally of high quality thus reducing the time taken in the coding and test phases.
3 Pascal programs are relatively easy to modify and should be machine independent.

The disadvantage of using a high-level language is that the compilation process is not 100% efficient and the resultant code will have larger memory requirements and execute more slowly than the equivalent assembly code written by a good programmer.

The advantage of using assembly languages is that at each stage in the program the most efficient combination of instructions and addressing modes can be used. Thus assembly language programs should be used where efficiency and speed of the executable code is a major requirement, e.g. in a real-time process control system.

The disadvantages of using an assembly language are:

1 Coding from the Structured English design can be difficult.
2 The error detection facilities are generally very poor.
3 The program will only be suitable for the particular microprocessor.

```
--------------------------------------------------------------
PROGRAM LIST_STOCK;
(* Pascal Program E.1a
          - to display total value of an item in stock,
            item data stored in records                    *)
TYPE STOCK_RECORD = RECORD
                          ITEM_NAME:STRING[8];
                               ITEM_COST,
                               ITEMS_IN_STOCK:INTEGER;
                       END;
VAR RECORD1,RECORD2:STOCK_RECORD;

PROCEDURE STOCK_VALUE(RECORDN:STOCK_RECORD);
 (* display total value of an ITEM in stock:
    RECORDN is the record which holds the data *)
 BEGIN
 WITH RECORDN DO
    BEGIN
    WRITE('total value of ',ITEM_NAME);
    WRITELN(' in stock is ',ITEM_COST*ITEMS_IN_STOCK);
    END;
 END;

 BEGIN (* OF MAIN PROGRAM *)
 (* define some data values to item records *)
 WITH RECORD1 DO
      BEGIN
      ITEM_NAME:='NUTS        ';
      ITEM_COST:=10;
      ITEMS_IN_STOCK:=25
      END;
 WITH RECORD2 DO
      BEGIN
      ITEM_NAME:='BOLTS       ';
      ITEM_COST:=15;
      ITEMS_IN_STOCK:=20
      END;
 (* display total item values *)
 STOCK_VALUE(RECORD1);
 STOCK_VALUE(RECORD2);
 END.
--------------------------------------------------------------
```

Program E.1a Pascal program using a record structure

```
----------------------------------------------------------------------------
 1                        * Program E.1, display total value of the stock of an item
 2                        *          data is stored in records
 3                        * see Pascal program for documentation
 4                        *
 5                                    XREF      WRLINE,WRTEXT,WRDECW,WRCHAR
 6 001000 207C 00001012 START:        MOVEA.L   #REC1,A0         first record address
 7 001006 6122                        BSR.S     SVALUE           display value
 8 001008 207C 0000101E               MOVEA.L   #REC2,A0         second record address
 9 00100E 611A                        BSR.S     SVALUE           display value
10 001010 4E4E                        TRAP      #14              !! STOP PROGRAM
11                        * stock records
12 001012 4E 55 54 53     REC1:       DC.L      'NUTS    '       item name
13 00101A 000A                        DC.W      10               value of item
14 00101C 0019                        DC.W      25               number in stock
15 00101E 42 4F 4C 54     REC2:       DC.W      'BOLTS   '       item name
16 001026 000F                        DC.W      15               value of item
17 001028 0014                        DC.W      20               number in stock
18                        *
19                        * subroutine to display total value of an item in stock
20                        *
21                        * on entry A0 holds the address of the record to processed
22                        * internal registers:-
23                        *   D0 holds characters and values for calculations
24                        *   D1 holds loop count for display of item name
25                        *   A1 holds address of name text
26                        *
27                        * Define the element offsets in the record
28        00000000        ITNAME:     EQU       0            item name (2 long words)
29        00000008        ITCOST:     EQU       8            item cost (word)
30        0000000A        ITSTOCK:    EQU       10           items in stock (word)
31                        *
32 00102A 48E7       C040 SVALUE:     MOVEM.L   D0-D1/A1,-(A7)
33 00102E 4EB9 0002802C               JSR       WRLINE
34 001034 4EB9 00028038               JSR       WRTEXT           display first  message
35 00103A 74 6F 74 61 6C              DC.W      'total value of $'
36 00104A 43E8       0000             LEA       ITNAME(A0),A1    get item name address
37 00104E 7207                        MOVEQ     #7,D1            8 characters in a name
38 001050 1019            SVALU1:     MOVE.B    (A1)+,D0         get character of name
39 001052 4EB9 00028028               JSR       WRCHAR            and display it
40 001058 51C9       FFF6             DBRA      D1,SVALU1
41 00105C 4EB9 00028038               JSR       WRTEXT           display next message
42 001062 20 69 6E 20 73              DC.W      ' in stock is $'
43 001070 3028       0008             MOVE.W    ITCOST(A0),D0    get cost per item
44 001074 C0E8       000A             MULU.W    ITSTOCK(A0),D0   item cost * in stock
45 001078 4EB9 0002806E               JSR       WRDECW           display total value
46 00107E 4CDF       0203             MOVEM.L   (A7)+,D0-D1/A1
47 001082 4E75                        RTS                        return
48                                    END
WRLINE    0002802C X    WRTEXT    00028038 X    WRDECW    0002806E X
WRCHAR    00028028 X    START     00001000 L    REC1      00001012 L
REC2      0000101E L    ITNAME    00000000 E    ITCOST    00000008 E
ITSTOCK   0000000A E    SVALUE    0000102A L    SVALU1    00001050 L
----------------------------------------------------------------------------
```

Program E.1b Assembly language program using a record structure

```
----------------------------------------------------------------
PROGRAM LIST_STOCK;
(* Pascal Program E.2a
      - to display total value of an item in stock,
        item data stored in an array of records   *)

TYPE STOCK_RECORD = RECORD
                                   ITEM_NAME:STRING[8];
                                   ITEM_COST,
                                   ITEMS_IN_STOCK:INTEGER;
                               END;
VAR ITEM: ARRAY [0..10] OF STOCK_RECORD;

PROCEDURE STOCK_VALUE(NUMBER:INTEGER);
 (* display total value of an ITEM in stock:
    ITEM is the array of item records
    NUMBER is the number of the record to process *)
 BEGIN
 WITH ITEM[NUMBER] DO
    BEGIN
    WRITE('total value of ',ITEM_NAME);
    WRITELN(' in stock is ',ITEM_COST*ITEMS_IN_STOCK);
    END;
 END;

 BEGIN (* OF MAIN PROGRAM *)
 (* define some data values to item records *)
 WITH ITEM[0] DO
     BEGIN
     ITEM_NAME:='NUTS          ';
     ITEM_COST:=10;
     ITEMS_IN_STOCK:=25
     END;
 WITH ITEM[1] DO
     BEGIN
     ITEM_NAME:='BOLTS         ';
     ITEM_COST:=15;
     ITEMS_IN_STOCK:=20
     END;
 (* display total item values *)
 STOCK_VALUE(0);
 STOCK_VALUE(1);
 END.
----------------------------------------------------------------
```

Program E.2a Pascal program using an array of records

```
--------------------------------------------------------------------------------
1                          * Program E.2, display value of the stock of an item
2                          *        data is stored in an array of records
3                          * see Pascal program for documentation
4                          *
5                                    XREF      WRLINE,WRTEXT,WRDECW,WRCHAR
6 001000 207C 00001010 START:        MOVEA.L   #ITEM,A0          address of start of array
7 001006 7200                        MOVEQ     #0,D1             display  first record
8 001008 611E                        BSR.S     SVALUE            display value
9 00100A 7201                        MOVEQ     #1,D1             display  second record
10 00100C 611A                       BSR.S     SVALUE
11 00100E 4E4E                       TRAP      #14               STOP PROGRAM !!!!
12                         * array of stock records
13 001010 4E 55 54 53      ITEM:     DC.L      'NUTS    '        first item name
14 001018 000A                       DC.W      10                value of item
15 00101A 0019                       DC.W      25                number in stock
16 00101C 42 4F 4C 54                DC.W      'BOLTS   '        second item name
17 001024 000F                       DC.W      15                value of item
18 001026 0014                       DC.W      20                number in stock
19                         *
20                         * subroutine to display total value of an item in stock
21                         * on entry:-
22                         *    A0 holds the address of the start of the array of records
23                           *    D1 holds the record number to be displayed (starts from
0)
24                         * internal registers:-
25                         *    D0 holds characters and values for calculations
26                         *    D2 holds loop count for display of item name
27                         *    A1 holds address of name text
28                         *
29                         * Define the element offsets in the record
30        00000000         ITNAME:   EQU       0                 item name (two long words)
31        00000008         ITCOST:   EQU       8                 item cost (word)
32        0000000A         ITSTOCK:  EQU       10                items in stock (word)
33                         *
34 001028 48E7      E040   SVALUE:   MOVEM.L   D0-D2/A1,-(A7)
35 00102C C2FC      000C             MULU.W    #12,D1                 record is 12 bytes
36 001030 4EB9 0002802C              JSR       WRLINE
37 001036 4EB9 00028038              JSR       WRTEXT
38 00103C 74 6F 74 61 6C             DC.W      'total value of $'
39 00104C 43F0      1000             LEA       ITNAME(A0,D1.W),A1     get item name address
40 001050 7407                       MOVEQ     #7,D2                  8 chars in a name
41 001052 1019              SVALU1:  MOVE.B    (A1)+,D0               get character of name
42 001054 4EB9 00028028              JSR       WRCHAR                  and display it
43 00105A 51CA      FFF6             DBRA      D2,SVALU1
44 00105E 4EB9 00028038              JSR       WRTEXT                 display next message
45 001064 20 69 6E 20 73             DC.W      ' in stock is $'
46 001072 3030      1008             MOVE.W    ITCOST(A0,D1.W),D0     get cost per item
47 001076 C0F0      100A             MULU.W    ITSTOCK(A0,D1.W),D0    item cost * in stock
48 00107A 4EB9 0002806E              JSR       WRDECW                 display total value
49 001080 4CDF      0207             MOVEM.L   (A7)+,D0-D2/A1
50 001084 4E75                       RTS                              RETURN
51                                   END
WRLINE     0002802C X    WRTEXT     00028038 X    WRDECW     0002806E X
WRCHAR     00028028 X    START      00001000 L    ITEM       00001010 L
ITNAME     00000000 E    ITCOST     00000008 E    ITSTOCK    0000000A E
SVALUE     00001028 L    SVALU1     00001052 L
--------------------------------------------------------------------------------
```

Program E.2b Assembly language program using an array of records

Appendix F

Test circuits for parallel interfaces

F.1 Parallel port usage

The MC6821 PIA has two parallel ports (called A and B) and the MC68230 PIT has three parallel ports (A, B and C). Although in practice any bit of any port can be used for input or output, the sample programs in this book use port A for input and port B for output.

F.2 The Bytronic Multi-application board (Bytronic 1989)

The Bytronic multi-application board (and similar boards) contains circuits equivalent to many of the following sample circuits. The following assumes that signals from the board (input to the PIA or PIT) are connected to port A and signals to the board (output from the PIA or PIT) are connected to port B.

F.3 Switches and lights circuit

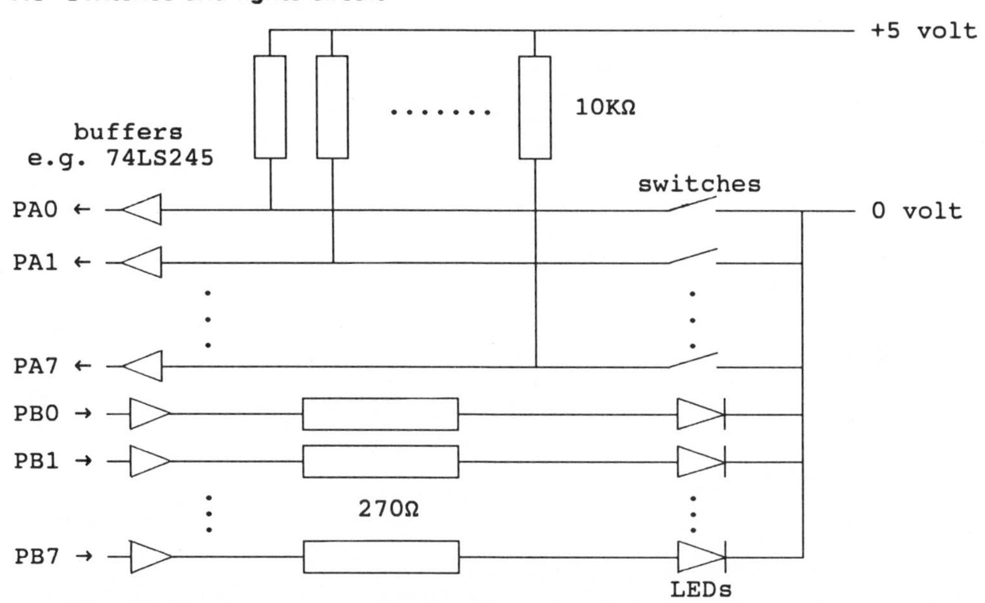

Fig. F.1 Switches and LEDs circuit for simple parallel I/O

When programming a parallel interface the first task is to attempt simple I/O (without handshaking) reading from and writing data to the device. Fig. F.1 shows a circuit which can be used to program simple I/O with the MC6821, MC68230 or any other parallel interface (the Bytronic multi-applications board has an equivalent circuit). The circuit contains:

1. A set of switches connected to port A for input. When a switch is:
   (a) open: + 5 volt is connected to port A as a binary 1 input;
   (b) closed: 0 volt is connected to port A as a binary 0 input.
2. A set of LEDs (light-emitting diodes) connected to port B for output. An LED lights when a binary 1 is output to the corresponding bit.

After programming simple data I/O the next thing is to program the handshaking or control lines. Care must be taken with signals input to the control lines in that they are generally sensitive to changes in logic level (not the levels themselves) and therefore such changes must have clean edges, e.g. use a Schmitt trigger or similar device to clean up noisy signals (Loveday 1984). Fig. F.2 shows a suitable test circuit:

1 A press switch is attached to CA1 & CB1 (6821) or H1 & H3 (68230) which enable high to low and low to high transitions to be input to the handshaking lines, thus:
   (a) Normally the switch is connected to + 5 volt and a binary 1 is input to CA1.
   (b) If the switch is pressed it is connected to 0 volts and binary 0 is input to CA1.
   (c) When it is released the switch reconnects to + 5 volt.
2 A yellow LED is connected to CA2 (6821) or H2 (68230) for output display.
3 A green LED is connected to CB2 (6821) or H4 (68230) for output display.

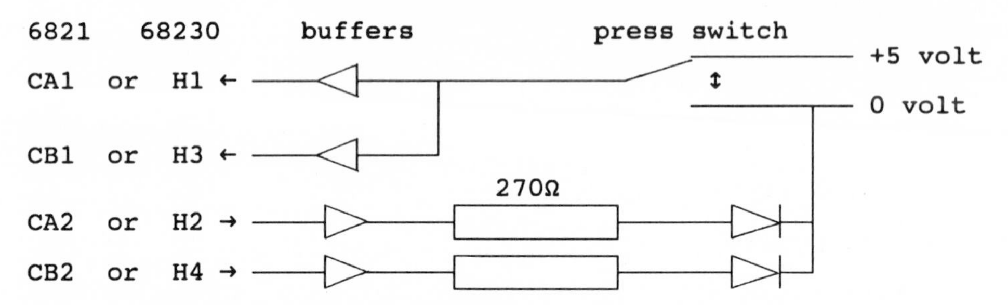

Fig. F.2 Circuit to test handshaking lines

The press switches of Fig. F.2 must be Hall effect or have debounce circuits to remove the contact bounce which occurs with normal mechanical switches.

The Bytronic multi-application board does not include the circuit of Fig. F.2. However, the motor speed sensor can be attached to CA1 or H1 to provide similar facilities to a switch.

F.4 Connecting a Centronics compatible parallel printer

A centronics compatible parallel printer may be connected to port B of a MC6821 or MC68230 as follows:

1 the MC6821 or MC68230 GND (ground) to printer GND;
2 the eight data lines of port B to the printer input data lines;
3 STROBE (data ready from interface) handshaking line to CB2 (6821) or H4 (68230);
4 ACKNLG (printer data acknowledge) handshaking line to CB1 (6821) or H3 (68230).

A parallel printer may be emulated using the switches & lights circuit described in section F.1. The data output and STROBE signal will be shown on the LEDs and a press switch serves as the ACKNLG signal, i.e. as the switch is pressed the next data should appear.

F.5 Connecting an Analogue to Digital converter

An A to D converter usually requires a signal from the computer to start conversion and returns a conversion finished signal. It may be connected to a MC6821 or MC68230 so:

1 the MC6821 or MC68230 GND (ground) to the A to D ground;
2 the eight A to D data lines to port A (a 16-bit A to D could use both ports);
3 START conversion (from interface) handshaking line to CA2 (6821) or H2 (68230);
4 Conversion FINISHED (from A to D) handshaking line to CA1 (6821) or H1 (68230).

The A to D converter may be emulated using the switches and lights circuits described in section F.2. The eight data switches emulate the converted data, an LED shows the conversion START signal and press switch emulates the conversion FINISHED signal.

F.6 Matrix keypad

Fig. F.3 shows a typical matrix keypad; the output and input lines can be connected to a single parallel I/O port or two separate ports as required. The circuit operates as follows:

1 Assuming that no keys are pressed the input lines (columns) are connected to +5 volts via the resistors and hence a logical 1 would be read by a parallel interface.
2 A 0 is written (by a parallel interface) to each output line (row) in turn.
3 When a key is pressed an input line (column) would read logical 0 when the corresponding row was 0.

By knowing which row had a 0 written to it and which column had a 0 read from it the key pressed can be determined. An outline algorithm would be:

```
REPEAT (* writing a 0 to each row in turn *)
    write 0 to next row (all other rows are 1)
UNTIL bit pattern read from columns <> 111
IF row_number <> 3
    keypad number = (row_number * 3) + column_number + 1
```

For example, if key 8 had been pressed:

1 row bit pattern written = 1011 thus row_number = 2
2 column bit pattern read = 101 thus column_number = 1
3 keypad number = (2 * 3) + 1 + 1 = 8

The algorithm would have to be extended to take account of the last row (* 0 #). In practice an electronic circuit would perform the decoding of a physical key press into the corresponding key value, e.g. into an ASCII character code.

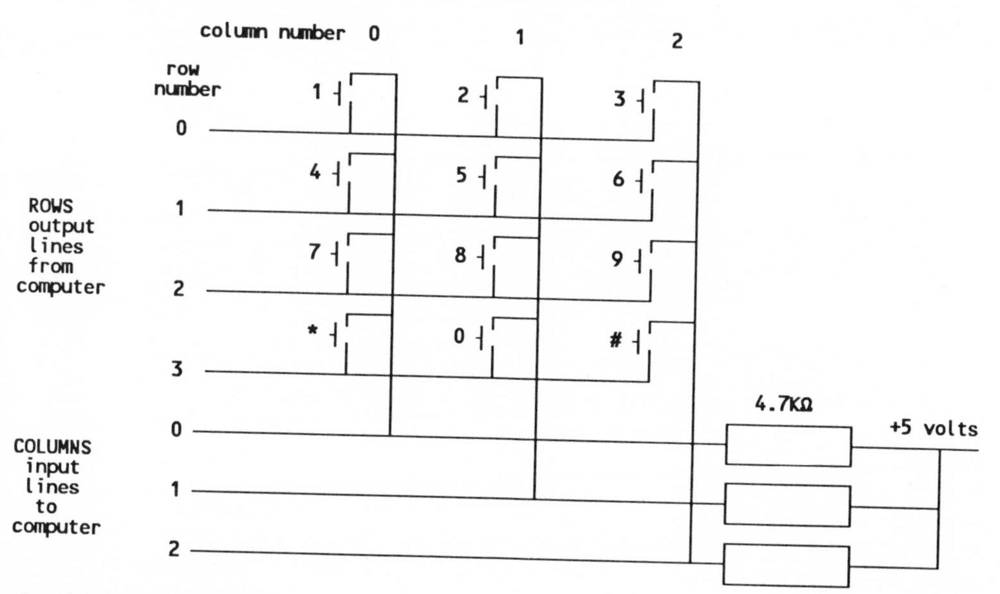

Fig. F.3 Matrix keypad

F.7 Motor/heater circuit

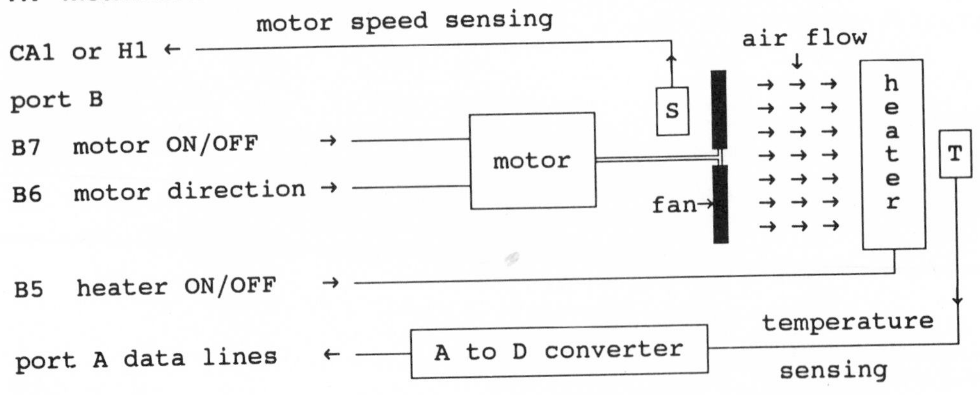

Fig. F.4 motor/header circuit (S and T are the speed and temperature sensors)

Fig. F.4 shows a motor/heater control circuit (the port connections are those used by the Bytronic multi-application board when used with Force, Bytronic and Kaycomp boards):

Motor ON/OFF signal (port B bit 7) turns the motor ON (logical 1) or OFF (logical 0)

Motor direction signal (port B bit 6) determines the motor forward (1) or reverse (0)

Heater ON/OFF signal (port B bit 5) switches the heater ON (logical 1) or OFF (logical 0)

Motor speed sensing The motor fan blades interrupts an infra red beam the changes in which can be detected by the signal input to CA1 (6821) or H1 (68230).

Temperature sensing A temperature sensor attached to the heater can be read via an analogue to digital converter connected to port A of a parallel interface. The Bytronic Multi-applications board A/D converter is *self clocking* (Bytronic 1989) and has no START and FINISHED conversion signals (i.e. it can be read at any time).

The fan of the motor is arranged to blow over the heater so it is possible to vary the heater temperature by switching the heater on and off and/or switching the motor on and off.

F.7.1 Motor speed control

The motor of Fig. F.4 is either ON or OFF and the simplest method of speed control is using PWM (pulse width modulation) in which the width of a sequence of pulses averaged over a period of time determines the voltage applied to the motor and hence speed, i.e.:

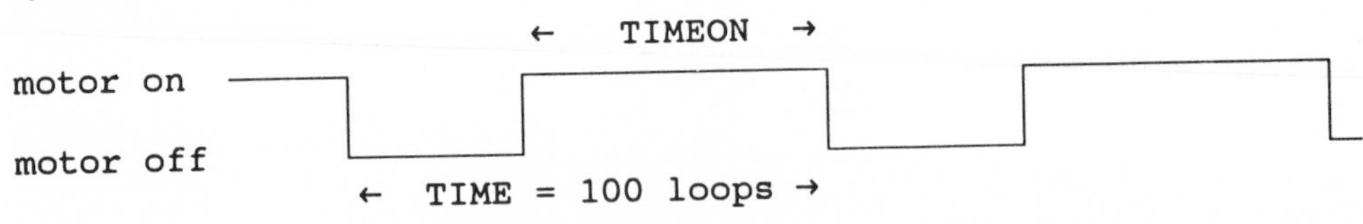

By increasing the width of the power ON pulse the voltage applied to the motor increases and hence the speed. The extreme cases are when TIMEON = 0 and no power is applied to the motor and when TIMEON = 100 and the motor runs at full speed. Note that the relationship between the width of the TIMEON pulse and the speed is not linear (in fact the pulse has to be a certain width to overcome friction and inertia to start the motor turning).

F.7.2 Problem: PWM motor speed control using a program loop

Implement a program for PWM control of the motor speed using a simple program loop. TIMEON may be read from switches or any other device (if an 8-bit number is read divide it by 2.55 to yield a value in the range 0 to 100). A PWM algorithm is:

```
LOOP for ever
     TIMEON = value read (8-bit) from switches / 2.55
     TIME = 100
     IF TIMEON <> 100
          switch motor power OFF
     REPEAT
        IF TIME = TIMEON
             switch motor ON
        TIME = TIME - 1
     UNTIL TIME = 0
ENDLOOP
```

The inner loop is executed 100 times. On each loop the counter (TIME) is decremented and when it equals the value read from the switches (TIMEON) the motor power is switched on and remains on until the count reaches 0.

F.7.3 Problem: Sensing motor revolutions

The motor fan blades interrupt an infra red beam the changes in which can be detected by the signal input to CA1 (MC6821) or H1 (MC68230). The speed can be determined by counting the signal changes on CA1 or H1 using polled or interrupt techniques. Extend the program implemented above to display an * every 50 revolutions. Note that other heat sources may upset IR detectors, i.e. a table lamp.

F.7.4 Problem: PWM motor speed control using timer interrupts

The problem with PWM motor speed control using a simple program loop is that it is not possible to do anything which may upset the PWM timing and hence the motor speed control, e.g. reading and/or writing information from/to a control console.

Implement a program to control the PWM motor/speed using interrupts generated by a timer chip (MC6840 PTM or MC68230 PIT). The switch values should be read within a loop in the main program and the value saved into a variable in memory, e.g:

```
LOOP for ever
   TIMEON = port A data register (switches or ADC) / 255
   WRITE(TIMEON)
ENDLOOP
```

The timer can be programmed to interrupt (e.g. every 100 microseconds) and the PWM line switched ON or OFF using a suitable algorithm, e.g. (TIME should initially be 100):

```
save registers used in routine
clear timer flag to remove cause of interrupt
IF TIME = 0  (* is end of loop reset TIME *)
   TIME = 100
   IF TIMEON <> 100
      switch off motor (* motor not on all the time *)
IF TIME = TIMEON
   switch on motor   (* time to switch on motor *)
TIME = TIME - 1
restore registers
RTE return from exception
```

Extend the program to:

1 Use interrupts to count the revolutions of the motor and display the RPM value on the screen; counting for 15 seconds gives a reasonable value.
2 On program start up select:
 (a) motor speed controlled via switches (as above);
 (b) to display a graph of speed against PWM pulse width.

Fig. F.5 Shows a sample run of the program. Remember:

1 Save and restore all registers used in the interrupt service routine.
2 Use named variables in memory to communicate between the main program and the interrupt service routine, NOT registers.
3 Take care that the timer interrupt service routine does not take longer to execute than the period of the timer interrupt, i.e. on RTE the interrupt service routine is immediately reentered and the main program is never executed to read a new switch value. In such cases recode the interrupt service routine to be more efficient or lengthen the timer interrupt period.

```
Display 50's of revolutions (each *) for 15 seconds against pulse width %

  0                                         0 RPM
  5                                         0 RPM
 10                                         28 RPM
 15 ****                                    1004 RPM
 20 **********                              2020 RPM
 25 **************                          3020 RPM
 30 *****************                       3460 RPM
 35 ******************                      3832 RPM
 40 ********************                    4136 RPM
 45 *********************                   4404 RPM
 50 **********************                  4600 RPM
 55 ***********************                 4772 RPM
 60 ************************                4920 RPM
 65 *************************               5044 RPM
 70 *************************               5148 RPM
 75 *************************               5236 RPM
 80 **************************              5308 RPM
 85 **************************              5392 RPM
 90 ***************************             5456 RPM
 95 ***************************             5520 RPM
100 ***************************             5604 RPM
```

Fig. F.5 Sample program run showing graph of motor speed against PWM pulse width

F.7.5 Problem: Closed loop control of motor speed

Design, code and test a program which reads the motor speed set point (from the switches or some other input device) and using a PID (proportional, integral and derivative) algorithm controls the motor speed. Display graphically, with various values of the PID control constants, system response to a step change in the set point.

F.7.6 Simple temperature control using heater and motor

The fan of the motor is arranged to blow air over the heater so it is possible to vary the heater temperature (read by the temperature sensor) by switching the heater on and off and/or switching the motor on and off.

Implement a program which, by switching the heater and motor on/off, maintains the temperature read by the sensor between two levels. A suitable algorithm is:

```
read temperature lower bound and upper bound
Start up system with heater on and motor off
LOOP for ever
     DELAY to allow system to settle, e.g. 1 second
     read temperature value
     display temperature value on the screen
     IF temperature < lower bound
          switch on heater, switch off fan motor
     ELSE
          if temperature > upper bound
               switch on fan motor, switch off heater
ENDLOOP
```

Extend the program to display a graph of the temperature value against time. Fig. F.6 shows a sample run of the program.

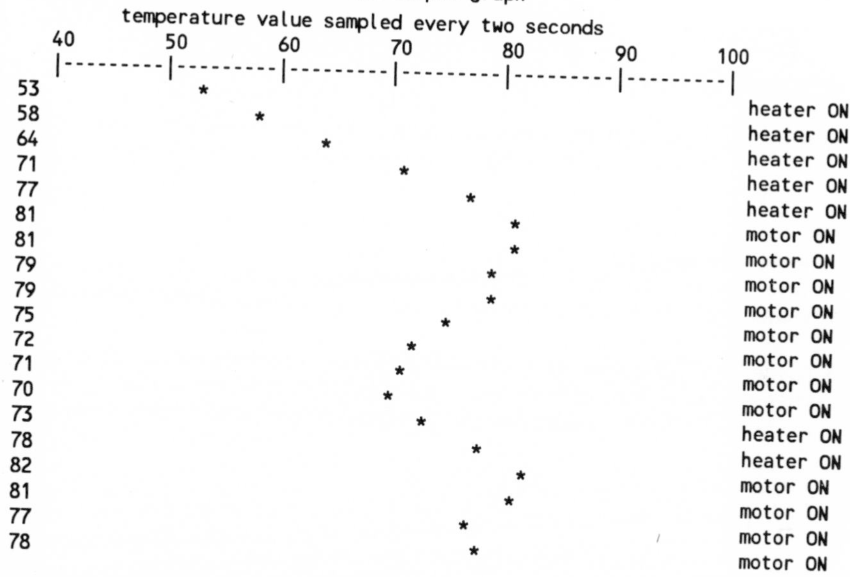

Fig. F.6 Sample run of program showing temperature variation against time

Note that the values displayed in Fig. F.6 are not the actual temperatures but the value read from the analogue to digital converter. In a real control application the system would have to be calibrated to determine the relationship between the physical temperature and A to D value read by the computer.

Appendix G

Answers to Exercises

Chapter 1

Exercise 1.1 (page 5) Calculations in decimal, binary and hexadecimal.

```
 16 00010000 10      45 00101101 2D     110 01101110 6E     110 01101110 6E
+32 00100000 20     +60 00111100 3C     -45 11010011 D3     +45 00101101 2A
 48 00110000 30     105 01101001 69      65 01000001 41    -101 10011011 9B
```

1 The third example (110 - 45) gave a carry out of bit 7 (the sign bit) which is ignored in the result.
2 In the fourth example overflow occurred when 110 + 45 gave a negative result. Overflow occurs when the number of bits available is too small to hold the result. A twos-complement signed 8-bit binary number can represent signed decimal values in the range -128 to +127. The addition of 100 and 45 gave a result greater than 127, thus setting the sign bit (bit 7) making the final result negative and incorrect.

Exercise 1.2 (page 5) Characters with ASCII codes in decimal, hex and binary.

Character	A	B	C	Z	a	b	c
Decimal	65	66	67	90	97	98	99
Hexadecimal	41	42	43	5A	61	62	63
Binary	01000001	01000010	01000011	01011010	01100001	01100010	01100011

Character	0	1	2	9	-	?	=
Decimal	48	49	50	57	45	63	61
Hexadecimal	30	31	32	39	2D	3F	3D
Binary	00110000	00110001	00110010	00111001	00101101	00111111	00111101

From Table A.2 it can be seen that the letter and digit character codes are in ascending numeric order, i.e. the characters '0' to '9' have the ASCII codes 48 to 57. This simplifies:

1 testing to see if a character is within a given range, e.g. a character read from a keyboard is a digit if its code is in the range 48 to 57 decimal;
2 the conversion between characters and numeric values, e.g. subtract the ASCII code for '0' (30 hexadecimal) from the character to give its equivalent numeric value.

Exercise 1.3 (page 8) Each model of processor has a different assembly language. Thus a programmer who is an expert in one assembly language can be limited in choice of job opportunities and/or have to learn a new language when changing employment.

Chapter 2

Exercise 2.1 (page 13)

1 The maximum size of primary memory is limited by the number of bits that the processor uses to address the memory, e.g. 24 bits can address 16Mbytes.
2 (a) RAM (Random Access (read/write) Memory) is random access memory that can be read/written; used for storage of programs and data during execution.
  (b) ROM is Read Only (random access) Memory: information can only be read; used for storage of permanent programs and data, e.g. the bootstrap loader.
3 Secondary memory (disks & tape) is used for long term program and data storage.

4 (a) Random Access: any data word can be accessed directly, e.g. primary memory.
(b) Sequential Access: data is accessed in sequence, e.g. a magnetic tape where to access data down the tape it is necessary to read over intermediate data.
5 A computer bus is the information or data highway which carries information between the various components of a computer system.

Exercise 2.2 (page 15)

1 High speed registers are used to store temporary information during processing.
2 The status register contains information about the result of the last calculation, e.g. if the result was zero or negative. The information can be used for program control.
3 A floating point co-processor would be required for applications where the majority of the calculations use real numbers, e.g. CAD design, mathematics, statistics, etc.

Chapter 3 The following listings are from a FORCE single board MC68000 microcomputer running the ROM based M68 Monitor (commands are invoked by single key hits and are self explanatory).

Exercise 3.1 (page 21) M68 monitor power up checks and command help.

```
--------------------------------------------------------------------------------
MC68000 microcomputer monitor
Copyright Brian Bramer, January 1990, Leicester Polytechnic
System hardware checks, hit RESET button to abort
FORCE microcomputer, CPU MC68000, 127K bytes RAM
MC6840 PTM @ 4CF41 interrupt test                                    OK
MC6850 ACIA @ 50040 write check OK, ACIA @ 50041 write check         OK
MC6821 PIA @ 5CEF1 data direction registers write/read check         OK
Monitor checksum C406 OK, Editor checksum 28E7                       OK
Test system RAM address 00000008 to 00000FFF                         OK
Test user RAM address 00001000 to 0001FFFF..                         OK
Test finished in 0 hours 0 minutes 3.2 seconds, 00000020 interrupts occurred
00000001 test sequence(s) executed, no errors found                all OK

MC68000 monitor V1.04b, please enter command (<ESC> to abort, ? for help)
M68> Help
 Valid commands and subcommands are:
 E - Edit/assemble a program
 D - Dissemble a program
 R - Register: initialise, set, display
 M - Memory:   display, set, modify, block
 X - Convert decimal/hexadecimal/text values
 L - Load S-record program from: Terminal
 S - System tests
 G - Go a program
 T or I - Trace program execution
 B - Breakpoint: display, set, clear
 C - Continue program execution from breakpoint or trace
 Data may be hexadecimal (default), decimal (prefix with .) or text in '...'
 Hit <ESC> to abort a command sequence
M68>
--------------------------------------------------------------------------------
```

On power up the M68 monitor performs a sequence of memory and I/O device checks and then prompts the user for command input with the prompt **M68>** . Commands are generally single keystrokes, e.g. in response to the key **H** the monitor displays the help screen which lists the commands available.

Chapter 4

Exercise 4.1 (page 25) Clear registers, sets D0 = 10 decimal, D1 = 10 hexadecimal, D2 = 'A' and displays the registers (user input is shown in heavy type):

```
-------------------------------------------------------------------------------
MC68000 monitor V1.04b, please enter command (<ESC> to abort, ? for help)
M68> Register initialise
M68> Register set
 Enter D or A (data/address registers) and number ? D0 new value ? .10
M68> Register set
 Enter D or A (data/address registers) and number ? D1 new value ? $10
M68> Register set
 Enter D or A (data/address registers) and number ? D2 new value ? 'A'
M68> Register display
Program Counter (PC) = 00001000  Status Register (SR) = 2700  CCR = -----
D0 = 0000000A  D1 = 00000010  D2 = 41000000  D3 = 00000000
D4 = 00000000  D5 = 00000000  D6 = 00000000  D7 = 00000000
A0 = 00000000  A1 = 00000000  A2 = 00000000  A3 = 00000000
A4 = 00000000  A5 = 00000000  A6 = 00000000  A7 = 00000FFC
M68>
-------------------------------------------------------------------------------
```

Exercise 4.2 (page 26) Load values into memory and check them.

```
-------------------------------------------------------------------------------
M68> Memory block fill from address $1000 to (or +offset) $2000
          with byte value $0
M68> Memory set starting at address $1000
Enter new values (<CR> for next, < for previous byte), <ESC> to exit
00001000 00 ? $12              12
00001001 00 ? $34              34
00001002 00 ? $5678            5678
00001004 00 ? $9ABCDEF0        9ABCDEF0
00001008 00 ? '01234567890'    01234567890
00001013 00 ? $<ESC> hit
M68> Memory display from address $1000 to (or +offset) +$10
00001000 12 34 56 78 9A BC DE F0 30 31 32 33 34 35 36 37 .4Vx....01234567
00001010 38 39 30 00 00 00 00 00 00 00 00 00 00 00 00 00 890.............
M68> Memory set starting at address $1000
Enter new values (<CR> for next, < for previous byte), <ESC> to exit
00001000 12 ? $10              10
00001001 34 ? $1234
  you are trying to load a word/long word starting at an odd address

00001001 34 ?
-------------------------------------------------------------------------------
```

1 **memory block fill command:** fills addresses $1000 to $2000 with 0

2 **memory set command:** load values into memory starting at address $1000
   (a) displays the address and its current value (byte)
   (b) the user enters a value (decimal ,hex or character) terminated by <CR>
   (c) the monitor works out from the size of the value if it is byte, word or long word
   (d) the value is written into memory, read back and displayed (as a write check)
   (e) the address is incremented by the length of the last data

3 **memory display command:** displays lines of memory contents with the start address, byte values of 16 ($10) locations followed by the ASCII character code equivalents

4 the final **memory set command** loads a byte value into location $1000 and then attempts to load a word value ($1234) starting at address $1001. The monitor displayed an error message and then prompts for input.

Exercise 4.3 (page 31) Load a machine code program and execute it.

```
--------------------------------------------------------------------------------
M68> Memory set starting at address $1000
Enter new values (<CR> for next, < for previous byte), <ESC> to exit
00001000 10 ? $20                20
00001001 34 ? $3C                3C
00001002 56 ? $00                00
00001003 78 ? $00                00
00001004 9A ? $01                01
00001005 BC ? $0B                0B
00001006 DE ? $04                04
00001007 F0 ? $40                40
00001008 30 ? $00                00
00001009 31 ? $17                17
0000100A 32 ? $D0                D0
0000100B 33 ? $40                40
0000100C 34 ? $06                06
0000100D 35 ? $40                40
0000100E 36 ? $00                00
0000100F 37 ? $07                07
00001010 38 ? $4E                4E
00001011 39 ? $4E                4E
00001012 30 ? $<ESC> hit
M68> Register initialise
M68> go/run program, from start address (<CR> for 00001000) ? $
TRAP #14 instruction at PC = 00001012, returning to monitor

M68> Register display
Program Counter (PC) = 00001012  Status Register (SR) = 2700  CCR = -----
D0 = 000001EF  D1 = 00000000  D2 = 00000000  D3 = 00000000
D4 = 00000000  D5 = 00000000  D6 = 00000000  D7 = 00000000
A0 = 00000000  A1 = 00000000  A2 = 00000000  A3 = 00000000
A4 = 00000000  A5 = 00000000  A6 = 00000000  A7 = 00000FFC

M68> Dissemble from address (<CR> for 00001012) ? $1000
Dissembling - hit <ESC> to abort
Address  object code              dissembled code
00001000 203C 0000010B            MOVE.L    #$0000010B,D0
00001006 0440      0017           SUBI.W    #$0017,D0
0000100A D040                     ADD.W     D0,D0
0000100C 0640      0007           ADDI.W    #$0007,D0
00001010 4E4E                     TRAP      #$000E
00001012 0000      0000           ORI.B     #$00,D0
00001016 0000      0000           ORI.B     #$00,D0
0000101A 0000      0000           ORI.B     #$00,D0
<ESC> hit
M68>
--------------------------------------------------------------------------------
```

1 **memory set command:** loads the machine code program
2 **go command:** starts program execution (at a particular address)
3 the **TRAP instruction** terminates program execution returning control to the M68 monitor
4 **register display command:** displays the result in D0 as $1EF (495 decimal)
5 **dissemble command:** dissembles the program (following the TRAP command in memory is whatever was left by the last program - 0's in this case)

Chapter 5 Exercise 5.1 (page 34)

SUB subtraction of binary integer numbers
OR logical OR operation
MULS multiply signed binary integer numbers
MULU multiply unsigned binary integer numbers
TST test a value
ROR rotate a bit pattern right
NEG negate a signed binary integer number
ASL shift a signed binary number left
BCLR clear a bit in a value
JMP jump (transfer control) to a location in memory, i.e. equivalent to a GOTO

Chapter 6 Exercise 6.1 (page 43)

You should have an assembler listing similar to Fig. 6.1 in Chapter 6.

Chapter 7 Exercise 7.1 (page 52)

```
--------------------------------------------------------------------------------
 1                          * Exercise 7.1
 2                          * program to calculate RESULT = 2*(355+4*(89*2-7)-13*8)
 3 001000 720D              START:   MOVEQ    #13,D1     initialize D1 with 13
 4 001002 D241                       ADD.W    D1,D1      13*2
 5 001004 D241                       ADD.W    D1,D1      13*4
 6 001006 D241                       ADD.W    D1,D1      13*8
 7 001008 7459                       MOVEQ    #89,D2     initialize D2 with 89
 8 00100A D442                       ADD.W    D2,D2      89*2
 9 00100C 5F42                       SUBQ.W   #7,D2      89*2-7
10 00100E D442                       ADD.W    D2,D2      2*(89*2-7)
11 001010 D442                       ADD.W    D2,D2      4*(89*2-7)
12 001012 203C 00000163              MOVE.L   #355,D0    initialize D0 with 355
13 001018 D042                       ADD.W    D2,D0      355+4*(89*2-7)
14 00101A 9041                       SUB.W    D1,D0      355+4*(89*2-7)-13*8
15 00101C D040                       ADD.W    D0,D0      2*(355+4*(89*2-7)-13*8)
16 00101E 4E4E                       TRAP     #14        !! STOP PROGRAM
17                                   END
--------------------------------------------------------------------------------
```

1 The MOVEQ (line 3 & 7) and MOVE.L (line 12) instructions are used to move initial values into registers and clear all the existing contents. Register values can be displayed using a monitor without being distracted with irrelevant information.

2 The above program can be written in a number of ways using a different order of calculations. The main point is that the program structure should be clear and that the program gives the correct results. If there is a clear, easy way to do something, use it, even if the program is slightly longer than it would otherwise be. A clever piece of code may take a few bytes less of memory and run a bit faster but you may not be able to understand how it works in six months time.

Exercise 7.2 (page 55)

```
------------------------------------------------------------------------------
 1                       * Exercise 7.2
 2                       * program to calculate RESULT = A*(355+4*(89*2-B)-13*8)
 3                       * on entry it is assumed that A is in D3 & B in D4
 4 001000 720D           START:   MOVEQ    #13,D1    initialize D1 with 13
 5 001002 D241                    ADD.W    D1,D1     13*2
 6 001004 D241                    ADD.W    D1,D1     13*4
 7 001006 D241                    ADD.W    D1,D1     13*8
 8 001008 7459                    MOVEQ    #89,D2    initialize D2 with 89
 9 00100A D442                    ADD.W    D2,D2     89*2
10 00100C 9444                    SUB.W    D4,D2     89*2-B
11 00100E D442                    ADD.W    D2,D2     2*(89*2-B)
12 001010 D442                    ADD.W    D2,D2     4*(89*2-B)
13 001012 203C 00000163           MOVE.L   #355,D0   initialize D0 with 355
14 001018 D042                    ADD.W    D2,D0     355-4*+89*2-B)
15 00101A 9041                    SUB.W    D1,D0     355+4*(89*2-B)-13*8
16 00101C C1C3                    MULS     D3,D0     A*(355+4*(89*2-B)-13*8)
17 00101E 4E4E                    TRAP     #14       !! STOP PROGRAM
18                                END
------------------------------------------------------------------------------
```

```
------------------------------------------------------------------------------
M68> Register set
 Enter D or A (data/address registers) and number ? D3 new value ? $2
M68> Register set
 Enter D or A (data/address registers) and number ? D4 new value ? $7
M68> go/run program, from start address (<CR> for 00001000) ? $

TRAP #14 instruction at PC = 00001020, returning to monitor
M68> Register display
Program Counter (PC) = 00001020  Status Register (SR) = 2700  CCR = -----
D0 = 0000074E  D1 = 00000068  D2 = 000002AC  D3 = 00000002
D4 = 00000007  D5 = 00000000  D6 = 00000000  D7 = 00000000
A0 = 00000000  A1 = 00000000  A2 = 00000000  A3 = 00000000
A4 = 00000000  A5 = 00000000  A6 = 00000000  A7 = 00000FFC
M68> Register set
 Enter D or A (data/address registers) and number ? D3 new value ? $4
M68> go/run program, from start address (<CR> for 00001000) ? $

TRAP #14 instruction at PC = 00001020, returning to monitor
M68> Register display
Program Counter (PC) = 00001020  Status Register (SR) = 2700  CCR = -----
D0 = 00000E9C  D1 = 00000068  D2 = 000002AC  D3 = 00000004
D4 = 00000007  D5 = 00000000  D6 = 00000000  D7 = 00000000
A0 = 00000000  A1 = 00000000  A2 = 00000000  A3 = 00000000
A4 = 00000000  A5 = 00000000  A6 = 00000000  A7 = 00000FFC
M68>
------------------------------------------------------------------------------
```

The above listing shows two runs of Exercise 7.2 with variable values A and B being loaded into D3 and D4 respectively:

1 A (in D3) = 2 and B (in D4) = 7 to give RESULT = 1870 ($74E) as Exercise 7.1;
2 A (in D3) = 4 and B (in D4) = 7 to give RESULT = 3740 ($E9C)

Chapter 8 Exercise 8.1 (page 62)

```
------------------------------------------------------------------------------
1                          * Exercise 8.1 -  RESULT = A*(355+4*(89*2-B)-13*8)
2                          *  RESULT, A and B are named variables in memory
3 001000 720D              START:    MOVEQ     #13,D1     initialize D1 with 13
4 001002 D241                        ADD.W     D1,D1      13*2
5 001004 D241                        ADD.W     D1,D1      13*4
6 001006 D241                        ADD.W     D1,D1      13*8
7 001008 7459                        MOVEQ     #89,D2     initialize D2 with 89
8 00100A D442                        ADD.W     D2,D2      89*2
9 00100C 9479 00001030               SUB.W     B,D2       89*2-B
10 001012 D442                       ADD.W     D2,D2      2*(89*2-B)
11 001014 D442                       ADD.W     D2,D2      4*(89*2-B)
12 001016 203C 00000163              MOVE.L    #355,D0    initialize D0 with 355
13 00101C D042                       ADD.W     D2,D0      355-4*+89*2-B)
14 00101E 9041                       SUB.W     D1,D0      355+4*(89*2-B)-13*8
15 001020 C1F9 0000102E              MULS      A,D0       7A*(355+4*(89*2-B)-13*8)
16 001026 33C0 00001032              MOVE.W    D0,RESULT save RESULT
17 00102C 4E4E                       TRAP      #14        !! STOP PROGRAM
18                         * data area
19 00102E 0002             A:        DC.W      2          initial value of A
20 001030 0007             B:        DC.W      7          initial value of B
21 001032 0000             RESULT:   DC.W      0
22                                   END
------------------------------------------------------------------------------
```

```
------------------------------------------------------------------------------
M68> go/run program, from start address (<CR> for 00001000) ? $
TRAP #14 instruction at PC = 0000102E, returning to monitor

M68> Register display
Program Counter (PC) = 0000102E  Status Register (SR) = 2700  CCR = -----
D0  = 0000074E  D1 = 00000068  D2 = 000002AC  D3 = 00000004
D4  = 00000007  D5 = 00000000  D6 = 00000000  D7 = 00000000
A0  = 00000000  A1 = 00000000  A2 = 00000000  A3 = 00000000
A4  = 00000000  A5 = 00000000  A6 = 00000000  A7 = 00000FFC

M68> Memory modify word values starting at address $102E
Enter word values (<CR> next value, < previous value, <ESC> to exit)
0000102E 0002 ? $4          0004
00001030 0007 ? $<ESC> key hit

M68> go/run program, from start address (<CR> for 00001000) ? $
TRAP #14 instruction at PC = 0000102E, returning to monitor

M68> Register display
Program Counter (PC) = 0000102E  Status Register (SR) = 2700  CCR = -----
D0  = 00000E9C  D1 = 00000068  D2 = 000002AC  D3 = 00000004
D4  = 00000007  D5 = 00000000  D6 = 00000000  D7 = 00000000
A0  = 00000000  A1 = 00000000  A2 = 00000000  A3 = 00000000
A4  = 00000000  A5 = 00000000  A6 = 00000000  A7 = 00000FFC

M68> Memory modify word values starting at address $102E
Enter word values (<CR> next value, < previous value, <ESC> to exit)
0000102E 0004 ? $
00001030 0007 ? $
00001032 0E9C ? $
------------------------------------------------------------------------------
```

The above listing shows two runs of Exercise 8.1:

1 Initial values defined using DC.W of A = 2 (line 20) and B = 7 (line 21) gives RESULT = 1870 ($74E) as Exercise 7.1;
2 A (at address $102E) is then modified to = 4, gives RESULT = 3740 ($E9C).

Chapter 9 Exercise 9.1 (page 70)

Subroutines are used:

1 To aid modular programming, i.e. to break large programs into manageable modules.
2 To save repeating the same code at each point where it is needed, i.e. write a subroutine once and call it as often as required.
3 To provide libraries of common modules, e.g. mathematical routines.

Exercise 9.2 (page 76)

```
------------------------------------------------------------------------------------
 1                              * Exercise 9.2 -  RESULT = A*(355+4*(89*2-B)-13*8)
 2                              *   RESULT, A and B are named variables in memory
 3                              *
 4                              * WRITE('Calculate  RESULT = A*(355+4*(89*2-B)-13*8)')
 5                              * WRITE(NEWLINE,'Enter value of A ? '); READ(A)
 6                              * WRITE(NEWLINE,'Enter value of B ? '); READ(B)
 7                              * Calculate  RESULT = A*(355+4*(89*2-B)-13*8)
 8                              * WRITE(NEWLINE,'RESULT of A*(355+4*(89*2-B)-13*8) = ',RESULT)
 9                              *
10                                       XREF     WRTEXT,RDDECW,WRDECW
11 001000 4EB9 00028038 START:           JSR      WRTEXT
12 001006 43 61 6C 63 75                 DC.B     'Calculate  A*(355+4*(89*2-B)-13*8)'
13 001031 0A 0D                          DC.B     10,13
14 001033 45 6E 74 65 72                 DC.B     'Enter value of A ? $'
15 001048 4EB9 00028088                  JSR      RDDECW    read A in decimal
16 00104E 33C0 000010DC                  MOVE.W   D0,A      and save to memory
17 001054 4EB9 00028038                  JSR      WRTEXT
18 00105A 0A 0D                          DC.B     10,13
19 00105C 45 6E 74 65 72                 DC.B     'Enter value of B ? $'
20 001070 4EB9 00028088                  JSR      RDDECW    read B in decimal
21 001076 33C0 000010DE                  MOVE.W   D0,B      and save to memory
22                              * read A and B, now perform calculation
23 00107C 720D                           MOVEQ    #13,D1    initialize D1 with 13
24 00107E D241                           ADD.W    D1,D1     13*2
25 001080 D241                           ADD.W    D1,D1     13*4
26 001082 D241                           ADD.W    D1,D1     13*8
27 001084 7459                           MOVEQ    #89,D2    initialize D2 with 89
28 001086 D442                           ADD.W    D2,D2     89*2
29 001088 9479 000010DE                  SUB.W    B,D2      89*2-B
30 00108E D442                           ADD.W    D2,D2     2*(89*2-B)
31 001090 D442                           ADD.W    D2,D2     4*(89*2-B)
32 001092 203C 00000163                  MOVE.L   #355,D0   initialize D0 with 355
33 001098 D042                           ADD.W    D2,D0     355-4*+89*2-B)
34 00109A 9041                           SUB.W    D1,D0     355+4*(89*2-B)-13*8
35 00109C C1F9 000010DC                  MULS     A,D0      7A*(355+4*(89*2-B)-13*8)
36 0010A2 33C0 000010E0                  MOVE.W   D0,RESULT save RESULT
37 0010A8 4EB9 00028038                  JSR      WRTEXT
38 0010AE 0A 0D                          DC.B     10,13
39 0010B0 52 45 53 55 4C                 DC.B     'RESULT of A*(355+4*(89*2-B)-13*8)= $'
40 0010D4 4EB9 00028074                  JSR      WRDECW    display RESULT in decimal
41 0010DA 4E4E                           TRAP     #14       !! STOP PROGRAM
42                              * data area
43 0010DC 0000                  A:       DC.W     0         initial value of A
44 0010DE 0000                  B:       DC.W     0         initial value of B
45 0010E0 0000                  RESULT:  DC.W     0
46                                       END
------------------------------------------------------------------------------------
```

```
------------------------------------------------------------------------------
M68> go/run program, from start address (<CR> for 00001000) ? $
Calculate  RESULT = A*(355+4*(89*2-B)-13*8)
Enter value of A ? 2
Enter value of B ? 7
RESULT of A*(355+4*(89*2-B)-13*8)= 1870
TRAP #14 instruction at PC = 000010DC, returning to monitor

M68> go/run program, from start address (<CR> for 00001000) ? $
Calculate  RESULT = A*(355+4*(89*2-B)-13*8)
Enter value of A ? 2 4
Enter value of B ? 7
RESULT of A*(355+4*(89*2-B)-13*8)= 3740
TRAP #14 instruction at PC = 000010DC, returning to monitor
------------------------------------------------------------------------------
```

Chapter 10 Exercise 10.1 (page 84)

```
------------------------------------------------------------------------------
 1                             * Exercise 10.1 - program to read a number and display
 2                             *  message if it was 0 or a positive or negative value
 3                             *
 4                             * LOOP FOR ever
 5                             *     WRITE(newline,'please enter number')
 6                             *     READ(NUMBER)
 7                             *     IF NUMBER=0
 8                             *          WRITE('number was 0')
 9                             *     ELSE
10                             *         if NUMBER > 0
11                             *               WRITE('number was positive ')
12                             *         ELSE
13                             *               WRITE('number was negative')
14                             *     END IF
15                             * END LOOP
16                             *
17                                      XREF      WRLINE,WRTEXT,RDDECW
18 001000 4EB9 0002802C START:          JSR       WRLINE
19 001006 4EB9 00028038                 JSR       WRTEXT      prompt user for input
20 00100C 70 6C 65 61 73                DC.W      'please enter number ? $'
21 001024 4EB9 00028088                 JSR       RDDECW      read a number
22 00102A 4EB9 0002802C                 JSR       WRLINE
23 001030 4A40                          TST.W     D0          test D0
24 001032 6618                          BNE.S     NZERO
25 001034 4EB9 00028038                 JSR       WRTEXT      is zero display message
26 00103A 6E 75 6D 62 65                DC.W      'number was zero$'
27 00104A 60B4                          BRA.S     START
28 00104C 6B1C            NZERO:        BMI.S     MINUS
29 00104E 4EB9 00028038                 JSR       WRTEXT      is a positive number
30 001054 6E 75 6D 62 65                DC.W      'number was positive$'
31 001068 6096                          BRA.S     START
32 00106A 4EB9 00028038 MINUS:          JSR       WRTEXT      is a negative number
33 001070 6E 75 6D 62 65                DC.W      'number was negative$'
34 001084 6000      FF7A                BRA.L     START
35                                      END
------------------------------------------------------------------------------
```

18-21 the user is prompted for input and a number read from the keyboard
23 TST is used to test the number and set the Condition Codes
24 branch to NZERO if value < > 0
25-27 executed if value was 0
28 branch to MINUS if value (tested in line 23) is negative
29-31 executed if value is > 0
32-33 executed if value is < 0

Chapter 11

Exercise 11.1 (page 87)

The sequence of execution of RDBINW (program listing on next page) is:

line

40 D1 (which will store the resultant value) is set to 0
42 a character is read into D0
43-46 IF the character is '0' ($30) or '1' ($31) execution continues to line 47
ELSE the loop terminates (line 53) and control returns to the main program (line 54 returns control to line 17) with the binary number in D0
47 echoes the character onto the screen
48 converts the ASCII code of the **character** in D0 into the binary value, by subtracting the ASCII character code for '0' (which is $30) from the character code read from the keyboard
49 the binary value entered so far is multiplied by two
50 the new bit value is added to the binary value
51 the loop returns to line 42

Labelling in subroutines can be difficult (i.e. when trying to avoid using the same labels that are in the main program or other subroutines). One simple way is shown in the program listing, in which labels start with the name of the subroutine, or a shortened form of it, and end in digits.

Exercise 11.2 (page 90)

Main program sequence (program listing on next page but one):

11 the loop count-1 is moved into D2
12 the subroutine LINE is called to display the line of characters
13 performs the loop control
14 Terminates the program after 5 loops

Sequence in subroutine LINE:

30 a newline is output to the screen
31 subroutine ASTX is called to display '**********'
32 the loop count-1 to display '?' characters is moved into D1
33 the character '?' is moved into D0
34 writes '?' to the screen
35 loop control for writing the '?' characters
36 subroutine ASTX is called to display '**********'
37 returns to calling program (to line 13)

Sequence in subroutine ASTX:

50 the loop count-1 is moved into D1
51 the character '*' is moved into D0
52 writes '*' to the screen
53 loop control for writing the '*' characters
54 returns to calling program (to line 32 or 37)

The character operands for MOVEQ are defined as byte length by use of EQU.B (lines 29 and 49). This program contains a number of loops within a loop (D2 being used to hold the outer loop count and D1 the inner loop counts).

Exercise 11.1 program listing

```
-------------------------------------------------------------------------------
 1                          * Exercise 11.1
 2                          * Test program for RDBINW that will loop indefinitely reading
 3                          *   binary values and displaying them in hexadecimal.
 4                          *
 5                          * LOOP for ever
 6                          *     WRITE(newline,'enter binary number (word) ? ')
 7                          *     CALL RDBINW to read a binary value into D0
 8                          *     WRITE(' equals in hexadecimal ')
 9                          *     CALL WRHEXW to display in hexadecimal
10                          * END LOOP
11                          *
12                                    XREF      WRLINE,WRHEXW,RDCHAR,WRCHAR,WRTEXT
13 001000 4EB9 0002802C START:        JSR       WRLINE      display a newline
14 001006 4EB9 00028038               JSR       WRTEXT      display prompt message
15 00100C 65 6E 74 65 72              DC.W      'enter binary number (word) ? $'
16 00102A 612A                        BSR.S     RDBINW        read a binary number
17 00102C 4EB9 00028038               JSR       WRTEXT        display result message
18 001032 20 20 20 20 65              DC.W      '    equals in hexadecimal $'
19 00104E 4EB9 00028056               JSR       WRHEXW        display number in hexadecimal
20 001054 61AA                        BSR.S     START         read next number
21                          *
22                          * subroutine to read a binary number from the keyboard into D0
23                          *
24                          * SUBROUTINE RDBINW
25                          * BINARY_VALUE=0
26                          * READ(character)
27                          * LOOP WHILE (character='0') OR (character='1')
28                          *     WRITE(character)
29                          *     BIT_VALUE=character - ASCII '0'
30                          *     BINARY_VALUE=2*BINARY_VALUE+BIT_VALUE
31                          *     READ(character)
32                          * END LOOP
33                          *
34                          * D1 (word) is used to store BINARY_VALUE during routine
35                          * D0 is used for character input and to store BIT_VALUE
36                          *
37                          * on exit:-
38                          * D0 holds the binary word value
39                          *
40 001056 4281              RDBINW:   CLR.L   D1              initialize the binary value
41                          *         loop WHILE character = '0' or = '1'
42 001058 4EB9 00028084 RDBIN1:       JSR     RDCHAR          read a character into D0
43 00105E 0C00      0030              CMPI.B  #'0',D0         is character a '0'
44 001062 6706                        BEQ.S   RDBIN2          if zero calculate value
45 001064 0C00      0031              CMPI.B  #'1',D0         is it a '1'
46 001068 6610                        BNE.S   RDBIN3          if not 0 or 1 is end of loop
47 00106A 4EB9 00028028 RDBIN2:       JSR     WRCHAR          echo character
48 001070 0400      0030              SUBI.B  #'0',D0         convert from ASCII to number
49 001074 D281                        ADD.L   D1,D1           multiply current value by 2
50 001076 D240                        ADD.W   D0,D1           and add new bit
51 001078 60DE                        BRA.S   RDBIN1          repeat loop again
52                          *         end of WHILE loop
53 00107A 2001              RDBIN3:   MOVE.L  D1,D0           move value into D0 and return
54 00107C 4E75                        RTS
55                                    END
-------------------------------------------------------------------------------
```

Exercise 11.2 program listing

```
--------------------------------------------------------------------------------
 1                          * Exercise 11.2 - display text pattern
 2                          *
 3                          * LOOP FOR 5 times
 4                          *     CALL subroutine LINE
 5                          * END LOOP
 6                          * END PROGRAM
 7                          *
 8                          * D2 contains the loop counter
 9                          *
10                                  XREF     WRLINE,WRCHAR
11 001000 7404              START:  MOVEQ    #4,D2          display 5 lines
12 001002 6106              LOOP1:  BSR.S    LINE           display a line
13 001004 51CA     FFFC             DBRA     D2,LOOP1       loop again if required
14 001008 4E4E                      TRAP     #14
15                          *
16                          * subroutine to draw a line so:-
17                          *      **********??????????**********
18                          * SUBROUTINE line
19                          * WRITE(newline)
20                          * CALL astx to display 10 * characters
21                          * LOOP FOR 10 times
22                          *     WRITE('?')
23                          * END LOOP
24                          * CALL astx to display 10 * characters
25                          * RETURN to calling program
26                          *
27                          * D1 contains the loop counter
28                          *
29        0000003F          CHARQU: EQU.B    '?'            define ? as a byte
30 00100A 4EB9 0002802C     LINE:   JSR      WRLINE         new line
31 001010 6112                      BSR.S    ASTX           display **********
32 001012 7209                      MOVEQ    #9,D1          display 10 ? characters
33 001014 703F                      MOVEQ    #CHARQU,D0     set up ? in D0
34 001016 4EB9 00028028     LINE1:  JSR      WRCHAR         display a ?
35 00101C 51C9     FFF8             DBRA     D1,LINE1       display 10 ?
36 001020 6102                      BSR.S    ASTX           display **********
37 001022 4E75                      RTS
38                          *
39                          * subroutine to display **********
40                          *
41                          * SUBROUTINE ASTX
42                          * LOOP FOR 10 times
43                          *     WRITE('*')
44                          * END LOOP
45                          * RETURN to calling program
46                          *
47                          * D1 contains the loop counter
48                          *
49        0000002A          CHARAS: EQU.B    '*'            define * character
50 001024 7209              ASTX:   MOVEQ    #9,D1          display 10 *
51 001026 702A                      MOVEQ    #CHARAS,D0     set up * in D0
52 001028 4EB9 00028028     ASTX1:  JSR      WRCHAR         display a *
53 00102E 51C9     FFF8             DBRA     D1,ASTX1       display 10 *
54 001032 4E75                      RTS
55                                  END
--------------------------------------------------------------------------------
```

Chapter 12

Exercise 12.1 (page 95)

```
 1                              * Exercise 12.1
 2                              * Test program for WRBINW that will loop indefinitely
 3                              *  reading  binary values and displaying them.
 4                              *
 5                              * LOOP for ever
 6                              *     WRITE(newline,'enter binary number (word) ? ')
 7                              *     CALL RDBINW to read a binary value into D0
 8                              *     WRITE(' number entered was ')
 9                              *     CALL WRBINW to display in binary
10                              * END LOOP
11                              *
12                                        XREF      WRLINE,WRHEXW,RDCHAR,WRCHAR,WRTEXT
13 001000 4EB9 0002802C START:    JSR       WRLINE       display a newline
14 001006 4EB9 00028038           JSR       WRTEXT       display prompt message
15 00100C 65 6E 74 65 72          DC.W      'enter binary number (word) ? $'
16 00102A 611C                    BSR.S     RDBINW       read a binary number
17 00102C 4EB9 00028038           JSR       WRTEXT       display result message
18 001032 20 20 6E 75 6D          DC.W      '  number was $'
19 001040 4EB9 00001070           JSR       WRBINW        display number in binary
20 001046 61B8                    BSR.S     START        read next number
21                              *
22                              * subroutine to read a binary number from the keyboard into D0
23                              *
24                              * SUBROUTINE RDBINW
25                              * BINARY_VALUE=0
26                              * READ(character)
27                              * LOOP WHILE (character='0') OR (character='1')
28                              *     WRITE(character)
29                              *     BIT_VALUE=character - ASCII '0'
30                              *     BINARY_VALUE=2*BINARY_VALUE+BIT_VALUE
31                              *     READ(character)
32                              * END LOOP
33                              *
34                              * D1 (word) is used to store BINARY_VALUE during routine
35                              * D0 is used for character input and to store BIT_VALUE
36                              *
37                              * on exit:-
38                              * D0 holds the binary word value
39                              *
40 001048 4281                  RDBINW:   CLR.L   D1         initialize the binary value
41                              *         loop WHILE character = '0' or = '1'
42 00104A 4EB9 00028084 RDBIN1:   JSR     RDCHAR     read a character into D0
43 001050 0C00      0030            CMPI.B  #'0',D0    is character a '0'
44 001054 6706                      BEQ.S   RDBIN2     if zero calculate value
45 001056 0C00      0031            CMPI.B  #'1',D0    is it a '1'
46 00105A 6610                      BNE.S   RDBIN3     if not 0 or 1 is end of loop
47 00105C 4EB9 00028028 RDBIN2:   JSR     WRCHAR     echo character
48 001062 0400      0030            SUBI.B  #'0',D0    convert from ASCII to number
49 001066 D281                      ADD.L   D1,D1      multiply current value by 2
50 001068 D240                      ADD.W   D0,D1      and add new bit
51 00106A 60DE                      BRA.S   RDBIN1     repeat loop again
52                              *         end of WHILE loop
53 00106C 2001                  RDBIN3:   MOVE.L  D1,D0      move value into D0 and return
54 00106E 4E75                            RTS
55                              *
```

**** continued on next page ****

```
----------------------------------------------------------------------------
56                          * subroutine to display a 16 bit word value in D0 in binary
57                          *  version using shift and rotate operations
58                          *
59                          * SUBROUTINE WRBINW
60                          * LOOP FOR 16 times
61                          *     rotate most significant bit of number into bit 0 of D0
62                          *     add character '0' to form ASCII character 0 or 1
63                          *     display bit value
64                          * END LOOP
65                          * RETURN
66                          *
67                          * on entry D0 (word) holds value to be displayed
68                          *
69                          * working values:-
70                          *   D2 (word) holds value being displayed
71                          *   D1 (word) holds loop counter
72                          *   D0 holds the most significant bit then the character 0/1
73                          *
74 001070 720F              WRBINW:  MOVEQ    #15,D1      16 bits to display
75 001072 3400                       MOVE.W   D0,D2       move value into D2
76 001074 E34A              WRBIN1:  LSL.W    #1,D2       shift bit 15 into X
77 001076 4200                       CLR.B    D0          clear D0 to receive that bit
78 001078 E310                       ROXL.B   #1,D0       rotate X into D0
79 00107A 0600     0030              ADDI.B   #'0',D0     add ASCII 0 to form character
80 00107E 4EB9 00028028              JSR      WRCHAR      display bit value
81 001084 51C9     FFEE              DBRA     D1,WRBIN1   LOOP again if required
82 001088 4E75                       RTS                  return to calling program
83                                   END
----------------------------------------------------------------------------
```

With the subroutine WRBINW each bit of the word value to be displayed (starting with the most significant bit) is moved via the X (extend) bit in the CCR into D0 where it is used to form the character '0' (numeric value $30) or '1' (numeric value $31) which is then displayed.

74 16 bits are to be displayed so the loop count-1 is moved into D1
75 the value to be displayed is moved from D0 into D2 (D0 will be used to output the characters)
76 the most significant bit of the word in D2 is shifted into X (and C)
77 D0 is cleared ready for the character to be formed (CLR does not affect the X bit)
78 using ROXL the X bit (set by line 76) is rotated left into bit 0 of D0
79 the ASCII character code for '0' is added to D0 (i.e. forming the character '0' or '1' depending upon the bit shifted in line 76)
80 the bit value is displayed
81 if the loop is not complete branch back to line 76 to display the next bit
82 return to calling program

Exercise 12.2 (page 96)

Main program and RDBINW as for Exercise 12.1 (to line 55).

```
--------------------------------------------------------------------------------
56                          * subroutine to display a 16 bit word value in D0 in binary
57                          *    subroutine using single bit operations
58                          *
59                          * SUBROUTINE WRBINW
60                          * BIT_NUMBER = 15
61                          * LOOP
62                          *     IF BIT_NUMBER of NUMBER = 0
63                          *         WRITE('0')
64                          *     ELSE
65                          *         WRITE('1')
66                          *     END IF
67                          *     BIT_NUMBER = BIT_NUMBER -1
68                          * UNTIL BIT_NUMBER < 0
69                          * RETURN
70                          *
71                          * on entry D0 holds value to be displayed
72                          *
73                          * working values:-
74                          *   D2 holds NUMBER being displayed
75                          *   D1 holds BIT_NUMBER
76                          *   D1 holds the ASCII character '0' or '1'
77          00000030        CHAR0:    EQU.B    '0'          define character 0
78 001070 720F              WRBINW:   MOVEQ    #15,D1       16 bits to test
79 001072 3400                        MOVE.W   D0,D2        move value into D2
80 001074 7030              WRBIN1:   MOVEQ    #CHAR0,D0    character '0' into D0
81 001076 0302                        BTST.L   D1,D2        test bit of number
82 001078 6704                        BEQ.S    WRBIN2       if bit i= 0 go
83 00107A 08C0      0000              BSET.L   #0,D0        bit = 1, form character '1'
84 00107E 4EB9 00028028 WRBIN2:      JSR      WRCHAR       display character
85 001084 51C9      FFEE              DBRA     D1,WRBIN1    display next bit or end loop
86 001088 4E75                        RTS
87                                    END
--------------------------------------------------------------------------------
```

Each bit of the word value to be displayed is tested (starting with the most significant bit 15), and the corresponding character '0' ($30) or '1' ($31) output. The sequence is:

78 15 is moved into D1 where it will be used for two things:
(a) in line 81 to test a bit number (the least significant is bit 0),
and (b) as the loop control counter in line 85 (which decrements D1 on each loop)
79 the value to be displayed is moved into D2 (D0 is to be used for character output)
80 the ASCII character code for '0' is moved into D0
81 a bit in D2 is tested; the bit number is specified by the value in D1 (this value is decremented on each loop by the DBRA in line 85)
82 if the bit was non-zero a branch is made to line 84 to display character '0'
83 the bit was 1, so BSET is used to form the character '1' in D0
84 the character '0' or '1' is displayed
85 loop control (also decrements number of bit to test in line 81)
86 return to calling program

This shows the use of BTST and BSET. It would be more efficient to replace line 83 with a MOVEQ instruction (to move '1' into D0).

Chapter 13 Exercise 13.1 (page 103)

```
-------------------------------------------------------------------------------
 1                          * Exercise 13.1: to test string compare subroutine
 2                          *
 3                          * CALL STRING with identical strings
 4                          * CALL STRING with non-identical strings
 5                          *
 6                                   XREF      WRLINE,WRTEXT
 7 001000 7005              START:   MOVEQ     #5,D0          number of chars in first test
 8 001002 207C 00001022              MOVEA.L   #S1,A0         address of strings
 9 001008 227C 00001028              MOVEA.L   #S2,A1
10 00100E 6132                       BSR.S     STRING         compare them
11 001010 700A                       MOVEQ     #10,D0         second pair of strings
12 001012 207C 0000102E              MOVEA.L   #S3,A0
13 001018 227C 00001038              MOVEA.L   #S4,A1
14 00101E 6122                       BSR.S     STRING         compare them
15 001020 4E4E                       TRAP      #14            !! STOP PROGRAM
16                          * test data: first string compare OK, second should fail
17 001022 41 42 43 44       S1:      DC.W      'ABCDE'
18 001028 41 42 43 44       S2:      DC.W      'ABCDE'
19 00102E 30 31 32 33       S3:      DC.W      '0123456789'
20 001038 30 31 32 33       S4:      DC.W      '0123456X89'
21                          *
22                          * SUBROUTINE STRING to compare two character strings
23                          *
24                          * LOOP FOR string length
25                          *     IF characters[index] are not same
26                          *         EXIT LOOP
27                          *     increment character string index
28                          * END LOOP
29                          * IF all characters were identical
30                          *      WRITE('strings are identical')
31                          * ELSE
32                          *      WRITE('strings are not identical')
33                          *
34                          * on entry:- D0 holds length of strings (word value)
35                          *            A0 holds start address of first string
36                          *            A1 holds start address of second string
37                          *
38                          * characters are compared using CMPM; compares two operands
39                          *  using autoincrement mode, i.e. after the CMPM the values
40                          *  of A0 & A1 are incremented to point at the next character
41                          *
42 001042 4EB9 0002802C STRING:  JSR       WRLINE
43 001048 5340                   SUBQ.W    #1,D0          decrement count for DBRA
44 00104A B308          STRNG1:  CMPM.B    (A0)+,(A1)+    compare two characters
45 00104C 6622                   BNE.S     STRNG2         if not equal is an error
46 00104E 51C8      FFFA         DBRA      D0,STRNG1      if D0>= do next characters
47 001052 4EB9 00028038          JSR       WRTEXT         string identical message
48 001058 73 74 72 69 6E         DC.W      'strings are identical$'
49 00106E 6020                   BRA.S     STRNG3
50 001070 4EB9 00028038 STRNG2:  JSR       WRTEXT         string not identical message
51 001076 73 74 72 69 6E         DC.W      'strings are not identical$'
52 001090 4E75          STRNG3:  RTS                      RETURN to calling program
53                               END
-------------------------------------------------------------------------------
```

In subroutine **STRING** each character in the first string is compared with the corresponding character in the second string. The strings are defined in the main program using the DC.W pseudo-operation (lines 17 to 20). Before calling subroutine STRING (at lines 10 and 14) the following information is set up in lines:

7 & 11 the length of the string is moved into D0
8 & 12 the address of the start of the first string is moved into A0
9 & 13 the address of the start of the second string is moved into A1

In subroutine STRING a newline is displayed, then at line:

43 one is subtracted from the number of characters in the strings (for use as a loop counter in the DBRA instruction at line 46)
44 `CMPM.B (A0)+,(A1)+` compares two characters where both operands are in memory and are addressed using postincrement addressing mode (after the compare the addresses are incremented to point at the next characters to be compared)
45 if the characters were not the same the loop EXITs to line 50 where a message 'strings are not identical' is displayed
46 loop control
47 if the loop is completed the message 'strings are identical' is displayed

Chapter 14 Exercise 14.1 (page 108)

```
---------------------------------------------------------------------------
 1                           * Exercise 14.1 - moving data to/from the stack
 2                           *
 3                           * Set A7 (Stack Pointer) to $1000
 4                           * Load D0 with some long word test data
 5                           * Push a byte, word and long word (D0) onto Stack
 6                           * Pop values off stack and use WRREGS to display them
 7                           *
 8                                        XREF      WRREGS
 9 001000 2E7C 00001000 START:            MOVEA.L   #$1000,A7       Set Stack Pointer
10 001006 203C 89ABCDEF                   MOVE.L    #$89ABCDEF,D0   put data in D0
11 00100C 1F3C     0041                   MOVE.B    #'A',-(A7)      Push byte
12 001010 3F3C     1234                   MOVE.W    #$1234,-(A7)    Push word
13 001014 2F00                            MOVE.L    D0,-(A7)        Push long word
14 001016 122F     0006                   MOVE.B    6(A7),D1        byte 'A' into D1
15 00101A 342F     0004                   MOVE.W    4(A7),D2        word $1234 into D2
16 00101E 2617                            MOVE.L    (A7),D3         long word into D3
17 001020 4EB9 00028094                   JSR       WRREGS
18 001026 4E4E                            TRAP      #14             !! STOP PROGRAM
19                                        END
---------------------------------------------------------------------------
```

The data that has been pushed onto the stack is accessed using address *register indirect with displacement addressing mode*. The top of the stack points to the last word pushed, i.e. the most significant half of the long word value. Thus the word value is stored four bytes above the top of the stack, and the byte value six bytes above.

Chapter 19 Exercise 19.1 (page 160)

The advantages of parallel communications are that it is simple to interface devices and can be very fast (limited by line bandwidth and speed of receiver/transmitter). Its disadvantages are that each data bit requires a separate communications line and handshaking is generally required. The distance over which parallel communications can be used is therefore relatively short (e.g. maximum hundred metres). It is used for local communications where speed and simplicity are an important requirement.

The main advantage of serial communications is that only a single communications line is required (two for simultaneous communications both ways). The disadvantage is, as the transmission of data is in serial form, the maximum data rates are limited. It is therefore used for communication over long distances, e.g. terminal to host computer.

Chapter 20 **Exercise 20.1** (page 165) Program 20.2, an extension of Exercise 20.1, rotates a pattern and reverses direction when a switch is hit.

Chapter 21 **Exercise 21.1** (page 181) Program 21.2, an extension of Exercise 21.1, rotates a pattern and reverses direction when a switch is hit.

Chapter 23

Exercise 23.1 (page 202) Force an address error

```
------------------------------------------------------------------------------
1                          * Program 23.1 - generate an address error
2                          *
3 001000 42B8      1001 START:     CLR.L      $1001
4 001004 4E4E                     TRAP       #14
5                                 END
------------------------------------------------------------------------------
```

```
------------------------------------------------------------------------------
M68> go/run program, from start address (<CR> for 00001000) ? $
Address error (exception vector 3) - fault address was 00001001
The PC may be up to five words in advance of the instruction being executed
Program Counter (PC) = 00001004  Status Register (SR) = 2700  CCR = -----
D0  = 00000000 D1  = 00000000 D2  = 00000000 D3  = 00000000
D4  = 00000000 D5  = 00000000 D6  = 00000000 D7  = 00000000
A0  = 00000000 A1  = 00000000 A2  = 00000000 A3  = 00000000
A4  = 00000000 A5  = 00000000 A6  = 00000000 A7  = 00000FFC
Instruction Register (IR) was 42B8; dissembling instruction (and a few more):
Address  object code             dissembled code
00001000 42B8      1001          CLR.L      $1001.W
00001004 4E4E                    TRAP       #$000E
00001006 04D8                    ILLEGAL_instruction
00001008 0002      06CA          ORI.B      #$CA,D2
MC68000 monitor V1.1, please enter command (<ESC> to abort, ? for help)
M68>
------------------------------------------------------------------------------
```

Attempting to clear a word operand starting from address $1001 causes an address error. An address exception occurs and the monitor displays an appropriate message, the contents of the CPU registers and then dissembles the instruction in question plus a few more (the values in locations 1006 and 1008 are garbage left from past programs).

Exercise 23.2 (page 210) HALT the processor

```
------------------------------------------------------------------------------
1                          * Program 23.2 - HALT the processor
2                          * e.g. load A7 with an odd address then execute RTS
3                          *
4 001000 2E7C 00001001 START:     MOVE.L     #$1001,A7           set up stack pointer
5 001006 4E75                     RTS
6                                 END
------------------------------------------------------------------------------
```

The program (executing in Supervisor Mode) loads A7 with an odd address. The execution of RTS causes an address error exception which, when it attempts to push information onto the stack, generates another address error and HALTs the processor. If the program was executed in User Mode the RTS would generate an address error exception which would switch the processor into Supervisor Mode and the Supervisor Stack Pointer (which would be OK) would be used to push the exception information.

References

Bramer, B, 1988, 'Problems of software portability with particular reference to engineering CAD systems', IEE Computer Aided Engineering Journal, Vol. 5 No. 6, December, pp. 233-236.

Bramer, B, 1989a, 'Selection of computer systems to meet end-user requirements', IEEE Computer Aided Engineering Journal, Vol. 6 No. 2, April, pp. 52-58.

Bramer, B., 1989b, 'IBM/PC MS-DOS full screen editor with program error analyser', Laboratory Microcomputer, December.

Bramer, B, 1990, 'Using a common host system to develop software products for a variety of target computer environments', IEE Computer-Aided Engineering Journal, Vol. 7 No. 5, October, pp. 129-134.

Bytronic 1989, 'The Bytronic multi-applications board', Bytronic Associates, 27b Cloeshill Road, Sutton Coldfield, B75 7AX, UK.

Cambel, J, 1984, 'The RS-232 Solution', Sybex.
This book describes in simple terms the RS232 standard and how to connect different devices.

Chikofsky, E J and Rubenstein, B L, 1988, 'CASE: reliability engineering for information systems', IEEE Software, Vol. 5 No. 2, March, pp. 11-16.
This paper discusses the need to use CASE tools in the implementation of complex software systems.

Clements, A, 1987, 'Microprocessor System Design: 68000 hardware, Software and Interfacing', PWS Computer Science.

Coats, R F, 1985/86, '68000 Board', Electronics and Wireless World, October pp. 51-54, November pp. 51-54, December pp. 36-38, January pp. 67-70, February pp. 72-74, May pp. 24-27.

Collard, D, 1986, 'IBM PCs as the ideal software engineering workstations', .EXE, Vol. 1 No. 3, May, pp. 19.

Edenfield, R W, Gallop, M G, Ledbetter, W B, McGarity, R C, Quintana, E E and Reininger, R A, 1990:
'The 68040 processor: Part 1, design and implementation', IEEE Micro, Vol. 10 No. 1, February, pp. 66-78.
'The 68040 processor: Part 2, memory design and chip verification', IEEE Micro, Vol. 10 No. 3, June, pp. 22-35.

EIA RS-232-C 1981, 'Interface between Data Terminal Equipment and Data Communications Equipment Employing Serial Binary Data Interchange', Washington DC: Electronic Industries Association Engineering Department.

Farrell III, J J, 1984, 'The Advancing Technology of Motorola's Microprocessors and Microcomputers', pp. 55-63, IEEE Micro, Vol. 4, No. 5, October.
This paper reviews the evolution of Motorola's microprocessor and microcomputer products from the MC6800 to the MC68020. It shows that with advancing technology modern microcomputers can be used for applications that needed large mainframes a few years ago.

Force 1984, '68000 Profi Kit Users Manual', Second Edition January, Force Computers Inc., 2041 Mission College Blvd., Santa Clara, CA 95054, USA.

Foster, C C, 'Computer Architecture', Van Nostrand Reinhold, 1976.

This book describes prefetch and other techniques used to enhance processor performance of the mainframes in the 1970's.

Hunt, J W, 1982, 'Programming Languages', pp. 70-88, IEEE Computer, Vol. 15 No. 4, April.

This paper reviews the development, design and application of programming languages in particular for engineering applications.

Huntsman, C and Cawthron, D, 1983, 'The MC68881 Floating-point Co-processor', pp. 44-54, IEEE Micro, Vol. 3 No. 4, December.

A paper that describes floating point standards and the use of the MC68881 as a co-processor to 16/32-bit systems.

Jelemensky, J, Goler, V, Burgess, B, Eifert, J and Miler, G, 1989, 'The MC68332 Microcontroller', IEEE Micro, Vol. 9 No. 4, August, pp. 31-50.

Loveday, G, 1984, 'Practical interface circuits for micros', Pitman.

This book contains an introduction to basic digital circuits and then goes on to describe various practical application circuits for ADCs, DACs, motor control, etc.

Loui, M. C., 1988, 'The case for assembly language programming', IEEE Transactions on Education, Vol. 31 No. 3, August, pp. 160-163.

MacCregor, D and Mothersole, D S, 1983, 'Virtual Memory and the MC68010', IEEE Micro, Vol. 3 No. 3, June, pp. 24-39.

MacGregor, D, Mothersole, D and Moyer, B, 1984,'The Motorola MC68020', IEEE Micro, Vol. 4 No. 4, August, pp. 101-118.

This paper describes the architecture of the MC68020 and compares its performance with the MC68000 and MC68010.

Motorola 1982, 'The MC68230 PIT provides an effective parallel interface', Application Note AN854, Motorola Inc.

Motorola 1983a, 'MC68230 Parallel Interface/Timer (PI/T)', MC68230/D, Motorola Inc.

Motorola 1983b, 'M68000 Family Resident Structured Assembler Reference Manual', M68KMASM/D7, Motorola Inc.

Motorola 1983c, 'MC6840 Programmable Timer', DS-9802-R2, Motorola Inc.

Motorola 1984a, 'A terminal interface, printer interface and background printing for a MC68000 based system using the MC68681 DUART', Application Note AN899, Motorola Inc.

Motorola 1984b, 'MC68008 minimum configuration system', Application Note AN987, Motorola Inc.

Motorola 1985a, 'MC6821 Peripheral Interface Adapter (PIA)', DS9435R4, Motorola Inc.

Motorola 1985b, 'MC6850 Asynchronous Communications Interface Adapter (ACIA', DS9493R4, Motorola Inc.

Motorola 1985c, 'MC68681 Dual Asynchronous Receiver/Transmitter (DUART)', MC68681/D, Motorola Inc.

Motorola 1987, 'A small-area network using the MC68681 DUART, Application Note AN990, Motorola Inc.

Motorola 1989, 'MC68000, MC68008, MC68010, MC68HC000, 8-/16-/32-bit Microprocessors User's Manual Seventh Edition', M68000UM/AD REV 6, Motorola Inc.

This *reference manual* provides details of the M68000, MC68008, MC68010 and MC68HC000 processor architectures and operations (it does not describe how to write programs).

Motorola 1991, 'MC68040 Benchmark Board, AN435/D, Motorola Inc.

Nelson, D and Leach, P, 1984, 'The architecture and applications of the Apollo Domain', IEEE CG&A, Vol. 4 No. 4, April, pp. 58-66.

Oman, P W, 1990, 'CASE analysis and design tools', IEEE Software, Vol. 7 No. 3, May, pp. 37-43.

Real Time Systems 1986, 'XA8 Users' manual', Version 1.6, Real Time Systems Ltd., 6th November.

Safavi, M, 1990, 'Single-chip micros - an overview', Electronics World, Vol. 96 No. 1654, August, pp. 665-676.

Smith, D B and Oman, P W, 1990, 'Software tools in Context', IEEE Software, Vol. 7 No. 3, May, pp. 15-19.

Steward, D, 1987, 'Software Engineering with systems analysis and design', Brooks/Cole.

Stritter, E and Gunter, T, 1979, 'The Motorola 68000', IEEE Computer, Vol. 12 No. 2, February, pp. 43-52.

Tanenbaum, A S, 1987, 'Operating Systems', Prentice-Hall.

Toong, H M and Gupa, A, 'An Architectural Comparison of Contemporary 16- bit Microprocessors', IEEE Computer, Vol. 1 No. 2, May 1981, pp. 26-37.

This paper provides a comparison of the Intel 8086, Zilog Z8000, Motorola 68000 and the National Semiconductor 16000 Series 16-bit microprocessors. Architectural details and application areas of microcomputers using these are discussed.

Wallace, D R and Fujii, R U, 1989, ' Software verification and validation: an overview', IEEE Software, Vol. 6 No. 3, May, pp. 10-17.

Whitesmiths 1986, 'Interface manual for the MC68000', Ver. 3.1, Whitesmiths Ltd., October.

Whitesmiths 1987a, 'C Language Manual', Ver. 3.2, August, Whitesmiths Ltd.

Whitesmiths 1987b, 'Compiler users' manual for DOS/86', Ver. 3.2, April, Whitesmiths Ltd.

Wilcox, A D, 1987, '68000 Microcomputer Systems: Designing and Troubleshooting', Prentice-Hall.

The IEEE is the Institute of Electrical and Electronic Engineers, 345 East 47th Street, New York, New York 10017, USA.

Information on Motorola Products and copies of application notes can be obtained from (or the nearest Motorola sales office):

Motorola Semiconductor Products Literature Distribution Centre, Motorola Inc., P.O.Box 20924, Phoenix, Arizona 85306, USA.

Technical Literature Distribution, Motorola Ltd., 88 Tanners Drive, Blakelands, Milton Keynes, MK14 5BP.

Index